Cases in Medical Microbiology and Infectious Diseases
THIRD EDITION

Cases in Medical Microbiology and Infectious Diseases

THIRD EDITION

Peter H. Gilligan, Ph.D.

Director, Clinical Microbiology-Immunology Laboratories
University of North Carolina Hospitals
Professor, Microbiology-Immunology and Pathology
University of North Carolina School of Medicine
Chapel Hill, North Carolina 27514

M. Lynn Smiley, M.D.

Senior Vice President of Clinical Research, Trimeris, Inc.
Durham, North Carolina
Clinical Professor, Medicine (Infectious Diseases)
University of North Carolina School of Medicine
Chapel Hill, North Carolina 27514

Daniel S. Shapiro, M.D.

Director, Clinical Microbiology and Molecular Diagnostics Laboratories
Boston Medical Center
Associate Professor, Medicine, Pathology and Laboratory Medicine
Boston University School of Medicine
Boston, Massachusetts 02118

ASM PRESS, WASHINGTON, D.C.

Copyright © 2003   ASM Press
American Society for Microbiology
1752 N St., N.W.
Washington, DC 20036-2904

Library of Congress Cataloging-in-Publication Data

Gilligan, Peter H., 1951–
   Cases in medical microbiology and infectious diseases / Peter H.
Gilligan, M. Lynn Smiley, Daniel S. Shapiro.--3rd ed.
       p. ; cm.
Includes bibliographical references and index.
   ISBN 1-55581-207-4 (alk. paper)
   1. Medical microbiology--Case studies. 2. Communicable diseases--Case
studies.
   [DNLM: 1. Communicable Diseases--microbiology--Case Report. 2.
Infection--microbiology--Case Report. 3. Diagnosis, Differential--Case
Report. WC 100 G481c 2003] I. Smiley, M. Lynn, 1952– II. Shapiro,
Daniel S., 1959– III. Title.
   QR46 .G493 2003
   616.9--dc21

                                                    2002011720

ISBN 1-55581-207-4

To our children

CONTENTS

ACKNOWLEDGMENTS

We would like to thank Charles Upchurch for updating the excellent glossary originally compiled by Susan Gibbs and Paul Walden. We thank Mary Ellen Mangum and Melissa Jones for collecting clinical materials that were used for the many new photographs in this edition. We thank Bob Bagnell for his excellent photographic work. We are grateful to Joan Barenfanger for the *Ehrlichia* photos; Thomas Bouldin for the Creutzfeldt-Jakob figure; Lynn Garcia for the *Trichomonas* figure; Jerome O. Klein for the measles rash photo; Frederick T. Koster for the hantavirus case; Krishnan Parayth for the photos of the coccidioidomycosis patient; Thomas Treadwell for the meningococcemia and dengue cases and patient photos; and Alison Holmes and Fiona Cooke for their contributions toward making the Table of Normal Values relevant to health care professionals who work with units that are not commonly in use in the United States.

We would like to thank Mary McKenney for her excellent copyediting. We would particularly like to thank Ellie Tupper, ASM Press, for overseeing this project with diligence and encouragement. Any shortcomings in this text are solely the responsibilities of the authors.

Finally, we would like to remember our colleague and friend Roy S. Hopfer, who passed away much too young in spring of 2001. Roy's photographs have graced all three editions of this text, including photos used on the covers of the first two editions. Roy had a remarkable fund of medical mycology knowledge. He taught us a great deal, and what we learned from him is reflected in these pages. He is greatly missed.

INTRODUCTION

TO THE THIRD EDITION

Since the publication of the second edition of this text, two events in the United States clearly have spotlighted infectious diseases. In the fall of 1999, an outbreak of encephalitis in New York City due to West Nile virus, newly introduced to the United States, brought the problem of emerging infectious disease into the public discourse. In the fall of 2001, bioterrorism in the form of "anthrax" letters sent through the U.S. mail resulted in five deaths. The specter of bioterrorism has resulted in a renewed interest, not only in the medical community but also in the general public, in the epidemiology, detection, pathogenesis, treatment, and prevention of infectious diseases, especially in those that have the potential to sicken and kill thousands to millions. The worldwide spread of HIV and AIDS as well as the ever-increasing problem of antimicrobial drug resistance continue to be major health problems that will not be solved anytime soon. In this setting, we have prepared the third edition of this text.

The goal of this text continues to be to challenge students to develop a working knowledge of the variety of microorganisms that cause infections in humans. Initially, this text was geared exclusively to medical students for use during their basic science curriculum. We have learned from many students that they found this text to be valuable in preparation for both Parts 1 and 2 Examinations of the National Board of Medical Examiners. We also see students using this text during their clinical infectious disease rotations. Although medical students continue to be our main target audience, we recognize that this text is used by a variety of different learners, including advanced undergraduate students, other professional students, physicians in training at both the resident and fellow stage, and faculty members at all levels from high schools to community colleges to colleges and universities and medical schools. We hope all readers of this new edition find it useful.

To try to make this book of greater value to all of our readers, we have included an extensive introductory chapter called "A Primer on the Laboratory Diagnosis of Infectious Disease." The goal of this primer is to explain in simple and concise language the strengths and weaknesses of different approaches that the clinical laboratory uses in assisting the physician in making the diagnosis of infectious diseases. All of the approaches described in this introductory section will later be encountered in the cases. It is our hope that this section will assist readers as they work through the cases that follow in understanding the strategies used in establishing an infectious disease diagnosis.

The cases are presented as "unknowns" and represent actual case histories of patients we have encountered during our professional duties at two university teaching hospitals. Each case is accompanied by several questions to test knowledge in four broad areas:

- the organism's characteristics and laboratory diagnosis
- pathogenesis and clinical characteristics
- epidemiology
- prevention, and in some cases, drug treatment and resistance

Whenever possible, we have tried to incorporate an exciting new area of study, cellular microbiology, into the case discussions. Cellular microbiology is the study of the manner in which microbes interact with their target host cell. As information gained from work within the Human Genome Project is translated into the area of microbial pathogenesis, the understanding of the interaction of the microbial invader with the target host cell will lead to exciting new targets for antimicrobial therapy. This is becoming of greater importance as resistance to conventional antimicrobial therapies becomes more widespread among an ever-increasing number of human pathogens.

There are 15 new cases in the third edition. As with the second edition, we have tried to find cases that represent contemporary trends in infectious diseases. This has led us to delete some cases. We have deemphasized opportunistic infections in HIV patients because the use of highly active antiretroviral therapy fortunately has resulted in a significant reduction in these types of complications. We have tried to increase the number of cases in transplant recipients, because the number of these patients is increasing, and opportunistic infections remain a major problem in this immunosuppressed population. We try to balance these tertiary care problems with typical cases of community-acquired infections that will be more frequently encountered by the majority of physicians. Because this text relies on actual case studies seen in our two institutions, we are happy to say that no cases due to bioterrorism are presented here, although organisms that are possible bioterrorism agents are represented and their potential for use is highlighted.

The book continues to be richly illustrated to familiarize students with the appearance of specific microorganisms in appropriate clinical settings. This is done in response to the decision by many medical schools to limit or eliminate hands-on laboratory experiences. Although photographs are an imperfect substitute for direct observation, we believe that students should develop an appreciation for the importance of microscopy and direct observation in infectious disease diagnosis.

This text has three secondary goals. One is to help students learn how to develop a differential diagnosis. The second is to help students begin to understand the language of medicine. An extensive glossary, originally compiled for the first edition by T. Paul Walden, M.D., and greatly expanded in the second edition by Susan Gibbs, M.D., has been updated for this edition by a current medical student, Charles Upchurch. The purpose of the glossary is to explain the medical terminology so the cases are accessible to readers, especially those who are not physicians, clinical scientists, or physicians-in-training.

The third goal of this text is to present interesting and challenging cases that make the study of medical microbiology and infectious diseases enjoyable. We hope that we have succeeded in this goal and, as always, are open to feedback from interested readers on how we can make future editions of the textbook reflect the fascination that we have for microbiology and infectious diseases.

TO THE STUDENT

This text was written for you. It is an attempt to help you better understand the clinical importance of the basic science concepts you learn either in your medical microbiology or infectious disease course or through your independent study. You may also find that this text is useful in reviewing for Part I of the National Board of Medical Examiners exam. It should be a good reference during your Infectious Disease rotations.

Below is a sample case, followed by a discussion of how you should approach a case to determine its likely etiology.

SAMPLE CASE

A 6-year-old child presented with a 24-hour history of fever, vomiting, and complaining of a sore throat. On physical examination, she had a temperature of 38.5°C, her tonsillar region appeared inflamed and was covered by an exudate, and she had several enlarged cervical lymph nodes. A throat culture plated on sheep blood agar grew many beta-hemolytic colonies. These colonies were small with a comparatively wide zone of hemolysis.

What is the likely etiologic agent of her infection?

The first thing that should be done is to determine what type of infection this child has. She tells you that she has a sore throat. On physical examination, she has sign of an inflamed pharynx with exudate, which is consistent with her symptoms. (Do you know what an exudate is? If not, it's time to consult the glossary in the back of this text.) She also has enlarged regional lymph nodes, which support your diagnosis of pharyngitis (sore throat).

What is the etiology of her infection? The obvious response is that she has a "strep throat," but in reality there are many agents which can cause a clinical syndrome indistinguishable from that produced by group A streptococci, the etiologic agent of "strep throat." For example, sore throats are much more frequently caused by viruses than streptococci. Other bacteria can cause pharyngitis as well, including *Mycoplasma* spp., various *Corynebacterium* spp., *Arcanobacterium* sp., and *Neisseria gonorrhoeae*. All of these organisms would be in the differential diagnosis, along with other perhaps more obscure causes of pharyngitis.

However, further laboratory information narrows the differential diagnosis considerably; small colonies that are surrounded by large zones of hemolysis are consistent with beta-hemolytic streptococci, specifically group A streptococci. On the basis of presenting signs and symptoms and the laboratory data, this child most likely has group A streptococcal pharyngitis.

Specific aids have been added to the book to assist you in solving the cases.

1. For this edition, a new section called "A Primer on the Laboratory Diagnosis of Infectious Diseases" has been added. The purpose of this section is to explain the application and effectiveness of different diagnostic approaches used in the clinical microbiology laboratory. We recommend that you read this primer before beginning your study of the cases.

2. At the beginning of each book section is a brief introduction and a list of organisms. Only organisms on this list should be considered when solving the cases in that section. These lists have been organized on the basis of important characteristics of the organisms.

3. A table of normal values is available inside the front cover of this book. If you are unsure whether a specific laboratory or vital sign finding is abnormal, consult this table.

4. A glossary of medical terms which are frequently used in the cases is available at the end of the text. If you do not understand a specific medical term, consult the glossary. If the term is not there, you will have to consult a medical dictionary or other medical texts.

5. Figures demonstrating microscopic organism morphology are presented in many of the cases, as are key radiographic, laboratory, clinical, or pathologic findings. They provide important clues in helping you determine the etiology of the patient's infection.

A FINAL THOUGHT

The temptation for many will be to read the case and its accompanying questions and then go directly to reading the answers. You will derive more benefit from this text by working through the questions and subsequently reading the case discussion.

Have fun and good luck!

A PRIMER

ON THE LABORATORY DIAGNOSIS OF INFECTIOUS DISEASES

The accurate diagnosis of infectious diseases often but not always requires the use of diagnostic tests to establish their cause. The utilization of diagnostic tests in the managed care environment is carefully monitored and is frequently driven by standardized approaches to care called "clinical pathways." These pathways include using a predefined set of diagnostic tests for patients who present with signs and symptoms characteristic of certain clinical conditions, such as community-acquired pneumonia. Currently, the Infectious Disease Society of America has published over 30 different "practice guidelines" dealing with various infectious diseases, including HIV (human immunodeficiency virus), tuberculosis, group A streptococcal pharyngitis, diarrheal disease, and pneumonia, from which clinical pathways can be derived. Clinical pathways and practice guidelines fall under the concept of "evidence-based medicine." "Evidence-based medicine" relies on review and interpretation of data in the medical literature as a basis for clinical decision-making.

In some patients, such as an otherwise healthy child with a rash typical of varicella (chicken pox), the etiology of the child's infection can be established with a high degree of certainty by physical examination alone. The use of diagnostic testing in this setting would be viewed as wasteful of the health care dollar. On the other hand, a 4-year-old who presents with enlarged cervical lymph nodes and a sore throat should have a diagnostic test to determine whether he or she has pharyngitis due to group A streptococci. The reason why such testing is necessary is that certain viral syndromes are indistinguishable clinically from group A streptococcal pharyngitis. Since group A streptococcal pharyngitis should be treated with an antibiotic to prevent poststreptococcal sequelae, and viral infections do not respond to antibiotics, determining the cause of the infection in this particular case is central to appropriate patient management. Far too often, antibiotics are given without diagnostic testing in a child with a sore throat. As a result, many children with viral pharyngitis are given antibiotics. This inappropriate use of antibiotics increases antibiotic selective pressure. This can result in greater antimicrobial resistance among organisms in the resident microflora of the throat, such as *Streptococcus pneumoniae*. In addition, patients may develop antibiotic-associated complications, such as mild to severe allergic reactions or gastrointestinal distress including diarrhea. One of the goals of the third edition of this text is to help you think in a cost-effective way about how best to use laboratory resources. As an introduction to this edition, we will present a general overview of the various laboratory approaches that are used in the diagnosis and management of infectious diseases.

ACCURACY IN LABORATORY TESTING

The clinical microbiology laboratory must balance the requirements of timeliness with those of accuracy.

As an example, consider the identification of a gram-negative rod from a clinical specimen. If the organism is identified with the use of a commercially available identification system, an identification and an assessment of the probability of that identification will be made on the basis of biochemical test results and a comparison of these results with a database. So, if the result states that the organism is *Enterobacter cloacae* with 92% probability, the laboratory may very well report this identification. Assuming that the 92% probability figure generated by the commercial system is on target (many commercial systems do a worse job with anaerobic bacteria), this means that there is a probability of 8%, or about 1 time in 12, that this identification will be incorrect.

Certainly, it would be possible for the laboratory to perform additional testing to be more certain of the identification. The problem is that by doing so there would be a delay, perhaps a clinically significant one, in the reporting of the results of the culture. In some cases such a delay is unavoidable (e.g., when the result of the identification in the commercial system is below an arbitrary acceptable probability and manual methods must be used) or clinically essential (e.g., when a specific identification is required and the isolate must be sent to a reference laboratory for identification; an example is *Brucella* spp., which require prolonged therapy and are potential agents of bioterrorism).

Similarly, the methods most commonly used in the clinical laboratory for susceptibility testing are imperfect. The worst errors, from the clinical point of view, are those in which the laboratory reports an organism as susceptible to a particular antibiotic to which, in fact, it is resistant. In some cases additional tests are employed to minimize the risk of this occurring. For example, in addition to standard testing using either an automated or a manual method, recommended susceptibility testing of *Staphylococcus aureus* includes the use of Mueller-Hinton agar, in which the antibiotic oxacillin is present at a known concentration. Even if the results of the standard susceptibility testing indicate susceptibility to oxacillin, if there is growth of the *S. aureus* isolate on the oxacillin-containing Mueller-Hinton plate, the organism is reported as resistant to oxacillin (as well as to all other beta-lactams).

Unfortunately, very few such checks exist to correct erroneous bacterial susceptibility assays. In addition to the oxacillin-containing plate, there are plates containing medium with vancomycin. These are used in the laboratory to screen for the possibility of vancomycin resistance in enterococci and the rarely seen increase in the minimum inhibitory concentration (MIC) in *S. aureus*. In general, there is a delay in the ability of automated susceptibility methods to reliably identify newly described mechanisms of antibiotic resistance. As a result, manual methods are often required. The performance of automated susceptibility testing methods varies, and certain combinations of organism and antibiotic have an unacceptably high error rate. In such cases, backup methods, such as disk diffusion or MIC testing, should be employed. Laboratories with a significant number of susceptibility tests to perform commonly use automated susceptibility methods because of the labor-intensive nature of manual testing and the speed with which automated systems are able to report results—often in a few hours as compared with overnight incubation, as is the case with manual methods.

Diagnostic tests vary in their **sensitivity** and **specificity**. As an example, consider a hypothetical STD (sexually transmitted disease) clinic in which the rapid plasma reagin (RPR) test, a screening test for syphilis, is being evaluated in 1,000 patients with genital ulcer disease who are suspected of having primary syphilis:

		PRIMARY SYPHILIS		
		PRESENT	ABSENT	
RPR TEST RESULT	**POSITIVE**	420	60	Positive predictive value = 420/(420 + 60) = 0.88 Positive predictive value = 88%
	NEGATIVE	220	300	Negative predictive value = 300/(300 + 220) = 0.58 Negative predictive value = 58%
		Sensitivity = 420/(420 + 220) = 0.66 Sensitivity = 66%	Specificity = 300/(300 + 60) = 0.83 Specificity = 83%	

On the basis of these data, the sensitivity of this test (the true-positive rate, calculated as true-positive results divided by the number of patients with disease) in primary syphilis is 66%. The specificity (1 minus the false-positive rate) is 83%. Note that in this high-prevalence population (the prevalence here is the total number of cases in which primary syphilis is present—640 divided by the total number of individuals, 1,000—and is thus 0.64 or 64%), the predictive value of a positive test is fairly good, at 88%. **The positive predictive value of an assay varies with the prevalence of the disease in the population.** This is a key point. An example of this in our syphilis serology example in a low-prevalence population will serve to illustrate the point.

The same RPR serologic assay is being used in a hypothetical population of octogenarian nuns, none of whom are or have been sexually active in at least 6 decades, and all of whom have been confined to their abbey since before the Second World War.

SYPHILIS

		PRESENT	ABSENT	
RPR TEST RESULT	**POSITIVE**	1	169	Positive predictive value = 1/170 = 0.006 Positive predictive value = 0.6%
	NEGATIVE	0	830	Negative predictive value = 830/830 = 1.00 Negative predictive value = 100%
			Specificity = 830/999 = 0.83 Specificity = 83%	

In this patient population, there is only one true case of syphilis, presumably acquired many years previously. The specificity of the test in this patient population is essentially the same as it is in the individuals attending the STD clinic (in reality, it is likely to be different in different populations and also in different stages of syphilis). Since there is one case of syphilis, and 169 of the positive RPR results are false-positive test results, the positive predictive value in this patient population is only 0.6%. Clearly, this is a patient population in which the decision to test for syphilis using the RPR assay is not cost-effective.

In making a decision to order a specific test, the physician should know what he or she will do with the test results—essentially, how the results will alter the care of the patient. In a patient who the physician is certain does not have a specific disease, if the test for that disease has an appreciable rate of false-positive results, a positive test result is likely to be false positive and should not alter clinical care. Conversely, if the physician is certain that a patient has a disease, there is no good reason to order a test with a low sensitivity, as a negative result will likely be false negative. Tests are best used when there is uncertainty and when the results will alter the posttest probability and, therefore, the management of the patient.

SPECIMEN SELECTION, COLLECTION, AND TRANSPORT

Each laboratory test has three stages.

1. **The preanalytical stage:** The caregiver selects the test to be done, determines the type of specimen to be collected for analysis, ensures that it is properly labeled with the patient's name, and facilitates rapid and proper transport of this specimen to the laboratory.

2. **The analytical stage:** The specimen is analyzed by the laboratory for the presence of specific microbial pathogens. The remaining sections of this chapter describe various analytic approaches to the detection of pathogens.

3. **The postanalytical stage:** The caregiver uses the laboratory results to determine what therapies, if any, to use in the care of the patient.

The preanalytical stage is the most important stage in laboratory testing! If the wrong test is ordered, if the wrong specimen is collected, if the specimen is labeled with the wrong patient's name, or if the correct specimen is collected but is improperly transported, the microbe causing the patient's illness may be not detected in the analytical stage. As a result, at the postanalytical stage, the caregiver may not have the appropriate information to make the correct therapeutic decision. The maxim frequently used in laboratory medicine is "garbage in, garbage out."

Specimen selection is important. A patient with a fever, chills, and malaise may have an infection in any one of several organ systems. If a patient has a urinary tract infection and if urine is not selected for culture, the etiology and source of the infection will be missed. Careful history taking and physical examination play an important role in selecting the correct specimen.

Continuing with the example of a patient with a fever due to a urinary tract infection, the next phase in the diagnosis of infection is the collection of a urine specimen. Because the urethra has a resident microflora, urine specimens typically are not sterile. A properly collected urine specimen is one in which the external genitalia are cleansed and midstream urine is collected. Collection of midstream urine is important because the initial portion of the stream washes out much of the urethral flora. Even with careful attention to detail, clean-catch urine can be contaminated with urethral flora, rendering the specimen uninterpretable at the postanalytical stage.

An important concept when considering the transport of clinical specimens for culture is to recognize that they contain living organisms whose viability is influenced by transport conditions. These organisms may be killed by changes in temperature, drying of the specimen, exposure to oxygen, lack of vital nutrients, or changes in specimen pH. Transport conditions that support the viability of any clinically significant organisms present in the specimen should be established. It should also be noted that the longer the transport takes, the less likely it is that viability will be maintained. Rapid transport of specimens is important for maximal accuracy at the analytical stage.

If the correct test is selected, the proper specimen is collected and transported, but the specimen is labeled with the wrong name, the test findings might be harmful to two different patients. The patient from whom the specimen came might not receive the proper therapy, while a second patient whose name was mistakenly used to label the specimen might receive a potentially harmful therapy.

DIRECT EXAMINATION

Macroscopic

Once a specimen is received in the clinical laboratory, the first step in the determination of the cause of an infection is to examine it. Frequently, infected urine, joint, or cerebrospinal fluid specimens will be "cloudy" because of the presence of microorganisms and white blood cells, suggesting that an infectious process is occurring. Occasionally, the organism can be seen by simply looking for it in a clinical specimen or by looking for it on the patient. Certain worms or parts of worms can be seen in the feces of patients with ascariasis or tapeworm infections. Careful examination of an individual's scalp or pubic area may reveal lice, while examination of the anal region may result in the detection of pinworms. Ticks can act as vectors for several infectious agents, including Rocky Mountain spotted fever, Lyme disease, and ehrlichiosis. When found engorged on the skin, physicians may remove and submit these ticks to the laboratory to determine their identity. This is done because certain ticks (deer ticks) act as a vector for certain infectious agents (*Borrelia burgdorferi*, the organism that causes Lyme disease). Knowing the vector may help the physician determine the patient's diagnosis.

Microscopic

Because most infectious agents are visible only when viewed with the aid of a light microscope, microscopic examination is central to the laboratory diagnosis of infectious diseases. Microscopic examination does not have the overall sensitivity and specificity of culture or the newer molecular diagnostic techniques. However, microscopic examination is very rapid, is usually relatively inexpensive (especially when compared with molecular techniques), is available around the clock in at least some formats in most institutions, and in many clinical settings but by no means all is highly accurate when done by highly skilled laboratorians. The organisms can be detected either unstained or by using a wide variety of stains, some of which are described below. Microbes have characteristic shapes that are important in their identification. Morphology can be very simple, with most clinically important bacteria generally appearing as either cocci (Fig. 1a) or rods (Fig. 1b). The rods can be very long or so short that they can be confused with cocci (coccobacilli); they can be fat or thin, have pointed ends, or be curved. The arrangement of cocci can be very helpful in determining their identity. These organisms can be arranged in clusters (staphylococci), pairs or diplococci (*S. pneumoniae*), or chains (various streptococcal and enterococcal species).

Fungi are typically divided into two groups based on morphology. One is a yeast (Fig. 2), which is a unicellular organism, and the other is a mold which is a multicellular organism with complex ribbon-like structures called hyphae (Fig. 3). Organisms that are referred to as parasites may be unicellular—the protozoans (Fig. 4)—or highly

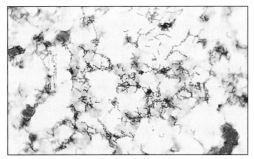

Figure 1a

Figure 1b

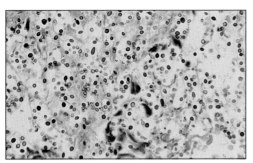

Figure 2

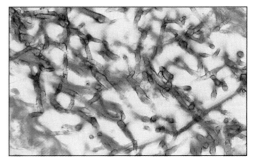

Figure 3

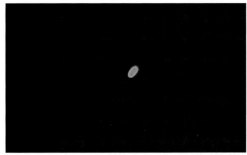

Figure 4

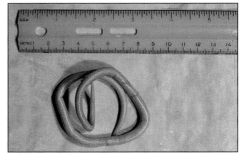

Figure 5

complex—the nematodes (Fig. 5) and cestodes. Parasites are typically identified on the basis of morphology alone.

Because of their small size, viruses cannot be visualized by light microscopy. Alternative approaches described below are needed to detect these microbes in clinical specimens.

Wet mounts

This technique is extremely simple to perform. As the name implies, the clinical specimen is usually mixed with a small volume of saline, covered with a glass coverslip, and examined microscopically. It is most commonly utilized to examine discharges from the female genital tract for the presence of yeasts or the parasite *Trichomonas vaginalis*. Wet mounts are also used to make the diagnosis of oral thrush, which is caused by the yeast *Candida albicans*. Using a special microscopic technique—dark-field microscopy—scrapings from genital ulcers and certain skin lesions can be examined for the spirochete *Treponema pallidum*, the organism that causes syphilis. This technique is not particularly sensitive but is highly specific in the hands of an experienced microscopist. It is typically done in STD clinics where large numbers of specimens are available, enabling the microscopist to maintain his or her skill in detecting this organism.

The wet mount can be modified by replacing a drop of saline with a drop of a 10% KOH solution to a clinical specimen. This technique is used to detect fungi primarily in sputum or related respiratory tract specimens, skin scrapings, and tissues. The purpose of the KOH solution is to "clear" the background by "dissolving" tissue and bacteria, making it easier to visualize the fungi.

Another modification of the wet mount is to mix a drop of 5% Lugol's iodine solution with feces. This stains any protozoans or eggs of various worms that may be present in the stool, making them easier to see and identify.

Gram stain

The most frequently utilized stain in the microbiology laboratory is the Gram stain. This stain differentiates bacteria into two groups. One is referred to as gram positive because of its ability to retain crystal violet stain, while the other is referred to as gram negative because it is unable to retain this stain (see Fig. 1). These organisms can be further subdivided based on their morphological characteristics.

The structure of the bacterial cell envelope determines an organism's Gram stain characteristics. Gram-positive organisms have an inner phospholipid bilayer membrane surrounded by a cell wall composed of a relatively thick layer of the polymer peptidoglycan. Gram-negative organisms also have an inner phospholipid bilayer membrane surrounded by a peptidoglycan-containing cell wall. However, in the gram-negative organisms, the peptidoglycan layer is much thinner. The cell wall in gram-negative organisms is surrounded by an outer membrane composed of a phospholipid bilayer. Embedded within this bilayer are proteins and the lipid A portion of a complex molecule called lipopolysaccharide. Lipopolysaccharide is also referred to as **endotoxin** because it can cause a variety of toxic effects in humans.

Because of their size or cell envelope composition, certain clinically important bacteria cannot be seen on Gram stain. These include all species of the genera *Mycobacterium, Mycoplasma, Rickettsia, Coxiella, Ehrlichia, Chlamydia,* and *Treponema.* Yeasts typically stain as gram-positive organisms, while the hyphae of molds may inconsistently take up stain but generally will be gram positive.

Gram stains can be performed quickly, but attention to detail is important to get an accurate Gram reaction. One clue to proper staining is to examine the background of the stain. The presence of significant amounts of purple (gram-positive) in the epithelial cells, red or white blood cells, or proteinaceous material, all of which should stain gram negative, suggests that the stain is under-decolorized and that the Gram reaction of the bacteria may not be accurate. This type of staining characteristic is frequently seen in "thick" smears. The detection of over-decolorization is much more difficult and is dependent on the observation skills of the individual examining the slide.

Staining of acid-fast organisms

Mycobacterium spp., unlike other bacteria, are surrounded by a thick mycolic acid coat. This complex lipid coat makes the cell wall of these bacteria refractory to staining by the dyes used in the Gram stain. As a result, bacteria within this genus usually cannot be visualized or, infrequently, may have a beaded appearance on Gram stain. Certain stains, such as carbol fuchsin or auramine-rhodamine, can form a complex with the mycolic acid. This stain is not washed out of the cell wall by acid-alcohol or weak acid solution, thus the term "acid-fast" bacterium.

Auramine and rhodamine are nonspecific fluorochromes. Fluorochromes are stains that "fluoresce" when excited by light of a specific wavelength. Bacteria that retain these dyes during the acid-fast staining procedure can be visualized with a fluorescent microscope (Fig. 6). In clinical laboratories with access to a fluorescent microscope, the auramine-rhodamine stain is the method of choice because the organisms can be visualized at a lower magnification. By screening at lower magnification, larger areas of the microscope slide can be examined more quickly, making this method more sensitive and easier to perform than acid-fast stains using carbol fuchsin.

Several other organisms are acid-fast, although they typically are not alcohol-fast. As a result, they are stained using a modified acid-fast decolorizing step whereby a weak acid solution is substituted for an alcohol-acid one. This technique is frequently used to distinguish two genera of gram-positive, branching rods from each other. *Nocardia* species are acid-fast when the modified acid-fast staining procedure is used, while *Actinomyces* species are not. *Rhodococcus equi* is a coccobacillus that may also be positive by modified acid-fast stain when first isolated. The modified acid-fast stain has also been effective in the detection of two gastrointestinal protozoan parasites, *Cryptosporidium* and *Cyclospora*. It should be noted that *Cyclospora* stains

Figure 6

inconsistently, with some organisms giving a beaded appearance while others do not retain the stain at all.

Trichrome stain

The trichrome stain is used to visualize protozoans in fecal specimens. This stain is particularly effective at staining internal structures, the examination of which is important in determining the identity of certain protozoans, such as *Entamoeba histolytica*. Modification of the trichrome stain is used in the detection and identification of microsporidia.

Direct fluorescent-antibody stains

The development of monoclonal antibodies has enhanced both the sensitivity and the specificity of staining techniques that use antibodies to detect microbes in clinical specimens. The most widely used staining technique that incorporates the use of antibodies is the direct fluorescent-antibody (DFA) stain. In this technique, a highly specific antibody is coupled to a fluorochrome, typically fluorescein, which emits an "apple-green" fluorescence. The antibody binds specifically either to antigens on the surface of the microbes or to viral antigens expressed by virally infected cells which can be visualized under the fluorescent microscope (Fig. 7). This technique is rapid, usually requiring 1 to 2 hours. In the hands of a skilled operator, the test is highly specific, although it frequently has a sensitivity of only 60 to 70% when compared with bacterial culture. Because of its rapidity, the test has been used to detect some relatively slow-growing or difficult-to-grow bacteria, such as *Bordetella pertussis* and *Legionella pneumophila*. For respiratory and herpes viruses, the sensitivity of this technique approaches 90% of the sensitivity of culture. The development of molecular amplification techniques, particularly real-time polymerase chain reaction (PCR) assays, for the detection of these bacterial and viral agents may render this application of DFA testing obsolete.

DFA staining is frequently used for the detection of microbes that cannot be cultured. DFA is the method of choice for detection of the nonculturable fungus *Pneumocystis carinii*, a common cause of pneumonia in people with AIDS. DFA is much more sensitive than other commonly used staining techniques, such as silver, Giemsa, or toluidine blue O staining. Likewise, for the gastrointestinal protozoans *Giardia lamblia* and *Cryptosporidium parvum*, DFA staining has been found to be much more sensitive than examination of wet mounts or the use of trichrome (for *Giardia*) or modified acid-fast stain (for *Cryptosporidium*).

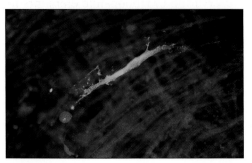

Figure 7

A novel application of the DFA technique is its use in detecting bloodstream infections with the cytomegalovirus (CMV). This virus is an important cause of infections in immunocompromised patients, especially in those who have received solid-organ or bone marrow transplant. In this technique, CMV antigens are detected on the surface of peripheral white blood cells. The number of infected cells per 100,000 white blood cells is determined. This technique, along with quantitative PCR, is currently believed to be the most accurate for diagnosing disseminated CMV infections and predicting those individuals who need to be receiving antiviral therapy.

Infectious disease diagnosis from peripheral blood smears and tissue sections

Not all staining used in the diagnosis of infectious disease is done in the microbiology laboratory. The hematologist and the anatomical pathologist can play important roles in the diagnosis of certain infectious diseases.

The peripheral blood smear is the method of choice for detection of one of the most important infectious diseases in the world, malaria, which is caused by protozoans within the genus *Plasmodium*. The various developmental stages of these parasites are detected in red blood cells. Other, less frequently encountered parasites seen in a peripheral blood smear include *Babesia* species, trypanosomes, and the microfilariae.

Bacterial and fungal pathogens may be seen in peripheral smears on occasion. The most likely of these is *Histoplasma capsulatum*, which is seen as small, intracellular yeasts in peripheral white blood cells. *Ehrlichia* spp. can produce a characteristic lesion, the morulae, which can be seen in peripheral white blood cells. There are two forms of the disease, one in which mononuclear cells are infected, and the other in which granulocytic cells are infected.

Examination of tissue by the anatomical pathologist is an important technique for detecting certain infectious agents. Tissue cysts due to toxoplasmosis can be detected in brain biopsy material from patients with encephalitis. The diagnosis of Creutzfeldt-Jakob disease is based on the finding of typical lesions on brain biopsy. The finding of hyphal elements in lung tissue is an important tool in the diagnosis of invasive aspergillosis and pulmonary zygomycosis. The observation of ribbon-like elements in a sinus biopsy is pathognomic for the diagnosis of rhinocerebral zygomycosis, a potentially fatal disease most frequently seen in diabetic patients.

Antigen detection

Visual examination of a clinical specimen is not the only means by which an infectious agent can be directly detected. A variety of tests have been developed that, like DFA, are dependent on the availability of highly specific antibodies to detect antigens of specific bacteria, fungi, viruses, and parasites. The most widely used antigen detection tests are various formats of the enzyme immunoassay (EIA) or the latex agglutination assay. These tests take anywhere from 10 minutes to 2 hours. The test most widely used is a 10- to 15-minute EIA for the detection of group A streptococci. The sensitivity of these various formats has been reported to be 80 to 90%, with a specificity usually greater than 95%. In the United States, there are over 50 different test formats marketed for the detection of this organism. The test is done in a wide variety of laboratories, clinics, and physicians' offices. Antigen detection tests are widely used in the United States to detect a variety of infectious agents, including *Cryptococcus neoformans*, *Clostridium difficile* toxin, respiratory syncytial virus (RSV), rotavirus, influenza virus, and *Giardia* and *Cryptosporidium* spp.

MOLECULAR DIAGNOSTICS

In addition to standard methods of culturing and identifying pathogenic microorganisms, there are now a number of molecular methods available that are able to detect the presence of the specific nucleic acid of these organisms. These methods are used in demonstrating the presence of the organism in patient specimens as well as in determining the identification of an isolated organism. In some cases, these methods are able to determine the quantity of the nucleic acid.

As an example, bacteria of a particular species will have a chromosomal nucleic acid sequence significantly different from that of another bacterial species. On the other hand, the nucleic acid sequence within a given species has regions that are highly conserved. For example, the base sequence of the *Mycobacterium tuberculosis* 16S ribosomal RNA gene differs significantly from the base sequence in the *Mycobacterium avium* complex 16S ribosomal RNA gene, yet the sequence of bases in this region among members of the *M. tuberculosis* complex is highly conserved. These properties form the basis for the use of genetic probes to identify bacteria to the species level. There are a number of commercially available genetic probes that can detect specific sequences in bacteria, mycobacteria, fungi, and viruses.

Nucleic acid hybridization is a method by which there is the in vitro association of two complementary nucleic acid strands to form a hybrid strand. The hybrid can be a DNA-RNA hybrid, a DNA-DNA hybrid, or, less commonly, an RNA-RNA hybrid. To do this, one denatures the two strands of a DNA molecule by heating to a temperature above which the complementary base pairs that hold the two DNA strands together are disrupted and the helix rapidly dissociates into two single strands. A second nucleic acid is introduced which will bind to regions that are complementary to its nucleic acid sequence. The stringency, or specificity, of the reaction can be varied by reaction conditions such as the temperature.

In addition to the direct demonstration of a nucleic acid sequence by hybridization, amplification assays (the process of making additional copies of the specific sequence of interest) are of increasing importance in clinical microbiology. The most

Figure 8 PCR. (A) In the first cycle, a double-stranded DNA target sequence is used as a template. (B) These two strands are separated by heat denaturation, and the synthetic oligonucleotide primers (solid bars) anneal to their respective recognition sequences in the 5' → 3' orientation. Note that the 3' ends of the primers are facing each other. (C) A thermostable DNA polymerase initiates synthesis at the 3' ends of the primers. Extension of the primer via DNA synthesis results in new primer-binding sites. The net result after one round of synthesis is two "ragged" copies of the original target DNA molecule. (D) In the second cycle, each of the four DNA strands in panel C anneals to primers (present in excess) to initiate a new round of DNA synthesis. Of the eight single-stranded products, two are of a length defined by the distance between and including the primer-annealing sites; these "short products" accumulate exponentially in subsequent cycles. (Reprinted from *Manual of Clinical Microbiology*, 7th ed., ©1999 ASM Press, with permission.)

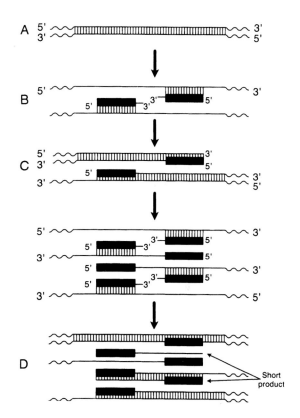

commonly used amplification assay is PCR (Fig. 8). PCR uses a DNA polymerase that is stable at high temperatures that would denature and inactivate most enzymes. This thermostable DNA polymerase most often is isolated from the bacterium *Thermus aquaticus*. Its stability at high temperature enables the enzyme to be used without the need for replacement after the high-temperature conditions of the DNA denaturization step that occurs during each cycle of PCR:

1. The target DNA sequence is heated to a high temperature that causes the double-stranded DNA to denature into single strands.

2. An annealing step follows, at a lower temperature than the denaturization step above, during which sets of primers, with sequences designed specifically for the PCR target sequences, bind to these target sequences.

3. Last is an extension step, during which the DNA polymerase completes the target sequence between the two primers.

Assuming 100% efficiency, the above three steps generate two copies of the target sequence. Multiple cycles (such as 30) in a thermal cycler result in a tremendous amplification of the number of sequences, so that the sequence is readily detectable using any of a variety of methods—colorimetric, chemiluminescent, or gel electrophoretic.

When the specific target nucleic acid is RNA rather than DNA, a complementary DNA (cDNA) sequence is made with the enzyme reverse transcriptase (RT) before PCR amplification in a procedure known as RT-PCR. Examples of pathogens for which RT-PCR is used include the RNA-containing viruses HIV-1 and hepatitis C virus (HCV).

An additional feature of PCR is that the amplified nucleic acid products can be directly sequenced. These sequences can be compared with sequences found in publicly accessible databases. This allows, for example, the identification of a bacterial organism to the level of species on the basis of a sequence of hundreds of bases in the 16S ribosomal RNA gene or, if the sequence is less closely related to sequences within the database, to the level of genus. In some cases, the organism may be an entirely new one. This method of PCR and sequencing of the product for the purposes of bacterial identification remains a research tool in clinical microbiology, but the rapidity with which it can be performed and the high quality of the databases makes it likely that this method will be increasingly used.

After the invention of PCR, a number of other amplification assays were developed, some of which have entered the clinical microbiology laboratory. These include ligase chain reaction (LCR), in which DNA ligase is used (as is a thermal cycler) to produce ligation products, and transcription-mediated amplification (TMA), which does not require a thermal cycler and relies on the formation of cDNA from a target single-stranded RNA sequence, the destruction of the RNA in the RNA:DNA hybrid by RNase H, and the formation of double-stranded cDNA (which can serve as transcription templates for T7 RNA polymerase). A similar procedure occurs during the nucleic acid sequence-based amplification (NASBA) assay. Strand-displacement amplification (SDA) does not require a thermal cycler and has two phases in its cycle: a target generation phase during which a double-stranded DNA sequence is heat denatured, resulting in two single-stranded DNA copies, and an exponential amplification phase in which a specific primer binds to each strand at the cDNA sequence. DNA polymerase extends the primer, forming a double-stranded DNA segment that contains a specific restriction endonuclease recognition site, to which a restriction enzyme binds, cleaving one strand of the double-stranded sequence and forming a nick, followed by extension and displacement of new DNA strands by DNA polymerase.

All of these assays—PCR, LCR, TMA, and SDA—have one thing in common: they amplify the target nucleic acid sequence, making many, many copies of the sequence. As you might imagine, there is the possibility that small quantities of the billions of amplified target nucleic acid sequences can contaminate a sample that will then undergo amplification testing, resulting in false-positive results. Steps are taken to minimize contamination, including physical separation of specimen preparation and amplification areas, positive displacement pipettes, and both enzymatic (in PCR) and nonenzymatic methods to destroy the amplified products.

An alternative method of demonstrating the presence of a specific nucleic acid sequence that does not require the amplification of the target is by amplification of the signal. One commonly used method that does so is the branched DNA assay (b-DNA) (Fig. 9), which is used particularly in quantitative assays, such as HIV and HCV viral load determinations. In this assay, specific oligonucleotides hybridize to the sequence of interest and capture it onto a solid surface. In addition, a set of synthetic enzyme-conjugated branched oligonucleotides hybridize to the target sequence. When an appropriate substrate is added, light emission is measured and compared with a standard curve. This permits quantitation of the target sequence. As there is no amplified sequence to be concerned about, the risk of contamination is dramatically reduced.

There are several commercially available molecular diagnostic assays for *Chlamydia trachomatis* and *Neisseria gonorrhoeae*. These include direct hybridization assays as well as tests that amplify the nucleic acid in specimens of cervical and urethral swabs and urine. The amplified assays are being used with increasing frequency both because of the convenience of using urine rather than a sample that requires more invasive specimen collection and because of the increased sensitivity of amplified as compared with

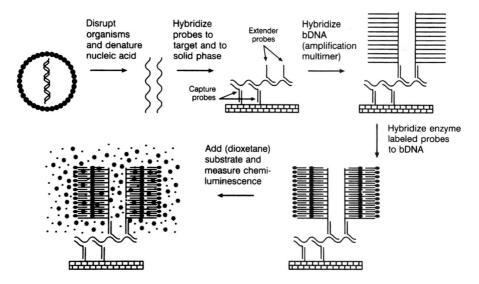

Figure 9   bDNA-based signal amplification. Target nucleic acid is released by disruption and is captured onto a solid surface via multiple contiguous capture probes. Contiguous extended probes hybridize with adjacent target sequences and contain additional sequences homologous to the branched amplification multimer. Enzyme-labeled oligonucleotides bind to the bDNA by homologus base pairing, and the enzyme-probe complex is measured by detection of chemiluminescence. All hybridization reactions occur simultaneously. (Reprinted from *Manual of Clinical Microbiology*, 7th ed., ©1999 ASM Press, with permission.)

nonamplified assays. Since *N. gonorrhoeae* is a fastidious organism that may not survive specimen transport, this test is of particular benefit in settings in which there may be a delay in the transport of the specimen to the laboratory; i.e., the viability of the organisms is not required to detect the presence of its nucleic acid. Similarly, the previous "gold standard" for the detection of *C. trachomatis*—tissue culture—was labor-intensive, required the use of living cell lines, and required rapid specimen transport on wet ice to ensure the viability of the organisms in the specimen. In many laboratories, *C. trachomatis* tissue culture has been replaced by amplification technologies, which have been shown to be more sensitive. As you might imagine, since these assays do not require the presence of living organisms, patients who have been treated with appropriate antibiotics may continue to have a positive assay for some time because of the presence of dead, and therefore noninfectious, organisms that contain the target nucleic acid.

Quantitative assays are now available for several different pathogens. These include tests that determine the level of HIV RNA in patients with HIV infection and are now recognized as one component of the standard clinical management of these patients. With the availability of highly active antiretroviral therapy (HAART) but the potential for antiviral drug resistance, it is important to be able to closely monitor the plasma level of HIV RNA, also known as the viral load. A clinical response to antiretroviral therapy can be demonstrated by a decrease in the viral load. Similarly, an increase in the viral load may indicate either the development of viral resistance to one or more of the antiviral agents being used to treat the patient or merely patient noncompliance with therapy. Modification of therapy may be made on the basis of a rising HIV viral load and the results of HIV genotyping studies.

HIV genotyping is a test that determines the specific nucleic acid sequence present in the virus infecting a patient. There are a number of ways that this test can be performed, and direct sequencing of amplified cDNA (using RT-PCR) is one example. These results are routinely compared with a database that contains nucleic acid sequences from viral strains that are known to be both sensitive and resistant to specific antiretroviral medications. This comparison permits the clinician to note what, if any, mutations are present in the virus infecting the patient and to predict with some reasonable degree of probability whether the viral isolate is resistant to antiretroviral medications, including those being taken by the patient. These data can help the physician make a rational choice of an antiretroviral regimen in a patient whose therapy is failing. One difficulty with this test is that patients are often infected with a mixture of different HIV viral populations, both because of the high frequency of mutation that occurs with HIV and because of the selection of resistant subpopulations while the patient receives antiretroviral therapy. As a result, there may be resistant subpopulations that are below the level of detection of the standard HIV genotyping assay and that could become clinically relevant under the selective pressure of continued antiretroviral therapy.

Detection of HCV RNA using PCR can be used both diagnostically and for following the effectiveness of therapy. The PCR product generated during the HCV RNA assay can be used for genotyping. Genotype is determined by a hybridization assay in which specific genes associated with specific genotypes are detected. Genotype 1 is more refractory to therapy than genotypes 2 and 3. Therefore, therapy is much more prolonged (48 versus 24 weeks) for genotype 1 than for 2 and 3.

CULTURE

Detection of bacterial and fungal pathogens by culture

Culture on artificial medium is the most commonly used technique for detecting bacteria and fungi in clinical specimens. Although not as rapid as direct examination, it is more sensitive and much more specific. For the majority of human pathogens, culture requires only 1 to 2 days of incubation. For particularly slow-growing organisms, such as *M. tuberculosis* and some fungi, the incubation period may last for weeks. By growing the organism, it is available for further phenotypic and genotypic analysis, such as antimicrobial susceptibility testing, serotyping, virulence factor detection, and molecular epidemiology studies.

Environmental and nutritional aspect of bacterial and fungal culture

Certain basic strategies are used to recover bacterial and fungal pathogens. These strategies are dependent upon the phenotypic characteristics of the organisms to be isolated and the presence of competing microflora in a patient's clinical specimen. Most human pathogens grow best at 37°C, human body temperature. Most bacterial and fungal cultures are performed, at least initially, at this temperature. Certain skin pathogens, such as dermatophytes and some *Mycobacterium* spp., grow better at 30°C. When seeking these organisms, cultures may be done at this lower temperature. A few clinically significant microorganisms will grow at low temperatures (4°C), while others prefer higher temperatures (42°C). These incubation temperatures may be used when attempting to recover a specific organism from specimens with a resident microflora, such as feces, since few organisms other than the target organism can grow at these temperature extremes.

Another important characteristic of human bacterial and fungal pathogens is the impact of the presence of oxygen on the growth of these organisms. Microbes can be divided into three major groups based on their ability to grow in the presence of oxygen. Organisms that can only grow in the presence of oxygen are called **aerobes.** Fungi and many bacteria are **aerobic** organisms. Organisms that can only grow in the absence of oxygen are called **anaerobes.** The majority of bacteria that make up the resident microflora of the gastrointestinal and female genital tracts are anaerobic organisms. Some bacteria can grow either in the presence or in the absence of oxygen. These organisms are called **facultative** organisms. A subgroup of facultative organisms is called **microaerophiles.** These organisms grow best in atmosphere with reduced levels of oxygen. *Campylobacter* spp. and *Helicobacter* spp. are examples of microaerophiles.

Besides temperature and oxygen, nutrients are an important third factor in the growth of microbes. Many bacteria have very simple growth requirements. They require an energy and carbon source, such as glucose; a nitrogen source, which may be ammonium salts or amino acids; and trace amounts of salts and minerals, especially iron. Some human pathogens have much more complex growth requirements, needing certain vitamins or less well defined nutrients such as animal serum. Organisms with highly complex growth requirements are often referred to as being **fastidious**. A fastidious bacterium which is frequently encountered clinically is *Haemophilus influenzae*. This bacterium requires both hemin, an iron-containing molecule, and NAD for growth.

Media

The selection of media to be used in isolation of pathogens from clinical specimens is dependent on several factors. First, the nutritional requirements of the specific pathogen must be met. For example, fastidious organisms require a medium which is enriched with specific nutrients, such as animal blood, serum, or other growth factors. If the clinical specimen is obtained from a site that has a resident microflora, certain strategies will be necessary to isolate a specific pathogen from the accompanying resident microflora. Often in this setting, a special type of medium called selective medium is used to recover these pathogens. This medium selects for the growth of a specific group of organisms. This is done by adding substances, such as dyes, antibiotics, or bile salts, that inhibit the growth of one group of organisms while permitting the growth of another. For example, MacConkey agar is a selective medium that contains bile salts and the dye crystal violet. These two substances are inhibitory for gram-positive organisms as well as some gram-negative ones. A wide variety of gram-negative rods grow on this medium. Some selective media are also differential. MacConkey agar is an example of a selective and differential medium. The gram-negative rods that grow on this agar can be differentiated from one another on the basis of the organism's ability to ferment the carbohydrate lactose. Organisms that ferment lactose are called lactose positive, and organisms that are unable to ferment lactose are called lactose negative (Fig. 10). When selecting media for culturing clinical specimens from sites with a resident microflora, typically both enriched and selective media are used. If gram-negative rods are a component of this microflora, then a selective-differential medium might be used as well.

Certain organisms will not grow on media commonly used to culture clinical specimens, because the media may not be enriched enough or may contain inhibitory substances. When these organisms are sought, the laboratory must be notified so that special isolation medium can be used. Two important respiratory tract patho-gens, *B. pertussis* and *L. pneumophila*, are examples of organisms that do not grow on standard laboratory media and require special media for their isolation.

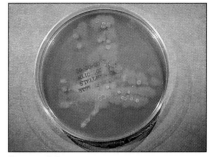

Figure 10

Organism identification and susceptibility testing

Once organisms are isolated, they may be identified, and in some cases susceptibility testing needs to be performed. Bacteria and fungi grow as colonies on agar plates. The appearance of these colonies is often useful in determining the identity of the organism. Colonies may appear flat or raised, smooth or rough, may pit the agar, or may hemolyze red blood cells in blood-containing agar. Molds, for example, have very characteristic "fuzzy" growth on agar. Colonies of organisms such as *S. aureus* may be pigmented or may secrete a diffusible pigment, as seen with *Pseudomonas aeruginosa*. Skilled microbiologists often have a very good idea of the identification of a microorganism based solely on its colonial appearance.

In specimens that come from an area of the body with a resident microflora, it is important to separate the colonies of organisms that may represent the resident microflora from the colonies of organisms that may be pathogens. Much of the time, this can be done on the basis of colonial appearance. However, some potential pathogens, such as *S. pneumoniae*, a common cause of bacterial pneumonia, cannot be readily differentiated from viridans group streptococci, a member of the resident oropharyngeal flora. In patients with suspected bacterial pneumonia, a sputum specimen may be obtained. Sputum consists of secretions coughed up from the lower airways that are expectorated through the oropharynx and submitted for culture. Because they pass through the oropharynx, sputum specimens almost always contain viridans group streptococci. The appearance of colonies produced by viridans group streptococci is very similar to that produced by *S. pneumoniae*. To determine whether or not these colonies are *S. pneumoniae*, one must do tests based on the phenotypic characteristics of the organism; these are referred to as **biochemical** tests. The bio-chemical test which is done most often to distinguish between these two organisms is the disc diffusion test, in which the organism's susceptibility to the compound

Figure 11   Left disk, optochin; right disk, oxacillin.

optochin is examined. *S. pneumoniae* (Fig. 11) is susceptible to optochin, while the viridans group streptococci are not. On the basis of this easily performed test, the identity of *S. pneumoniae* can be determined from a sputum specimen.

Bacteria are typically identified on the basis of colonial morphology, Gram stain reaction, the primary isolation media on which the organism is growing, and biochemical and serologic tests of various degrees of complexity. Figures 12 and 13 are flow charts that give fairly simple means of distinguishing commonly encountered human pathogens. Yeasts are identified in much the same way that bacteria are, while molds are generally identified on the basis of the arrangement of microscopic reproductive structures called conidia. It is important to accurately identify bacteria and fungi because certain organisms (e.g., *B. pertussis*) are the cause of certain clinical syndromes (in this case, whooping cough). Other bacteria (e.g., *Staphylococcus epidermidis*) may represent contamination with resident microflora in a clinical specimen (e.g., a wound culture). The accurate identification of a bacterium or fungus may help determine what role a particular microbe may be having in the patient's disease process.

Antimicrobial susceptibility typically is performed on rapidly growing bacteria if the organism is believed to play a role in the patient's illness and if the profile of antimicrobial agents to which the organism is susceptible is not predictable. Let's take three clinical scenarios to explain this concept.

A patient with a "strep throat" has group A streptococci recovered from his throat. Although the organism is clearly playing a role in the illness of this patient, antimicrobial susceptibility testing is not warranted. This organism is uniformly susceptible to first-line therapy—penicillin—and is susceptible more than 98% of the time to second-line therapy—the macrolide antibiotics such as erythromycin—although recent reports suggest that erythromycin resistance is becoming more frequent in this organism.

A patient presents with a leg abscess from which *S. aureus* is recovered. Susceptibility testing is indicated because some strains are resistant to the first-line drugs used to treat this infection—semisynthetic penicillins, including oxacillin and dicloxacillin—and the second-line drug, clindamycin. In this situation, the patient may be started on **empiric** antimicrobial therapy until the susceptibility of the organism is known. If the organism is resistant to the agent used for empiric therapy, then the patient should be treated with an alternative antimicrobial agent to which the organism is susceptible.

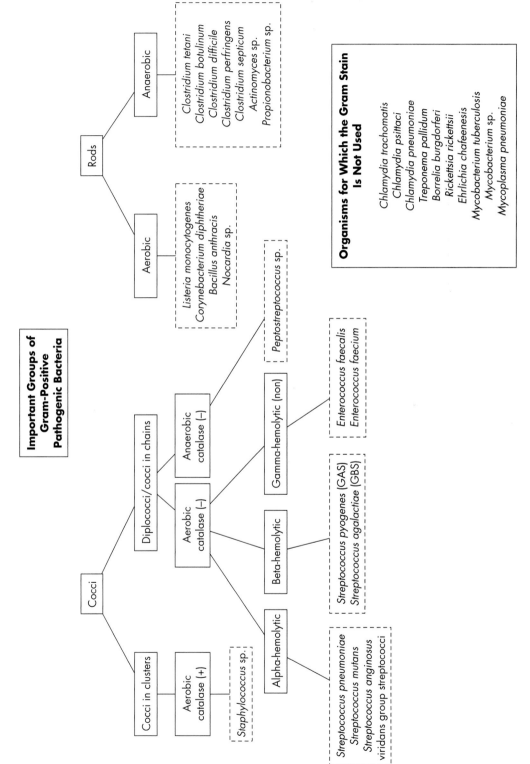

Figure 12

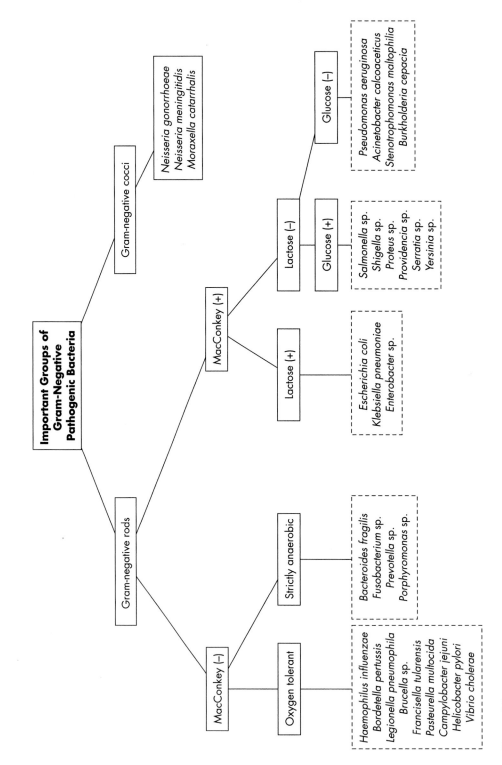

Figure 13

The third scenario is more subtle. A patient comes to the hospital with a high fever. He has two sets of blood cultures drawn in the emergency room. Two days later, *S. epidermidis* is recovered from one of these blood culture sets. As with *S. aureus*, this organism may show resistance to a variety of antimicrobial agents that are used to treat infected patients. However, no susceptibility testing is done by the laboratory, and this practice is acceptable to the clinician caring for the patient. Why? *S. epidermidis* is a component of skin microflora and may have contaminated the culture. If the laboratory had performed the susceptibility testing without considering that this isolate was a potential contaminant, they would be validating that the isolate was clinically significant. In this setting, the laboratory should only do susceptibility testing if instructed to by the caregiver, who is in a better position to know if this organism is clinically important.

There are several approaches to antibacterial susceptibility testing. All the approaches are highly standardized to ensure that the susceptibility results will be consistent from laboratory to laboratory. Screening of selected organisms for resistance to specific antimicrobial agents is one strategy that is frequently used, especially with the emergence of resistance in three organisms: *S. aureus* to oxacillin, *S. pneumoniae* to penicillin, and *Enterococcus faecium* and *Enterococcus faecalis* to vancomycin. Other strategies are to determine susceptibility to a preselected battery of antimicrobial agents using automated or manual systems that determine the MIC of antibiotics to the organism being tested or by using the disk diffusion susceptibility testing technique.

A novel approach to susceptibility testing is to perform MIC determinations using the E-test. The E-test is a plastic strip that contains a gradient of a specific antimicrobial agent. This strip is applied to a lawn of bacteria on an agar plate. Where the zone of inhibition intersects with the strip is the MIC value of that antibiotic for the organism tested. This test has many applications but is used most frequently for determining penicillin MIC values for *S. pneumoniae* isolates that show resistance to penicillin in the screening test previously described (Fig. 14).

Figure 14

Susceptibility testing is not routinely performed on fungal isolates. Because of their slow growth, special susceptibility testing techniques are used for the mycobacteria.

Tissue culture for Chlamydia and viruses

Both *Chlamydia*, a bacterium, and the viruses are obligate intracellular parasites. As such, they do not grow on artificial media, as fungi and other bacteria do. Rather, they can only grow by parasitizing living animal cells (including human cells) that are maintained by continuous tissue culture. Animals such as mice or chicken eggs can be inoculated in an attempt to isolate certain viruses, but this approach is rarely done. Tissue culture for *Chlamydia* may still be attempted, especially in situations where the detection of *C. trachomatis* is at issue in a legal proceeding, such as a case of sexual abuse of a child. However, molecular detection has become the standard method for diagnosis of *C. trachomatis* infection.

Tissue culture is still an important technique for the detection of viruses. Herpes simplex virus can be isolated from skin and genital tract lesions, often within the first 24 hours of incubation. Another herpesvirus, varicella-zoster virus, the etiologic agent of chicken pox and zoster, can also be isolated from skin lesions, but it typically requires 3 to 7 days to grow. The enteroviruses are the major etiologic agents of aseptic meningitis and can be isolated from cerebrospinal fluid.

A modification of the tissue culture technique is done to detect cytomegalovirus and several respiratory viruses in clinical specimens. In this method, the specimen is centrifuged onto tissue culture cells that are growing on a round glass coverslip inside a vial referred to as a shell vial. The cells are incubated for a brief period of time (24 to 72 hours) and then stained with fluorescent antibodies to detect the virus. This technique is much more rapid and sensitive than conventional tissue culture.

SEROLOGY

It is not always possible to isolate a microorganism by culture, visualize it microscopically, or detect it by antigenic or molecular detection techniques. In those situations, an alternative approach is to determine if the patient has mounted an immune response against a specific agent as evidence that he or she has been infected with that agent. The immune response is generally measured by detecting antibodies in the serum of patients. Thus the name serology.

Serology has both advantages and disadvantages. (i) Specimens for testing are readily available; (ii) antibodies are relatively stable molecules, so transport is not a major concern as it is with culture; and (iii) tests have been designed that can detect most known agents, such as HIV and HCV, which are difficult to detect by other means. Depending upon the target antigen against which the immune response is measured, the test can show both high sensitivity and high specificity. Compared with other techniques, these tests are relatively inexpensive and easy to perform, in part because they have been automated. As a result, they can be used to screen large numbers of specimens for selected infectious agents. For example, this approach is used to screen blood products used for transfusions to ensure that the transfused patient does not receive blood contaminated with hepatitis B and C viruses, HIV, or *T. pallidum*, the agent of syphilis.

Serologic tests also have several disadvantages and should be interpreted with some caution. To have a positive test, the patient must have mounted an immune response. Serum obtained from an acutely ill patient may have been taken during the window period in an infection before the patient had time to mount an immune response. Therefore, to get the most accurate result, **acute** and **convalescent** specimens should be obtained. The convalescent specimen should show a significant increase (or, in some cases, decrease) from the antibody level of an acute specimen. This is usually a fourfold change in the titer. Because the convalescent specimen should be obtained a minimum of 2 weeks after the acute specimen, serologic diagnosis is often retrospective. Because obtaining a convalescent specimen is often difficult logistically, the only value that may be available is that from the acute specimen. Patients may have relatively high antibody levels because of previous infection with the test organism and, as a result, may have a false-positive result. Antigenic cross-reactions between the test organism and other antigens may also lead to false-positive results. Some immunocompromised patients are unable to mount a response and may never have a positive serologic test.

Serologic tests can be done in combination using a **screening** test followed by a **confirmatory** test. This approach is used most commonly in the diagnosis of syphilis, HIV infection, HCV infection, and Lyme disease. The screening test should be highly sensitive so that all true-positive results will be detected. This test may not be highly specific, meaning that some results may be false positives. It should also be easily performed so that large numbers of specimens can be tested fairly inexpensively. The confirmatory test needs to be highly specific so that the correct diagnosis can be applied to the patient who screens positive for the infectious agent. It tends to be much more expensive and technically complex than the screening test. Western blotting or an equivalent technique is used in the confirmatory tests for Lyme disease, HIV infection, and HCV infection. In this technique, a patient is considered to be positive for the agent only if the patient has antibodies to multiple specific antigens.

ONE
Genitourinary Tract Infections

INTRODUCTION TO SECTION I

We begin this text with a discussion of infections of the genitourinary tract for two reasons. First, the number of microorganisms that frequently cause infection in these organs is somewhat limited. Second, **urinary tract infections (UTIs)** and **sexually transmitted diseases (STDs)** are two of the most common reasons why young adults, particularly women, consult a physician. UTIs are examples of **endogenous** infections, i.e., infections that arise from the patient's own microflora. In the case of UTIs, the microbes generally originate in the gastrointestinal tract and colonize the periurethral region before ascending the urethra to the bladder. STDs are **exogenous** infections; i.e., the infectious agent is acquired from a source outside the body. In the case of STDs, these agents are acquired by sexual contact.

UTIs are much more common in women than in men for a number of reasons. The urethra is shorter in women than in men, and straight rather than curved as in men, making it easier for microbes to ascend to the bladder. Prostatic secretions have antibacterial properties, which further protects the male. The periurethral epithelium in women, especially women with recurrent UTIs, is more frequently colonized with microorganisms that cause UTIs. It should also be noted that the incidence of UTIs is higher in sexually active women, as coitus can introduce organisms colonizing the periurethral region into the urethra. The incidence of nosocomial UTIs, however, is similar in women and men. In these infections, catheterization is the major predisposing factor.

The incidence of STDs is similar in both heterosexual men and women; however, the morbidity associated with these infections tends to be much greater in women. In particular, irreversible damage to reproductive organs, caused by both *Chlamydia trachomatis* and *Neisseria gonorrhoeae*, is all too common. Infections with these two organisms are almost always symptomatic in males, though the few men who do not have symptoms can be responsible for infecting many partners. By contrast, a significant number of women may be infected asymptomatically at first. They may manifest signs and symptoms of infection only when they develop **pelvic inflammatory disease,** which can result in sterility. Fetal loss or severe perinatal infection may be caused by two other STD agents, herpes simplex virus and *Treponema pallidum*, the etiologic agent of syphilis.

Important agents of genitourinary tract infections are listed in Table 1. Only organisms in this table should be considered in your differential diagnosis for the cases in this book. You should note that not all organisms that can be spread sexually, such as hepatitis B virus and *Entamoeba histolytica*, are listed. This is because these infections do not have genitourinary tract manifestations.

TABLE 1 SELECTED GENITOURINARY TRACT PATHOGENS

ORGANISM	GENERAL CHARACTERISTICS	SOURCE OF INFECTION	DISEASE MANIFESTATION
Bacteria			
Actinomyces spp.	Anaerobic, gram-positive bacilli	Endogenous	PID[a] associated with intrauterine device usage
Bacteroides fragilis	Anaerobic, gram-negative bacillus	Endogenous	Pelvic abscess
Chlamydia trachomatis	Obligate intracellular pathogen (does not Gram stain)	Direct sexual contact	Urethritis, cervicitis, PID
Enterobacter spp.	Lactose-fermenting gram-negative bacilli	Endogenous	Community or nosocomial UTI[b]
Enterococcus spp.	Catalase-negative, gram-positive cocci	Endogenous	Nosocomial UTI
Escherichia coli	Lactose-fermenting gram-negative bacillus	Endogenous	Community or nosocomial UTI
Haemophilus ducreyi	Fastidious pleiomorphic gram-negative bacillus	Direct sexual contact	Chancroid (painful genital ulcer)
Klebsiella pneumoniae	Lactose-fermenting gram-negative bacillus	Endogenous	Community or nosocomial UTI
Morganella morganii	Lactose-nonfermenting gram-negative bacillus	Endogenous	Community or nosocomial UTI
Neisseria gonorrhoeae	Gram-negative intracellular diplococcus	Direct sexual contact	Urethritis, cervicitis, PID
Proteus mirabilis	Lactose-nonfermenting, swarming gram-negative bacillus	Endogenous	Community or nosocomial UTI
Pseudomonas aeruginosa	Lactose-nonfermenting gram-negative bacillus	Catheterization	Nosocomial UTI
Staphylococcus saprophyticus	Catalase-positive, gram-positive coccus	Endogenous	Community-acquired UTI
Treponema pallidum	Spirochete (does not Gram stain)	Direct sexual contact; vertical, mother to child	Chancre (painless genital ulcer); primary, secondary, tertiary syphilis; neonatal syphilis

= STD

(continued next page)

TABLE 1 SELECTED GENITOURINARY TRACT PATHOGENS *(continued)*

ORGANISM	GENERAL CHARACTERISTICS	SOURCE OF INFECTION	DISEASE MANIFESTATION
Fungi			
Candida spp.	Yeast with pseudohyphae	Endogenous	Vaginitis, nosocomial UTI, balanitis
Parasites			
Phthirus pubis	Crab lice	Direct sexual contact	Pubic hair infestation
Trichomonas vaginalis	Protozoan	Direct sexual contact	Vaginitis
Viruses			
Herpes simplex viruses (HSV-1 and -2)	Enveloped DNA virus	Direct sexual contact; vertical, mother to child	Recurrent genital ulcers, fetal/neonatal infections, encephalitis
Human immunodeficiency viruses (HIV-1 and -2)	Retrovirus	Direct sexual contact; blood and body fluids; vertical, mother to child	AIDS, neonatal infection, dementia
Human papillomavirus	Nonenveloped DNA virus	Direct sexual contact	Genital warts, cervical carcinoma

[a]PID, pelvic inflammatory disease.

[b]UTI, urinary tract infection.

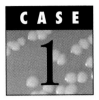

CASE 1 The patient was a 19-year-old female with a history of a urinary tract infection (UTI) 4 months prior to admission for which she was treated with oral ampicillin without complications. Five days prior to this admission she began to note nausea without vomiting. One day later she developed left flank pain, fevers, and chills and noted increased urinary frequency. She noted foul-smelling urine on the day prior to admission. She presented with a temperature of 38.8°C, and physical examination showed left costovertebral angle tenderness. Urinalysis of a clean-catch urine sample was notable for >50 white blood cells per high-power field, 3 to 10 red blood cells per high-power field, and 3+ bacteria. Urine culture was subsequently positive for >100,000 CFU of an organism per ml (seen growing on culture in Fig. 1 [sheep blood agar] and Fig. 2 [MacConkey agar]). Note that the organism is beta-hemolytic.

1. What do the urinalysis findings indicate? Explain your answer.

2. Why were the numbers of organisms in her urine quantitated on culture? How would you interpret the culture results in this case?

3. Which gram-negative rods are lactose fermenters? Which one is also beta-hemolytic?

4. This bacterium was resistant to ampicillin. What in this patient's history might explain this observation?

5. UTIs are more frequent in women than men. Why?

6. Did this woman have cystitis or pyelonephritis? Why is it important to differentiate the two?

7. Briefly explain the evolution of the organism causing this infection in terms of its ability to infect the urinary tract. What virulence factors have been shown to play a pathogenic role in this infection?

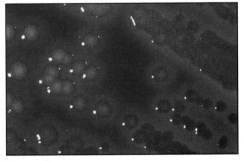

Figure 1

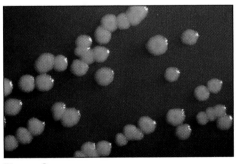

Figure 2

CASE DISCUSSION

1. Urine from normal individuals usually has fewer than 10 white blood cells per high-power field. Pyuria (the presence of >10 white blood cells per high-power field in urine) and hematuria (the presence of red blood cells in urine), as seen in this patient, are reasonably sensitive but not always specific indicators of UTI. The presence of bacteriuria (bacteria in urine) in this patient further supports this diagnosis. However, the presence of bacteriuria on urinalysis should always be interpreted with caution. Clean-catch urine, which is obtained by having the patient cleanse her external genitalia, begin a flow of urine, and then "catch" the flow of urine in "midstream," is rarely sterile because the distal urethra is colonized with bacteria. Urine is an excellent growth medium. Therefore, if urine is not analyzed fairly quickly (within 1 hour), the organism colonizing the urethra can divide (two to three generations per hour) if the urine specimen is left at room temperature. Organisms colonizing the urethra may be present in sufficient numbers to be visualized during urinalysis even though the patient is not infected.

2. In a normal individual, urine within the bladder is sterile. As it passes through the urethra, which has a resident microflora, it almost always becomes contaminated with a small number (<10^3 CFU/ml) of organisms. As a result of urethral contamination, essentially all clean-catch urine samples will contain a small number of organisms, so culturing urine nonquantitatively will not allow differentiation between colonization of the urethra and infection of the bladder. It should be noted that only a small number of clinical specimens other than urine are cultured quantitatively.

Patients in whom the bladder is infected tend to have very large numbers of bacteria in their urine. These organisms usually, but not always, are of a single species. Studies have shown that most individuals with true UTIs have greater than 10^5 CFU/ml in clean-catch urine specimens. There are exceptions to this generalization. In a woman with symptoms consistent with UTIs, bacterial counts as low as 10^2 CFU/ml of a uropathogen—e.g., *Escherichia coli*, *Klebsiella pneumoniae*, *Enterobacter* spp., *Proteus* spp., *Staphylococcus saprophyticus*—may indicate that she has a UTI. Colony counts of 10^2 CFU/ml of a uropathogen are highly sensitive for diagnosing UTIs but are of low specificity; colony counts of >10^5 CFU/ml are highly specific as well as being reasonably sensitive.

3. The lactose fermenters that are most commonly isolated from urine are the KEE organisms (*Klebsiella* spp., *E. coli*, and *Enterobacter* spp.). *E. coli* is recovered from approximately 80 to 85% of outpatients and 40 to 50% of inpatients with UTI. The observation that the organism is beta-hemolytic indicates that, in all likelihood, the organism is *E. coli*. Approximately 55% of *E. coli* isolates recovered from urine of patients with pyelonephritis are beta-hemolytic, whereas *K. pneumoniae* and *Enterobacter* spp. are rarely, if ever, beta-hemolytic. Another common gram-negative rod that is frequently beta-hemolytic is *Pseudomonas aeruginosa*. This organism is incapable of fermenting carbohydrates and should not be confused with lactose-fermenting isolates of *E. coli*. A spot indole test was done on the patient's isolate and it was positive, confirming the identity of this organism as *E. coli*.

4. The patient had a previous UTI, at which time she received oral ampicillin. One of the deleterious effects associated with the use of antimicrobial agents is the development of resistance. This occurs with some degree of frequency to gut flora, where plasmids coding for resistance may be mobilized in response to antimicrobial pressure, leading to the transfer of resistance to previously susceptible organisms, such as in this *E. coli* isolate. Not only may resistance to the agent supplying the pressure result, but also the plasmid may contain genes which code for resistance to other antimicrobial agents, the end result being a multidrug-resistant organism. Since the gut has been shown to be an important reservoir of organisms causing UTI, resistance to ampicillin in the face of prior ampicillin therapy is not surprising.

In this era of managed care, many patients with symptoms of UTI are no longer having urine cultures performed but rather are receiving empiric antimicrobial therapy. As a result, knowledge of resistance trends in *E. coli* is important to ensure proper selection of antimicrobial agents for empiric therapy. One study showed that resistance to ampicillin was found in approximately one-third of *E. coli* isolates recovered from the urine of women with cystitis, making this drug a relatively poor choice for empiric therapy. Trimethoprim-sulfamethoxazole is another antimicrobial agent that is used empirically for the treatment of UTI. Drug resistance in UTI isolates of *E. coli* is increasing, making this drug less useful as an empiric agent. Fluoroquinolone antimicrobial agents for now seem to be the empiric agent of choice for treating uncomplicated *E. coli* UTI. However, widespread use of these agents is sure to result in increased resistance among *E. coli* isolates recovered from patients with cystitis.

5. In adults, 90% of uncomplicated UTIs occur in women. It is one of the most common reasons why adolescent and adult women seek health care, resulting in approximately 10 million physician visits annually in the United States. The simplistic view of why women have more UTIs than men is that the shorter urethra in women results in a greater likelihood that organisms will ascend the urethra and enter the bladder. Sexual activity is thought to play a significant role in the introduction of uropathogens into the urethra. In addition, the use of diaphragms and spermicides has been shown to predispose women to UTIs. However, other factors which may play a role in this gender difference have been identified. It has been observed that prostatic fluid inhibits the growth of common urinary tract pathogens in urine, providing a unique defense mechanism for men. It has also been observed that specific uropathogens bind to vaginal and periurethral epithelial cells. Binding in the periurethral region by these organisms is often seen in women prior to the development of UTI, as well as in women who have recurrent UTIs. Binding of uropathogens to the periurethral epithelium is highest when estrogen levels reach their peak during the menstrual cycle. These observations may further explain why a preponderance of UTIs are seen in women.

6. The clinical presentation in this patient is consistent with acute pyelonephritis. Pyelonephritis is an infection of the kidney, whereas cystitis is an infection of the bladder. The findings of fever, chills, and left flank pain, with corresponding costovertebral angle tenderness, are all consistent with pyelonephritis. If white blood cell casts were seen in the patient's urinalysis, that finding would support the diagnosis of pyelonephritis. Culture results would not be useful in differentiating between the two types of infections. Radiographic or cystoscopic studies would be necessary to make a definitive diagnosis of pyelonephritis, but clinical judgment is usually sufficient. The reason why it is important to distinguish between pyelonephritis and cystitis is that antimicrobial treatment strategies differ. Cystitis therapy is usually brief, typically 3 days, while pyelonephritis therapy may be more prolonged, typically lasting 7 days to 2 weeks. The outcome of antimicrobial therapy is dependent upon the susceptibility of the *E. coli* strain. If patients are treated empirically with an antimicrobial agent to which their isolate is resistant, their outcome will be less favorable than in those who receive an antimicrobial agent to which their isolate is susceptible.

7. "Pathogenicity islands" are an exciting new concept for understanding the evolution of human microbial pathogens. They are relatively large pieces of DNA that encode virulence factors that have been inserted by recombination into chromosomal regions that appear to more readily allow "foreign" DNA. What that means practically is that organisms such as *E. coli* can quickly evolve from harmless gastrointestinal tract commensals to agents capable of causing UTI by incorporating DNA which encodes for virulence factors. Acquisition of virulence factors by gene transfer is a common theme in *E. coli* pathogenicity, not only in strains causing UTI but also in strains that cause diarrheal disease. Two virulence factors known to be important in the pathogenesis of *E. coli* pyelonephritis, P fimbrae and hemolysin, have been found on pathogenicity islands. Pathogenicity islands are found much more frequently in *E. coli* strains that cause cystitis and pyelonephritis than in fecal isolates.

The fimbriae are the major means of adhesion of uropathogenic *E. coli*, allowing them to bind to the various types of epithelial cells that line the urinary tract. Two different fimbriae found on the surface of uropathogenic *E. coli*, types P and 1, have been well studied. The P fimbriae are so designated because they agglutinate red blood cells possessing the P blood group antigen. They bind to uroepithelial cells and are resistant to phagocytosis. Over 80% of *E. coli* isolates causing pyelonephritis have pathogenicity islands that encode for these fimbriae. Type 1 fimbriae are distinct from the P fimbriae. Both agglutination of red blood cells and binding to uroepithelial cells by *E. coli* possessing type 1 fimbriae can be blocked by preincubating the organism with mannose, while binding of type P-fimbriated *E. coli* is not blocked by mannose. Type 1-fimbriated *E. coli* are thus said to be "mannose sensitive," while type P strains are said to be mannose insensitive. Type 1 fimbriae are found more frequently in patients with cystitis and less frequently in patients with pyelonephritis. Our patient likely had a P-fimbriated *E. coli* strain because she had pyelonephritis.

Another important virulence factor of uropathogenic *E. coli* is hemolysin. Hemolysin production is detected in approximately 55% of *E. coli* recovered from patients with pyelonephritis. Studies with renal tubular cells in primary culture have shown them to be quite sensitive to the cytotoxic activity of this virulence factor.

Aerobactin is a third virulence factor, found in approximately 75% of *E. coli* strains causing pyelonephritis. Aerobactin is a siderophore. Siderophores are molecules produced by bacteria and scavenge iron, an essential nutrient for bacteria, from the host. Strains of *E. coli* that produce aerobactins have been shown to grow faster in urine than nonproducing strains, although how important this is in the pathogenesis of urinary tract infection is unclear.

REFERENCES

1. **Finlay, B. R., and S. Falkow.** 1997. Common themes in microbial pathogenicity revisited. *Microbiol. Mol. Biol. Rev.* **61:**136–169.

2. **Gupta, K., D. Scholes, and W. E. Stamm.** 1999. Increasing prevalence of antimicrobial resistance among uropathogens causing acute uncomplicated cystitis in women. *JAMA* **281:**736–738.

3. **Guyer, D. M., J.-S. Kao, and H. L. T. Mobley.** 1998. Genomic analysis of a pathogenicity island in uropathogenic *Escherichia coli* CFT073: distribution of homologous sequences among isolates from patients with pyelonephritis, cystitis, and catheter-associated bacteriuria and from fecal samples. *Infect. Immun.* **66:**4411–4417.

4. **Talan, D. A., W. E. Stamm, T. H. Hooton, G. J. Moran, T. Burke, A. Iravani, J. Reuning-Scherer, and D. A. Church.** 2000. Comparison of ciprofloxacin (7 days) and trimethoprim/sulfamethoxazole (14 days) for acute uncomplicated pyelonephritis in women. *JAMA* **283:**1583–1590.

5. **Warren, J. W.** 1995. Clinical presentation and epidemiology of urinary tract infections, p. 3–27. *In* H. L. T. Mobley and J. W. Warren (ed.), *Urinary Tract Infections: Molecular Pathogenesis and Clinical Management.* ASM Press, Washington, D.C.

CASE 2

The patient was a 15-year-old male who was brought to the emergency room by his sister. He gave a 24-hour history of dysuria and noted some "pus-like" drainage in his underwear and on the tip of his penis. Urine appeared clear, and urine culture was negative although urinalysis was positive for leukocyte esterase and multiple white cells were seen on microscopic examination of urine. He gave a history of being sexually active with five or six partners in the past 6 months. He claimed that he and his partners had not had any sexually transmitted diseases. His physical exam was significant for a yellow urethral discharge and tenderness at the tip of the penis. (A Gram stain done in the emergency room is shown in Fig. 1.) He was given antimicrobial agents and scheduled for a follow-up visit 1 week later. He did not return.

1. Based on the Gram stain results, with what organism is this patient infected? What is the reliability of the Gram stain for establishing the diagnosis in this patient? How reliable is the Gram stain for detection of this organism in cervical specimens from infected women? What other direct detection technique is now available for laboratory diagnosis of the organism causing this patient's infection?

2. Are his urinalysis and urine culture findings consistent with his illness? Explain.

3. Why did his partners have a negative history for sexually transmitted diseases? For what complications are his sexual partners (whom he may have infected and/or who infected him) at increased risk?

4. What virulence factor(s) made by this organism is responsible for his symptoms?

5. Given his history, for what organisms is he at increased risk? Why do you think this patient was asked to return for a follow-up visit?

6. What antimicrobial agent(s) was he given in the emergency room? How has antimicrobial therapy for this infection evolved over the past 15 years and why was that evolution necessary?

7. Why is there no reliable vaccine against the organism causing this individual's infection?

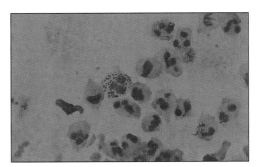

Figure 1

CASE DISCUSSION

1. The organism seen on Gram stain is a gram-negative, intracellular diplo-coccus consistent with *Neisseria gonorrhoeae*. In males with symptomatic ure-thritis, a Gram stain of a urethral discharge is a highly reliable test for diagnosis of *N. gonorrhoeae* urethral infection. The Gram stain will be positive for gram-negative, intracellular diplococci in approximately 95 to 100% of infected male patients. Gram stains of cervical specimens are positive in only 50 to 60% of females, making direct Gram stain a fairly unreliable tool for women suspected of having a gonococcal infec-tion. A number of molecular amplification techniques, including the polymerase chain reaction (PCR), the ligase chain reaction (LCR), and transcription-mediated amplifi-cation (TMA), are now commercially available. In males, these assays can be done on either urine or urethral swabs. In females, the assays can be performed on cervical swabs or urine. Less is known about the value of these methods in throat or rectal spec-imens. These methods have a sensitivity approaching that of culture, but false-positive results have been reported in some of these assays for closely related but saprophytic *Neisseria* spp. As clinical laboratories become more centralized in the era of managed care, these molecular methods are replacing *N. gonorrhoeae* culture. The reason for this changing diagnostic approach is that maintaining the viability of this fastidious organ-ism for culture is difficult when specimens have to travel significant distances to a cen-tral laboratory. Bacterial nucleic acid, on the other hand, is comparatively stable, making transport of these specimens for molecular amplification much easier and the detection of gonococci theoretically more accurate. Given the potential implications of a false-positive assay due to the presence of saprophytic *Neisseria* species, it is impor-tant for health care providers to understand the issues surrounding the specificity of the particular amplification assay that is being used in the diagnostic laboratory.

2. In patients with gonococcal urethritis, white blood cells wash from the urethra dur-ing urination. The white blood cells can be detected in urine by dipstick testing for leukocyte esterase (an enzyme produced by leukocytes) or by microscopic examination. *N. gonorrhoeae* is generally not recovered on urine culture because of the media and incubation conditions used (usually sheep blood agar and media selective for enteric gram-negative rods, with incubation times usually <48 hours and incubation under ambient air). *N. gonorrhoeae* requires an enriched medium such as chocolate agar and incubation times of at least 36 to 48 hours in 5% CO_2 for growth to be detected visi-bly. Therefore, a negative urine culture is consistent with the patient's disease.

3. Obtaining an accurate sexual history, especially from adolescents, may be difficult. The individual may not recognize signs and symptoms of sexually transmitted diseases or may be too embarrassed or ashamed to admit to them. However, given an incubation time of approximately 2 to 5 days for *N. gonorrhoeae* and an acute symptomatic history of 24 hours, it is most likely that this patient was recently infected. If the patient was "serially monogamous" (that is, sexually active exclusively with only one partner for varying lengths of time), it is likely that he was infected by one of his recent partners and that his previous partners had not been infected. A significant percentage of women may be infected asymptomatically, and it is possible that the sexual partner who infected him was asymptomatic.

Complications of *N. gonorrhoeae* infection are more common in women because of increased rates of asymptomatic infections. These complications tend to be severe. The major complication seen in women infected with *N. gonorrhoeae* is pelvic inflammatory disease (PID). PID can cause fallopian tube scarring and obstruction, which may result in infertility. Ectopic pregnancy is also more common in women with a history of PID. Though it is uncommon, both men and women can have disseminated gonococcal infection, which can present with a rash and septic arthritis.

4. *N. gonorrhoeae* induces an intense inflammatory response, which is manifested clinically in males as exudate from the urethra. Two virulence factors are important in this process: pili and lipo-oligosaccharide (LOS). Pili mediate attachment and stimulate nonspecific phagocytosis by epithelial cells in the urethra. LOS (endotoxin) can stimulate an inflammatory reaction to these phagocytized organisms.

5. This individual is at increased risk for a number of sexually transmitted diseases. Coinfections with *Chlamydia trachomatis* are common. Less frequent but still problematic would be syphilis (*Treponema pallidum*), herpes simplex virus, human papillomavirus, and HIV. Because of his history of multiple sexual partners and the diagnosis of a sexually transmitted disease, this individual is at increased risk for becoming infected with HIV. Sexually active teenagers are one of the populations in which HIV is most rapidly spreading in the United States. Emergency rooms are often hectic, with physicians needing to see many patients as rapidly as possible. This physician did not feel he could adequately counsel and get consent for HIV serologic testing in such an environment. The physician asked the patient to return to the clinic so appropriate counseling and HIV testing could be done. This patient chose not to return.

6. The current Centers for Disease Control and Prevention (CDC) guidelines for treating uncomplicated gonococcal urethritis are to administer a single dose of either an oral cephalosporin (cefixime) or an oral quinolone antimicrobial agent (ofloxacin or ciprofloxacin) to treat gonococcal infections, plus doxycycline or azithromycin to treat a presumed coinfection with *C. trachomatis*. Alternatively, an intramuscular injection of ceftriaxone is commonly given to treat gonococcal infections. Many centers, especially in areas of high HIV incidence, have abandoned intramuscular administration of antimicrobial agents for treatment of gonococcal disease in favor of oral therapy. The reason is concern among health professionals over needlestick injuries after injection of patients who are at high risk for HIV infection. Why not treat both the gonococcal and *C. trachomatis* infections with doxycycline? There are two reasons. First, CDC surveillance data in 1997 showed that 26% of gonococcal isolates were resistant to doxycycline. Second, compliance when antimicrobial agents must be taken twice daily for 7 days is often poor.

In addition to resistance to the tetracyclines, gonococcal resistance to penicillin therapy has become so widespread in the past 15 years that penicillin is no longer a reasonable therapeutic option for treating infections with this organism. Initially, penicillin resistance was due to a plasmid-encoded β-lactamase; β-lactamase is an enzyme that degrades the beta-lactam ring in penicillin, inactivating the drug. Subsequently, isolates were recovered that had chromosomal mutations that encoded modification in penicillin-binding proteins, making the binding of penicillin to the gonococci much less efficient. This decreased binding resulted in resistance to penicillin. Finally, surveillance studies are showing an increased number of gonococci resistant to quinolones. This is not surprising, since single mutations resulting in quinolone resistance have been reported in other organisms. How long it will be before there is widespread quinolone resistance depends on a variety of factors and is difficult to predict. Until quinolone resistance is >1.0%, the quinolones remain a reasonable therapeutic choice for gonococcal infection.

One of the issues to consider is will we know when we reach this threshold of quinolone resistance in the gonococci? The current molecular methods that are becoming increasingly used for diagnosis of gonococcal infections do not determine the antimicrobial resistance pattern of these organisms. Therefore, the CDC surveillance studies of gonococcal resistance are critical to the recognition of when that threshold is crossed.

7. The most successful bacterial vaccines elicit an immune response against either toxins produced by the organism (tetanus and diphtheria) or surface components of the bacteria (*Haemophilus influenzae* type b capsular polysaccharide or filamentous hemagglutinin in the acellular pertussis vaccine). Since the gonococcus does not produce a conventional exotoxin, the obvious target would be a surface component. Unfortunately, surface components of gonococci such as pili can undergo rapid antigenic variation because of frequent rearrangement of the pilin genes, making it impossible to produce a reliably protective vaccine antigen. Conserved and phenotypically stable determinants on the surface of the gonococcus have not yet been used in vaccine development. Whether they will be efficacious in providing mucosal immunity is beyond the scope of this discussion.

REFERENCES

1. **Centers for Disease Control and Prevention.** 2002. Sexually transmitted disease treatment guidelines. *Morb. Mortal. Wkly. Rep.* **51**(RR-6):1–80.

2. **Farrell, D.** 1999. Evaluation of AMPLICOR *Neisseria gonorrhoeae* PCR using cppB nested PCR and 16S rRNA PCR. *J. Clin. Microbiol.* **37**:386–390.

3. **Finlay, B. R., and S. Falkow.** 1997. Common themes in microbial pathogenicity revisited. *Microbiol. Mol. Biol. Rev.* **61**:136–169.

4. **Knapp, J. S., K. K. Fox, D. L. Trees, and W. L. Whittington.** 1997. Fluoroquinolone resistance in *Neisseria gonorrhoeae. Emerg. Infect. Dis.* **24**:142–148.

C A S E 3 This 16-year-old female presented to the emergency room of an urban medical center with complaints of crampy abdominal pain for days and vaginal bleeding. She denied symptoms of urinary tract infection and abnormal vaginal discharge and had not noted any chills or fever. She had no nausea or vomiting. The pain increased in the 24 hours prior to presentation, and at the time of examination she also noted pain in the right upper quadrant. She was sexually active with one male partner in the preceding 3 months, and claimed to use condoms as a method of birth control. She is the mother of one child.

On examination, her temperature was 38.3°C, and there was exquisite tenderness in the right upper quadrant as well as the left lower quadrant. No rebound tenderness or guarding was noted. On pelvic exam, cervical motion tenderness was present, as well as right and left adnexal tenderness. No masses were palpated.

1. Clinically, this patient was believed to have pelvic inflammatory disease (PID) and was admitted to the hospital for antibiotic treatment. What bacteria have been associated with PID?

2. An endocervical swab was obtained from the patient and used to inoculate McCoy cells. After 48 hours of incubation, the McCoy cells were stained with an immunofluorescent reagent that demonstrated the presence of inclusions (Fig. 1). This finding established the presence of what infectious agent? Briefly describe the life cycle of this organism in McCoy cells.

3. In addition to PID, in which this organism may well have a significant role, in what other clinical situations might this organism be expected to be recovered?

4. Tissue culture is one of several methods used to detect this organism in clinical specimens. What other methodologies currently exist? How do they compare with tissue culture with respect to sensitivity?

5. How effective are beta-lactam antibiotics in treating infections caused by this organism? What is the rationale for using a beta-lactam in addition to doxycycline in this patient's therapy? What else should be done epidemiologically in cases of PID?

6. What type of screening strategy has been used successfully to prevent PID? What populations have a high prevalence of chlamydial infection? What are the potential consequences of PID?

Not effective. — chlamydia = β-lactam resistant.

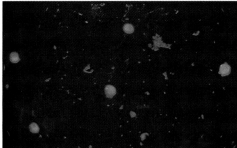

Figure 1

CASE DISCUSSION

1. PID has been associated with the sexually transmitted bacterial agents *Neisseria gonorrhoeae* and *Chlamydia trachomatis*. In addition, normal vaginal flora, including anaerobes and facultative aerobes, may be isolated from patients with PID who either have no documented gonococcal or chlamydial infection or have an infection documented with one of these pathogens. Knowledge about the role of *Mycoplasma* and *Ureaplasma* species in the pathogenesis of PID is evolving.

2. McCoy cells are used to culture *C. trachomatis*, an obligate intracellular pathogen. Because it is an intracellular pathogen, it cannot be cultured on enriched agar media, as can, for example, *N. gonorrhoeae*. After the infectious elementary body infects the McCoy cells, the organism is taken into the cell by a process called receptor-mediated endocytosis. The bacterium develops into a reticulate body within a membrane-bound structure called an inclusion. Reticulate bodies, the reproductive form of the organism, multiply by binary fission. The reticulate bodies then condense to form elementary bodies. Elementary bodies are released from the cell by lysis, release of intact inclusions, or exocytosis. The presence of chlamydial inclusions is demonstrated by staining these cells with a fluorescein-tagged monoclonal antibody that binds specifically to the chlamydial antigens present within the infected McCoy cells. These can then be viewed with a fluorescent microscope, where they will give a characteristic apple-green fluorescence, and the etiologic diagnosis can be established.

3. *C. trachomatis*, the most common sexually transmitted bacterial pathogen in the United States, is also an etiologic agent of both nongonococcal urethritis and epididymitis in males and cervicitis, endometritis, and salpingitis in women, and it can cause pneumonia and conjunctival disease in neonates if they have passed through an infected birth canal. It is worth noting that many patients are minimally symptomatic or asymptomatic with genital infection due to *C. trachomatis* and may not seek medical attention. *C. trachomatis*, as one of the causes of PID, is associated with infertility in women. Other serotypes of *C. trachomatis*, found rarely in the United States, cause lymphogranuloma venereum. Lymphogranuloma venereum is a genital tract infection characterized by enlarged, tender, and erythematous inguinal lymph nodes and is frequently accompanied by systemic symptoms of fever, headache, and malaise. Still other serotypes of *C. trachomatis* cause trachoma, a leading cause of blindness in the developing world.

4. Tissue culture, as described in the answer to question 2, has been regarded as the "gold standard" for the diagnosis of *C. trachomatis* infection. A number of other methodologies are approved by the U.S. Food and Drug Administration (FDA) and have been commercially available for many years. These include enzyme immunoassay, direct fluorescent antibody staining of the specimen with monoclonal antibodies (as seen in Fig. 1), and the use of a genetic probe that hybridizes specifically with nucleic acid of *C. trachomatis* in the specimen. Unfortunately, the sensitivity of each of these assays is less than that of well-performed tissue culture.

Tissue culture, however, is a labor-intensive technique. In addition, the sensitivity of tissue culture for *C. trachomatis* may be affected by a number of factors that may vary from laboratory to laboratory, including the type of swab used (some swabs are toxic), the type of transport medium used, whether the specimen undergoes a freeze/thaw cycle prior to tissue culture (which may decrease by 1 log the number of viable organisms), and several other factors. As a result, many laboratories have used the methodologies noted above rather than tissue culture. Over the past several years, however, with the FDA approval of PCR (the polymerase chain reaction), ligase chain reaction, and transcription-mediated amplification for the diagnosis of *C. trachomatis*, many laboratories now use amplification technology. These methodologies compare favorably in sensitivity with tissue culture and are called by some a "platinum standard" to emphasize their superiority in comparison to the "gold standard" of tissue culture. These amplification assays also can be used for urine specimens so that the patient's discomfort from obtaining a urethral or cervical swab can be avoided. Disadvantages of these assays, as of the time of this writing, include their higher cost, the potential for contamination resulting in a positive result in a patient without an infection, the possible nonspecific inhibition of the assays by blood or other components of cervical secretions and by compounds present in urine, and the lack of a confirmatory assay. An advantage of these assays besides their speed, sensitivity, and ease of use is that they are now also able to establish the diagnosis of gonococcal infection using the same methodology from a single specimen.

Tissue culture for the diagnosis of *C. trachomatis* has now been largely replaced by these assays, though for medicolegal cases, such as sexual abuse and rape, tissue culture remains the procedure of choice.

5. Empiric therapy for sexually active young women and others at risk for sexually transmitted diseases in whom PID is suspected includes, most commonly, a beta-lactam antimicrobial agent to treat *N. gonorrhoeae* and anaerobes, plus doxycycline to treat *C. trachomatis*. The combination is necessary because of the poor activity of beta-lactams against *C. trachomatis*. Beta-lactams characteristically have poor intracellular penetration. The intracellular location of the replicative phase of *C. trachomatis* (the reticulate bodies) protects it from the activity of beta-lactam antibiotics. Tetracyclines, including doxycycline, are the therapy of choice for *C. trachomatis* infection. Other combinations of antibiotics have been used with success in the treatment of PID, including intravenous clindamycin and gentamicin, intravenous ofloxacin and metronidazole, and others. Oral treatment of PID can be used in those patients who are able to be managed as outpatients.

Ofloxacin, a fluoroquinolone, is active against *C. trachomatis* and is often given in uncomplicated gonorrhea since treatment with ofloxacin is clinically effective against both *C. trachomatis* and *N. gonorrhoeae*, which can coinfect the same patient. It is worth noting, however, that resistance to fluoroquinolones has been reported in isolates of *N. gonorrhoeae*.

In addition, it is important for sex partners of patients who have PID to be evaluated because of the high risk of infection with *C. trachomatis* and *N. gonorrhoeae* even if these pathogens have not been isolated from the affected woman.

6. The use of criteria to identify women among a low-prevalence population who are at increased risk for chlamydial infections, to test these women for cervical chlamydial infections, and to treat those who are found to be infected has significantly reduced the incidence of PID in a low-prevalence population. Adolescent inner-city females are a very high-prevalence population for *C. trachomatis*, and, on the basis of a prospective longitudinal study, the screening of all sexually active adolescent females every 6 months has been advocated. Similarly, the high prevalence of both chlamydial and gonococcal infection in women entering jails and adolescents entering juvenile detention centers suggests that screening of these women may be worthwhile.

Untreated lower genital tract infections in women may lead not only to PID but to complications of PID, including infertility, ectopic pregnancy, and chronic pelvic pain.

REFERENCES

1. **Burstein, G. R., C. A. Gaydos, M. Diener-West, M. R. Howell, J. M. Zenilman, and T. C. Quinn.** 1998. Incident *Chlamydia trachomatis* infections among inner-city adolescent females. *JAMA* **280**:521–526.

2. **Centers for Disease Control and Prevention.** 2002. Sexually transmitted disease treatment guidelines. *Morb. Mortal. Wkly. Rep.* **51**(RR-6):1–80.

3. **Centers for Disease Control and Prevention.** 1999. High prevalence of chlamydial and gonococcal infection in women entering jails and juvenile detention centers—Chicago, Birmingham, and San Francisco, 1998. *Morb. Mortal. Wkly. Rep.* **48**:793–796.

4. **Scholes, D., A. Stergachis, F. E. Heidrich, H. Andrilla, K. K. Holmes, and W. E. Stamm.** 1996. Prevention of pelvic inflammatory disease by screening for cervical chlamydial infection. *N. Engl. J. Med.* **334**:1362–1366.

 **CASE 4** The patient was a 20-year-old female who presented to the emergency room with a 4-day history of fever, chills, and myalgia. Two days prior to this she had noted <u>painful genital lesions</u>. On the day of admission she developed <u>headache, photophobia, and a stiff neck</u>. Previously, she had been in good health. She admitted to being sexually active but had no history of sexually transmitted diseases (STDs).

On physical examination, she was alert and oriented. Her vital signs were normal except for a temperature of 38.5°C; pulse rate was 80 beats/min, and blood pressure was 130/80 mm Hg. A general examination was unremarkable except for slight nuchal rigidity. Her throat was clear, and there was no lymphadenopathy. A pelvic examination revealed extensive <u>vesicular and ulcerative lesions on the left labia minora and majora with marked edema</u>. The cervix had exophytic (outward-growing) necrotic ulcerations. Specimens were taken to culture for *Neisseria gonorrhoeae*, viruses, and *Chlamydia trachomatis*.

General laboratory tests were unremarkable. The VDRL (Venereal Disease Research Laboratory) test was negative. A lumbar puncture was done. The opening pressure was normal. The cerebrospinal fluid (CSF) showed a mild pleocytosis with a leukocyte count of 41/µl with 21% polymorphonuclear leukocytes and 79% mononuclear cells, a glucose level of 46 mg/dl, and a protein level of 68 mg/dl (slightly elevated). The CSF VDRL test was negative. A rapid diagnostic test was done on the genital lesion and gave positive results. Cultures from the genital lesions and CSF verified the diagnosis 2 days later. By that time the patient's condition had improved after 2 days of intravenous therapy. She was discharged home on oral medication.

1. What is the differential diagnosis of ulcerative genital lesions? Which rapid test was used so that specific therapy could be started?

2. Which complication of her underlying illness did she develop?

3. If she had been pregnant at the time of her infection, for what would her fetus be at risk?

4. Briefly describe the natural history of this infection.

5. Briefly describe the epidemiology of the agent causing her infection.

6. There are two different serotypes of the agent causing her infection. What similarities do they share and what are the differences between these agents?

CASE DISCUSSION

1. The most likely diagnosis is genital herpes. In studies of patients with genital lesions in the industrialized world, herpes simplex virus (HSV) is the most frequently recovered agent. Other agents that are common causes of genital lesions include *Treponema pallidum, Haemophilus ducreyi* (the etiologic agent of chancroid), human papillomavirus (genital warts), and the lymphogranuloma venereum-causing variant of *C. trachomatis.* Genital herpes lesions are painful, whereas lesions due to *T. pallidum* are usually painless. Genital infections such as chancroid or lymphogranuloma venereum can result in painful or painless ulcers, respectively, but they often result in suppurative lymphadenopathy. The diagnosis of HSV infection can be confirmed by isolating the virus in tissue culture cells. Using a shell vial culture technique, the virus can usually be detected within 24 hours. However, detection of HSV antigen or DNA in scrapings from the lesion by immunofluorescence or PCR is both more rapid than culture and, in the case of PCR, more sensitive as well. Tzanck preparations, in which smears taken from the edge of the lesion are examined for the presence of cells showing pathologic changes consistent with HSV infection, are also used in the diagnosis of genital lesions. This technique, although inexpensive, lacks both the sensitivity and specificity of culture, immunofluorescence, or PCR. HSV was detected in this patient by immunofluorescent staining of her genital lesion.

2. Among women with primary genital herpes, approximately one in three will have self-limited, aseptic meningitis. These patients typically have a pleocytosis with a lymphocytic predominance and an elevated protein level, as was seen in this case. It is unusual to grow HSV from cerebrospinal fluid of individuals with primary genital herpes because CSF cultures are rarely done in this clinical setting.

3. Her fetus would be at risk for neonatal herpes. Neonatal herpes is a relatively infrequent infection, occurring in between 1 in 2,000 and 1 in 5,000 births. However, it is estimated that 30 to 50% of women who have vaginal deliveries and primary HSV genital lesions at the time of birth will transmit the disease to their child, whereas ≤3% with recurrent infection will do so. Other factors that increase the likelihood of infection are prolonged rupture of membranes, a mother who is seronegative for HSV-2, and the use of fetal scalp monitors. It is estimated that approximately 6 to 14% of infected neonates are infected in utero. Neonates infected in utero typically show symptoms in the first 2 days of life. Approximately 80% are infected during passage through an infected birth canal. The rest are infected postpartum.

There are three forms of neonatal HSV infection. Most neonatal HSV infections occur in the second to third week of life. The most benign form, which is seen in 40% of cases, causes infection localized to the skin, eyes, and mouth. If recognized, it can be effectively treated with antiviral agents such as acyclovir. Encephalitis can be seen in 35% of cases. These children will have nonspecific central nervous symptoms not unlike those of neonatal bacterial meningitis, including seizures, lethargy, high fevers, poor feeding, and irritation. Mortality approaches 50% in untreated children, and long-term neurologic sequelae are seen in half of the survivors. The most severe manifestation of disease is disseminated infection, which occurs in approximately 25%. In this infection, multiple organs, including the brain, may be infected. These individuals typically have a viral exanthem in the setting of encephalitis and/or multiorgan failure. If the infection is untreated, mortality is very high, reaching 80%; those who survive often have profound psychomotor retardation.

4. HSV, like all herpesviruses, causes a lifelong, latent infection. In genital tract infections, the virus enters a latent state in the sacral nerve ganglia. Recurrences occur when the virus replicates in the neuron and is carried along the peripheral nerves to the epithelium. Symptomatic recurrences are common in genital herpes and may occur as frequently as 8 to 10 times per year, although the majority of individuals will have significantly fewer episodes. Recurrences are generally milder than the primary episode of disease. It should be noted that HSV-infected individuals can shed HSV in the absence of symptoms, so the use of condoms by these individuals is encouraged to prevent the spread of this agent. One of the major problems with the control of genital herpes is that at most only one-fourth of infected individuals know that they are infected.

5. HSV-2 infects approximately 20% of individuals in the United States. Infections are more common in females than in males and are more common in nonwhite populations. HSV-2 is a disease of poverty and is also associated with cocaine use. The higher the number of sexual partners, the greater the likelihood of HSV-2 infection. Infection rates among commercial sex workers may approach 100%. Over the past 2 decades, HSV-2 infection rates increased by 30% in the United States. The populations with the highest rate of increase were teenagers and individuals in their 20s. This increase occurred despite efforts to prevent STDs, especially infections with HIV.

6. There are two distinct serotypes of HSV—HSV-1 and HSV-2. HSV-1 is an infection primarily of the oropharyngeal mucosa, with latent infection occurring in the trigeminal ganglion, while HSV-2 primarily infects the genital mucosa. HSV-1 infections are typically acquired in early childhood, while HSV-2 infections occur after the individual becomes sexually active. HSV-1 has been reported to be responsible for approximately 10% of genital herpes cases, and HSV-2 can cause herpes labialis. HSV-1 appears to cause more severe central nervous system (CNS) infection outside the neonatal period. In contrast to aseptic meningitis associated with genital HSV-2 infection, herpes encephalitis in adults and older children is a more severe illness and is most often due to HSV-1 infection. Herpes encephalitis is a rare, sporadic CNS viral infection. It is believed to be the most common cause of nonepidemic viral encephalitis in adults in the United States. Patients present with fever, headache, and encephalopathic findings such as altered consciousness, behavioral and speech disturbances, and focal or diffuse neurologic signs. These neurologic changes are consistent with infection of the frontal lobe, the usual site of infection in the brain. The diagnosis can be confirmed by detecting HSV directly using fluorescent antibody staining of tissue obtained by brain biopsy. Because brain biopsy is dangerous and not 100% sensitive, alternative means of making this diagnosis have been sought. Detection of HSV DNA in the patient's CSF by PCR (polymerase chain reaction) has been found to be a reliable means of making this diagnosis. It is not clear why certain patterns of CNS infection with either HSV-1 or HSV-2 result in different CNS manifestations. The age of the patient, the route of viral dissemination (e.g., neural versus hematogenous), preexisting immunity, and/or specific viral properties may be factors.

REFERENCES

1. **Ashley, R. L., and A. Wald.** 1999. Genital herpes: review of epidemic and potential use of type-specific serology. *Clin. Microbiol. Rev.* **12:**1–8.

2. **Fleming, D. T., G. M. McQuillan, R. E. Johnson, A. J. Nahmais, S. O. Aral, F. K. Lee, and M. E. St. Louis.** 1997. Herpes simplex virus type 2 in the United States, 1976 to 1994. *N. Engl. J. Med.* **337:**1158–1159.

3. **Tang, Y.-W., P. S. Mitchell, M. J. Espy, T. F. Smith, and D. H. Persing.** 1999. Molecular diagnosis of herpes simplex virus infections in the central nervous system. *J. Clin. Microbiol.* **37:**2127–2136.

4. **Whitley, R. J., D. W. Kimberlin, and B. Roizman.** 1998. Herpes simplex viruses. *Clin. Infect. Dis.* **26:**541–555.

C A S E 5 This 26-year-old woman was referred to a public health clinic as a result of contact tracing in a case of gonorrhea. The woman, who had recently had unprotected sexual intercourse, had no symptoms. Physical examination was normal. Pelvic examination demonstrated a white vaginal discharge but was otherwise unremarkable. Cervical culture was obtained for *Neisseria gonorrhoeae*, and a specimen was submitted for *Chlamydia trachomatis* testing by ligase chain reaction. Examination of a wet mount of the vaginal discharge revealed the presence of a protozoan with a characteristic jerky motility. Figure 1 shows a Giemsa stain of the organism.

1. What organism did the wet preparation demonstrate? What other organism can cause vaginitis and can be detected by wet mount?

2. What other methodologies are available for detection of this organism?

3. How is infection with this organism most commonly acquired? What clinical presentations occur in women infected with this organism? In men infected with this organism?

4. This patient was asymptomatic when examined. She had had sexual contact with a partner who had a positive culture for *N. gonorrhoeae*. What would be appropriate antimicrobial therapy for this patient?

5. Why is infection with this organism of special concern in pregnant women? Would therapy be any different if this woman were pregnant?

6. What else should be done to prevent this patient from becoming reinfected with the organism identified on the wet preparation?

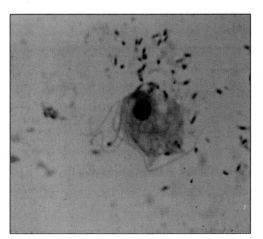

Figure 1

CASE DISCUSSION

1. The wet preparation demonstrated the trophozoites of the protozoan *Trichomonas vaginalis*. Examination of freshly prepared wet mounts of vaginal fluid, prostatic secretions, or urine from infected patients will reveal the organism in 40 to 80% of infected individuals. The organism is 7 to 23 μm in size, with a typical jerky motility. Microscopic examination for *T. vaginalis* is highly specific because its unique morphology makes it unlikely to be confused with any other organism that might typically be seen in genital tract secretions. This technique is the most widely used for laboratory diagnosis because it is inexpensive, rapid, easily performed, and requires relatively simple equipment (light microscope). Wet mounts can also be used to diagnose *Candida* vaginitis. In this form of vaginitis, yeast and pseudohyphae will be seen on wet mount. *Candida* vaginitis is frequently seen during or following a course of antimicrobial agents that alter the vaginal flora.

2. Culture and PCR (polymerase chain reaction) techniques have been developed to detect this organism. Culture is done by growing the organism in enriched broth. Recently, a specially designed pouch that allows the direct examination of the broth microscopically for trophozoites has been developed. Culture is more sensitive than direct examination, but because of its complexity, expense, and length of time to result, it is typically used only in research settings.

PCR for *T. vaginalis* has been found to be more sensitive than direct examination and both more rapid and sensitive than culture. False-positive reactions with PCR are of concern. As with culture, this technique is restricted to the research setting.

3. *T. vaginalis* is typically transmitted via sexual contact. Since *Trichomonas* infection is not a reportable disease, the number of cases that occur annually is unknown. However, it is estimated that 3 million women are infected annually in the United States, making this parasite an important health issue. Women can be asymptomatically infected, but most infections result in a vaginal discharge. Symptoms of itching or burning are frequently associated with this discharge. The infection can also involve the urethra, resulting in symptoms of dysuria. In men, most cases are asymptomatic, though some men have symptoms of urethral involvement, including a urethral discharge. Involvement of the prostate or seminal vesicles may occur as well.

4. Clearly, this woman must be treated for the *T. vaginalis* infection, the diagnosis having been established on the basis of a microscopic examination of her discharge. The drug of choice for this infection is metronidazole (Flagyl). It should be noted that there are an increasing number of reports of treatment failures due to metronidazole-resistant strains of *T. vaginalis*, but the frequency with which resistant isolates are recovered is unknown. Even though she was asymptomatic, she was at a very high risk for a coinfection with *N. gonorrhoeae* because that organism had been detected in her male sexual partner. This finding prompted her visit to the clinic. Her gonococcal infection was treated with an intramuscular injection of ceftriaxone. In addition, since gonococcal infections are often associated with infection by *C. trachomatis*, she was given oral doxycycline. Her cervical culture was subsequently positive for *N. gonorrhoeae*, and the ligase chain reaction was positive for *C. trachomatis*. Remember that patients can be simultaneously infected with multiple sexually transmitted disease agents and that both *C. trachomatis* and *N. gonorrhoeae* frequently cause asymptomatic infections in women but not men.

The patient was also offered testing for HIV infection. Recent studies have shown that *T. vaginalis* infection increases the likelihood of HIV transmission.

5. *T. vaginalis* has been associated with pre-term labor, premature rupture of membranes, and low-birth-weight babies. The use of metronidazole during pregnancy has been controversial because this drug has been shown to be mutagenic in bacteria and carcinogenic in laboratory animals. Retrospective studies have shown that women treated with metronidazole during pregnancy do not have a higher rate of delivery of children with birth defects than those women who did not receive this drug during pregnancy. Nevertheless, some experts would caution against using metronidazole during the first trimester.

6. The patient's partner, who had been treated for gonorrhea and chlamydia, had not been treated for infection with *T. vaginalis*. As with other sexually transmitted diseases, treatment of both people within a sexual relationship is necessary to prevent reinfection by the untreated person. Treatment of only the person presenting and not the partner can result in a "ping pong ball" phenomenon, where the infection "bounces" back and forth between the two partners. In addition, the patient was advised on the risks of unprotected sex and informed that condom use may help to prevent disease transmission.

REFERENCES
1. **Anonymous.** 1999. National guidelines for the management of *Trichomonas vaginalis*. *Sex. Transmit. Infect.* **75**(Suppl. 1): S21–23.

2. **McKee, K. T., Jr., P. R. Jenkins, R. Garner, R. A. Jenkins, E. D. Nannis, I. F. Hoffman, J. L. Schmitz, and M. S. Cohen.** 2000. Features of urethritis in a cohort of male soldiers. *Clin. Infect. Dis.* **30**:736–741.

3. **Petrin, D., K. Delgaty, R. Bhatt, and G. Garber.** 1998. Clinical and microbiological aspects of *Trichomonas vaginalis*. *Clin. Microbiol. Rev.* **11**:300–317.

TWO

Respiratory Tract Infections

INTRODUCTION TO SECTION II

Respiratory tract infections are a major reason why children and the elderly seek medical care. These infections are more common in cold-weather months in locales with temperate climates. Respiratory tract infections are primarily spread by inhalation of aerosolized respiratory secretions from infected hosts. Some respiratory tract pathogens, such as rhinoviruses, can also be spread by direct contact with mucous membranes, but this mode of transmission is much less common than inhalation. Organisms that are part of the endogenous flora of the oropharynx may, under certain conditions, such as aspiration of oropharyngeal secretions, be able to cause clinical disease. Animal exposure may result in some of the less common but more severe bacterial causes of respiratory infection, including inhalation anthrax, pneumonic plague, and tularemia pneumonia. These zoonotic agents are also potential agents of bioterrorism. For the purposes of our discussions, we will divide these types of infections into two groups, upper tract and lower tract infection.

The most common form of upper respiratory tract infection is pharyngitis. Pharyngitis is seen most frequently in children from 2 years of age through adolescence. The most common etiologic agents of pharyngitis are viruses, particularly adenoviruses, and group A streptococci. Pharyngitis due to group A streptococci predisposes individuals to the development of the poststreptococcal sequelae rheumatic fever and glomerulonephritis. Because rheumatic fever can be prevented by penicillin treatment of group A streptococcal pharyngitis, aggressive diagnosis and treatment of pharyngitis due to this organism is needed.

Otitis media is a common infectious problem in infants and young children. The most frequently encountered agents of this infection are the bacteria *Streptococcus pneumoniae*, *Haemophilus influenzae*, and *Moraxella catarrhalis*. These organisms, along with selected viruses and anaerobic bacteria from the oral cavity, are the most important pathogens in sinusitis.

S. pneumoniae, *H. influenzae*, and *M. catarrhalis*, adenoviruses, and *Chlamydia trachomatis* (in neonates) are the common etiologic agents of conjunctivitis. External otitis, a common problem in swimmers, is more common in warm-weather months. *Staphylococcus aureus* and *Pseudomonas aeruginosa* are the most common agents of this relatively benign condition. Malignant external otitis is a serious medical condition seen primarily in diabetics, the elderly, and the immunocompromised. The infection can spread from the ear to the temporal bone. The most common etiology of malignant otitis externa is *P. aeruginosa*. Two other life-threatening infections of the upper respiratory tract are rhinocerebral mucormycosis (zygomycosis) and bacterial epiglottitis. Rhinocerebral mucormycosis is most common in diabetics, especially those with ketoacidosis. In this infection of the sinuses, fungi within the zygomycetes, such as *Mucor* and *Rhizopus* spp., invade blood vessels, resulting in necrosis of bone and thrombosis of the cavernous sinus and internal carotid artery. Treatment of this infection requires aggressive surgical debridement of the infected tissue in addition to antifungal therapy. Epiglottitis is most commonly caused by *H. influenzae* type b. In this disease, the airway may become compromised because of swelling of the epiglottis, with death due to respiratory arrest. With the widespread use of *H. influenzae* type b vaccine, this rare disease should essentially disappear.

Three childhood infections with respiratory manifestations or complications that were common in the early part of the 20th century—diphtheria, whooping cough, and measles—are now rare diseases in the developed world. This is due to the development and use of vaccines in children that are effective against the etiologic agents of these diseases, *Corynebacterium diphtheriae*, *Bordetella pertussis*, and measles virus.

Viruses play an important role in upper respiratory tract infections. The common syndrome of cough and "runny" nose is usually due to rhinoviruses. More severe upper respiratory infections such as the "croup" are due to respiratory syncytial virus and influenza and parainfluenza viruses. These viruses can also cause lower tract infection and are an important cause of morbidity and mortality in the very young and very old.

When discussing lower respiratory tract infections, it is important to look at four different groups of patients: patients with community-acquired infections; patients with nosocomial infections; patients with underlying lung disease; and immunocompromised individuals, especially those with AIDS.

Common agents of community-acquired lower respiratory tract infections include *S. pneumoniae*; *Klebsiella pneumoniae*, especially in alcoholics; *Mycoplasma pneumoniae*, especially in school-age students through young adulthood; *Mycobacterium tuberculosis*, especially in individuals born in countries with a high prevalence of tuberculosis; respiratory syncytial virus in infants and young children: and influenza A virus. The dimorphic fungi *Histoplasma capsulatum* and *Coccidioides immitis* usually cause mild, self-limited diseases in patients residing in specific geographic locales. *S. pneumoniae*, *H. influenzae*, *S. aureus*, and *M. catarrhalis* may cause bronchitis and/or pneumonia in adults following viral pneumonia. Aspiration due to seizure disorders, semiconscious states from excessive consumption of alcohol or other drugs, or impairment of the gag reflex, as may occur following a stroke, may result in aspiration pneumonia or lung abscess caused by the organisms residing in the oral cavity. The anatomic location of the lung process depends upon the patient's position at the time of aspiration.

Nosocomial infections due to the organisms listed above certainly occur. Particular emphasis is placed on preventing the spread of *M. tuberculosis* in all patient populations and on preventing the nosocomial spread of respiratory syncytial virus in pediatric patients. Nosocomial pneumonia due to methicillin-resistant *S. aureus* and multidrug-resistant gram-negative bacilli, such as *P. aeruginosa*, are common problems in intubated patients. Because of their ability to survive within hospital water and air conditioning systems, the potential for outbreaks of pneumonia due to *Legionella* spp. is a constant threat.

Patients with chronic obstructive pulmonary disease brought on by smoking frequently develop bronchitis. *S. pneumoniae*, *M. catarrhalis*, *H. influenzae*, and *P. aeruginosa* are frequent causes of this type of infection. Patients with cystic fibrosis have chronic airway infections which are primarily responsible for their premature death. *S. aureus* and mucoid strains of *P. aeruginosa* are the most important agents of such chronic airway disease. Both of these patient populations have an increased risk of developing allergic bronchopulmonary aspergillosis. Patients with cavitary lung disease, frequently due to prior *M. tuberculosis* infection, are at increased risk for another type of infection, an aspergilloma or fungus ball caused by *Aspergillus* spp. This fungus grows in the form of a ball in the preformed cavity. A distinction between actual tissue invasion with this fungus and noninvasive disease is clinically difficult but is important.

The diagnosis of the etiology of lung infection in immunocompromised patients is one of the most daunting in clinical microbiology and infectious disease. It has been greatly facilitated by the development of the flexible bronchoscope, which provides a relatively noninvasive means to sample the airways and alveoli. Immunocompromised patients are typically at risk for essentially all recognized respiratory tract pathogens. However, a distinction must be made between different types of immunosuppression—defects in cell-mediated immunity, humoral immunity, and neutrophil number or function—because different types of immunosuppression predispose patients to infection with different pathogens. The most common immunosuppressed state is the result of cigarette smoking, which causes impaired removal of pathogens due to defective mucociliary clearance. Although smoking results in a significantly increased rate of both bronchitis and pneumonia, smokers are not normally described as immunosuppressed.

In AIDS patients, *Pneumocystis carinii*, *S. pneumoniae*, and multidrug-resistant *M. tuberculosis* are all seen more frequently than in other patient populations. Solid-organ transplant recipients have a greatly increased risk for pneumonia with cytomegalovirus, herpes simplex virus, *Legionella* spp., *P. carinii*, and *Nocardia* spp. Prophylactic antibiotics are frequently taken by these patients to prevent pulmonary infections with *P. carinii*. Prophylactic therapies are not as widely used for other agents for a variety of reasons, including expense, questionable efficacy of the prophylactic measures, or the rarity with which the organism is encountered. Profoundly neutropenic patients, especially those in whom the duration of neutropenia is prolonged, not only have a risk of infection with routine bacteria but have a very high risk of invasive aspergillosis and other invasive fungal infections.

TABLE 2 SELECTED RESPIRATORY TRACT PATHOGENS

ORGANISM	GENERAL CHARACTERISTICS	PATIENT POPULATION	DISEASE MANIFESTATION
Bacteria			
Actinomyces spp.	Branching gram-positive bacilli, usually anaerobic	Adults with aspiration	Lung abscess
Bacillus anthracis	Spore-forming, gram-positive bacillus	Victims of bioterrorism due to exposure to spores; woolsorters in endemic areas	Inhalation anthrax with widened mediastinum, high-grade bacteremia
Bordetella pertussis	Fastidious gram-negative bacillus	Children and adults	Whooping cough, atypical whooping cough
Chlamydia pneumoniae	Obligate intracellular bacterium; does not Gram stain	Children and adults	Pneumonia, bronchitis
Chlamydia psittaci	Obligate intracellular bacterium; does not Gram stain	Children and adults with exposure to birds	Pneumonia; ornithosis (psittacosis)
Chlamydia trachomatis	Obligate intracellular bacterium; does not Gram stain	Neonatal	Conjunctivitis, pneumonia
Corynebacterium diphtheriae	Catalase-positive, gram-positive, club-shaped bacillus	Unvaccinated adults and children; improperly vaccinated adults	Diphtheria
Enterobacter spp., *Escherichia coli*	Lactose-fermenting, gram-negative bacilli	Adults	Nosocomial pneumonia
Group A streptococci (*Streptococcus pyogenes*)	Catalase-negative, gram-positive cocci in chains	Children >2 years; adults	Pharyngitis, pneumonia with empyema
Group B streptococci (*Streptococcus agalactiae*)	Catalase-negative, gram-positive cocci in chains	Neonates	Pneumonia
Haemophilus influenzae	Pleomorphic gram-negative bacillus	Children; adults, especially with COPD[a]	Otitis media, conjunctivitis, epiglottitis, bronchitis, pneumonia

(continued next page)

TABLE 2 SELECTED RESPIRATORY TRACT PATHOGENS *(continued)*

ORGANISM	GENERAL CHARACTERISTICS	PATIENT POPULATION	DISEASE MANIFESTATION
Klebsiella pneumoniae	Lactose-fermenting, gram-negative bacillus	Adults	Community-acquired and nosocomial pneumonia
Legionella pneumophila	Poorly staining, fastidious, gram-negative bacillus	Adults, especially immunocompromised	Pneumonia
Moraxella catarrhalis	Oxidase-positive, gram-negative diplococcus	Children; adults with COPD	Otitis media, conjunctivitis, bronchitis
Mycobacterium tuberculosis	Acid-fast bacillus	Children and adults, especially HIV-infected	Tuberculosis
Mycoplasma pneumoniae	Fastidious; does not Gram stain	Children, adolescents, adults	Walking pneumonia
Neisseria gonorrhoeae	Oxidase-positive, gram-negative diplococcus	Individuals with oral-genital contact, neonates	Pharyngitis, conjunctivitis
Neisseria meningitidis	Oxidase-positive, gram-negative diplococcus	Adults	Pneumonia
Nocardia spp.	Partially acid-fast, aerobic, branching, gram-positive bacilli	Adults, especially with defects in cell-mediated immunity	Pneumonia with brain abscess
Nontuberculous mycobacteria (many species)	Acid-fast bacilli	Adults with chronic lung disease; CF[b] patients	Granulomatous lung disease
Prevotella sp.; *Porphyromonas* spp.	Anaerobic gram-negative bacilli	Adults with aspiration	Lung abscess
Pseudomonas aeruginosa	Glucose-nonfermenting, gram-negative bacillus	Adults and children; diabetic adults; nosocomial; CF patients	External otitis (swimmer's ear), malignant external otitis, ventilator-associated pneumonia, chronic bronchitis with mucoid strains
Staphylococcus aureus	Catalase-positive, gram-positive cocci in clusters	Nosocomial	Pneumonia, pneumonia superinfections

Organism	Morphology	Population	Clinical manifestation
Stenotrophomonas maltophilia	Glucose-nonfermenting, gram-negative bacillus	Nosocomial	Ventilator-associated pneumonia
Streptococcus pneumoniae	Catalase-negative, gram-positive diplococcus	Children and adults	Otitis media, sinusitis, conjunctivitis, pneumonia

Fungi

Organism	Morphology	Population	Clinical manifestation
Aspergillus spp.	Acute-angle-branching, septate hyphae in tissue; mold	Children and adults with chronic lung disease; adults with cavitary lung lesions; neutropenic individuals	Allergic bronchopulmonary aspergillosis; aspergilloma (fungus ball); invasive pneumonia
Blastomyces dermatitidis	Broad-based budding yeast; dimorphic	Adults	Pneumonia
Coccidioides immitis	Spherules in tissue; mold with arthroconidia at 30°C	Children and adults, especially in desert southwest of U.S. and northern Mexico	Flu-like illness with pneumonia; can disseminate
Cryptococcus neoformans	Encapsulated, round yeast	Adults with defects in cell-mediated immunity, especially with AIDS	Pneumonia, often asymptomatic, preceding meningitis
Histoplasma capsulatum	Very small, intracellular yeast; dimorphic	Adults, primarily with AIDS, especially in Missouri and Ohio River Valleys and Caribbean	Pneumonia, mediastinal fibrosis
Pneumocystis carinii	Clusters of 4- to 6-μm cysts in tissue and secretions	Immunocompromised individuals, especially with AIDS	Pneumonia
Rhizopus sp., *Mucor* sp.	Ribbon-like, nonseptate hyphae in tissue; rapidly growing mold	Diabetics, neutropenic individuals	Rhinocerebral zygomycosis, invasive pneumonia

Parasites

Organism	Morphology	Population	Clinical manifestation
Ascaris lumbricoides	Larvae	Children and adults	Usually asymptomatic, incidental finding
Echinococcus granulosus	Tapeworm (cestode)	Exposure to dogs in areas with sheep	Cyst in lung growing over the course of years; rupture from liver may lead to pleural space
Entamoeba histolytica	Amoeba	Children and adults with amoebic liver abscess	Empyema, hepatobronchial fistula, lung abscess

(continued next page)

61

TABLE 2 SELECTED RESPIRATORY TRACT PATHOGENS (continued)

ORGANISM	GENERAL CHARACTERISTICS	PATIENT POPULATION	DISEASE MANIFESTATION
Hookworm (*Necator americanus* and *Ancylostoma duodenale*)	Larvae	Children and adults	Usually asymptomatic, incidental finding
Paragonimus westermani	Fluke (trematode)	Children and adults in endemic areas	Hemoptysis, chronic bronchitis, bronchiectasis
Schistosoma spp.	Fluke (trematode); granulomas form around eggs	Children and adults in endemic areas	Pulmonary hypertension due to trapping of eggs in pulmonary capillaries
Strongyloides stercoralis	Rhabditiform larvae	Immunocompromised individuals	Wheezing, cough, pneumonia
Viruses			
Adenovirus	Enveloped, dsDNA[c]	Children and adults	Pharyngitis, bronchiolitis, pneumonia, conjunctivitis
Cytomegalovirus	Enveloped, dsDNA	Immunocompromised individuals	Pneumonia
Hantaviruses	Enveloped, ssRNA[d]	Children and adults	Adult respiratory distress syndrome, pneumonia
Herpes simplex virus	Enveloped, dsDNA	Immunocompromised individuals	Pneumonia
Influenza viruses	Enveloped, ssRNA	Children and adults, particularly elderly	Influenza, pneumonia
Parainfluenza virus types I, II, III	Enveloped, ssRNA	Infants and young children	Croup, bronchiolitis, pneumonia, laryngitis
Respiratory syncytial virus	Enveloped, ssRNA	Infants and young children; elderly	Cough, wheezing, bronchiolitis, pneumonia
Rhinoviruses	Nonenveloped, ssRNA	Children and adults	Common cold
Varicella-zoster virus	Enveloped, dsDNA	Immunocompromised individuals, pregnant women	Pneumonia

[a] COPD, chronic obstructive pulmonary disease.

[b] CF, cystic fibrosis.

[c] dsDNA, double-stranded DNA.

[d] ssRNA, single-stranded RNA.

CASE 6

The patient was a 5-year-old male who awoke on the day prior to evaluation with a sore throat and fever. His mother had him stay home from kindergarten and treated him symptomatically with Tylenol. He slept well but the next day awoke still complaining of sore throat and fever, as well as headache and abdominal pain. He was an only child and neither parent was ill.

On physical examination, he was noted to have a fever of 38.4°C. His physical examination was significant for a 2+ (on a scale of 1 to 4+) red anterior pharynx, tonsillar region, and soft palate. His anterior cervical lymph nodes at the angle of the mandible were slightly enlarged and tender. No skin lesions or rashes were seen. A culture of the organism causing this patient's infection is shown in Fig. 1.

1. What is the most likely organism causing this patient's infection?

2. Why is detection of this organism important in management of this infection? How can this organism be detected? What are the strengths and weaknesses of these methods?

3. This patient is at risk for two noninfectious sequelae. What are they? Briefly describe our current understanding of the pathogenesis of these two disease processes.

4. What antimicrobial resistance problems have been observed with this organism?

5. Sore throat associated with a maculopapular rash is frequently seen with this organism. What is this usually benign condition called? What virulence factor is believed to be responsible for production of this rash?

6. What potentially fatal infection has been seen with increasing frequency with this organism? What virulence factors produced by this organism are believed to be important in the pathogenesis of this infection?

7. What is the current status of vaccine development for this organism?

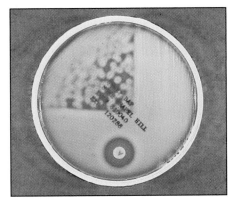

Figure 1

CASE DISCUSSION

1. This patient had *Streptococcus pyogenes* or group A streptococcal (GAS) pharyngitis. In Fig. 1, the growth of the beta-hemolytic organism is seen to be inhibited by a disc containing the antimicrobial agent bacitracin. Bacitracin inhibits the growth of approximately 95% of GAS strains and so it is a presumptive but highly accurate identification test for this organism. Definitive identification can be accomplished by serotyping the organism but is not necessary in most clinical situations.

GAS pharyngitis is most commonly seen in school-age children between 5 and 15 years of age. It has its highest incidence in the industrialized world in the winter and early spring but can occur year round.

This child's presentation was typical of GAS pharyngitis. He complained of a sore throat with fever, headache, and abdominal pain. His physical examination was also typical for a child with GAS pharyngitis: low-grade fever, a red throat, and cervical adenopathy. He did not have a peritonsillar exudate, which is commonly seen in these patients. His presentation was also consistent with viral pharyngitis.

2. Pharyngitis is the most common infectious disease reason for individuals, especially children between the ages of 5 and 15, to seek medical attention. Because the two major causes of pharyngitis, viruses (adenovirus, rhinovirus, coronavirus, and respiratory syncytial virus) and GAS, cannot be accurately distinguished on clinical grounds alone, it is important to distinguish the two in order to treat them properly. This can be best accomplished by the detection of GAS with throat swabs. Viral pharyngitis is treated symptomatically with antipyretics and warm salt-water gargles, while GAS pharyngitis should be treated both symptomatically and with antibiotics. Antibiotics are given primarily to prevent poststreptococcal sequelae (see answer to question 3 for further explanation). They also will prevent suppurative complications of GAS pharyngitis, such as peritonsillar and retropharyngeal abscesses, and decrease the infectivity of the infected individual. In school-age children, this is important so that they are less likely to infect their classmates and siblings, both at-risk populations.

There are two ways to detect GAS, by direct detection of group A polysaccharide antigen in throat swabs and by culture on a blood-containing agar plate. Direct antigen detection is accomplished by extracting the group A polysaccharide antigen from the throat swab and then performing an immunoassay on the extract. The test is very rapid, taking 10 to 15 minutes, and is highly specific (>95%), but when compared with culture it has a sensitivity of 80 to 90%, meaning that GAS will not be detected by this test in 10 to 20% of patients with GAS in their throats. The advantage of the "rapid strep test," as it is called, is that a swab can be obtained in the office or clinic and a result can be obtained while the patient waits, i.e., a "real-time" microbiology test. For patients who have a positive test, antibiotics can be prescribed on the spot. The problem is more complex with patients who have a negative test. Should they be considered negative for GAS and be advised to treat their pharyngitis symptomatically, with the physician explaining that no antibiotics are necessary since over half of pharyngitis episodes are due to viral infection? Or should the physician follow the recommendations of the Infectious Disease Society of America and the American Academy for Pediatrics, which both recommend that the decision to use antibiotics in patients with negative antigen tests should be based on culture results which are highly accurate but take 24 to 48 hours? Or should the physician give antibiotics without regard to antigen or culture results even though the majority of patients they see have viral pharyngitis and it cannot be distinguished from GAS pharyngitis on clinical grounds? Since it is estimated that 70% of individuals with pharyngitis get an antibiotic but GAS causes pharyngitis in, at most, 35% of children and 15% of adults, this third approach is probably used quite frequently. This approach probably contributes to the increasing rates of antimicrobial resistance that we are seeing with respiratory pathogens such as *Streptococcus pneumoniae*. The expert opinion is to back up negative antigen tests with culture and treat the patient on the basis of those results.

3. The patient was at risk for two poststreptococcal sequelae, rheumatic fever and glomerulonephritis. Because he received antimicrobial therapy, his risk of rheumatic fever is essentially zero. The likelihood of an untreated, infected person developing either one of these complications is rare but is dependent on the serotype of the organism with which he is infected. Serotyping of GAS is based on the M protein, a surface protein that is anchored in the organism's cell wall. There are over 80 different types of this antiphagocytic protein. Certain M types, such as M1 and M3, are associated with rheumatic fever and are said to be "rheumatogenic." Other strains, such as M12 and M49, are considered "nephritogenic" and are associated with glomerulonephritis. Glomerulonephritis is seen following both pharyngitis and skin infections (pyoderma or impetigo) whereas rheumatic fever is believed to occur only following pharyngitis.

These noninfectious poststreptococcal sequelae occur after an acute GAS infection. Rheumatic fever occurs 1 to 5 weeks after infection, while glomerulonephritis following pharyngitis occurs at 1 to 2 weeks and 3 to 6 weeks following pyoderma. Both sequelae are believed to be immune-mediated diseases whereby antibodies made in response to GAS react with tissues in the target organ.

In rheumatic fever, antibodies directed against the M protein are believed to cross-react with a variety of tissue components in the heart, including myosin, laminin, and tropomyosin. This can result in damage to heart valves and muscle and produce the carditis and heart murmurs that are manifestations of this syndrome.

In glomerulonephritis, streptococcal antibodies that cross-react with the glomerular basement membrane are believed to be important in the disease process as well as the deposition in the glomeruli of circulating immune complexes containing streptococcal antigens. Clinically, individuals present with edema, hypertension, and hematuria.

4. Despite the use of penicillin to treat GAS infections for over 50 years, this organism continues to be uniformly sensitive to this antimicrobial. In penicillin-allergic patients, erythromycin and the newer macrolide antimicrobials, clarithromycin and azithromycin, are recommended therapeutic agents for GAS pharyngitis. A study in Finland showed that GAS resistance to erythromycin was associated with increasing use of this antimicrobial. In 1993, almost 20% of GAS isolates were resistant to erythromycin. Following a national education effort, use of erythromycin and related antimicrobials declined. By 1996, the percentage of erythromycin-resistant strains of GAS declined to 8.6%. The important lesson here is that once resistance is present in an organism, reducing specific antimicrobial pressure will only result in a reduction in the number of resistant strains, not an elimination of them. A recent study of schoolchildren in Pittsburgh was notable for a very high rate of resistance of GAS to erythromycin of 48%.

5. Streptococcal pyrogenic exotoxins (Spe) A through C were once referred to as erythrogenic or scarlet fever toxins. Scarlet fever is considered to be a benign complication of pharyngitis caused by a pyrogenic exotoxin-producing strain of GAS. The skin rash seen in scarlet fever is believed to be superantigen mediated. (For further information concerning superantigens, see answer to question 6.)

6. Since the mid-1980s there have been an increasing number of cases of invasive disease due to GAS. The most severe manifestation of invasive disease is the streptococcal toxic shock syndrome. These patients are usually bacteremic and in the most severe cases have necrotizing fasciitis. Patients with the streptococcal toxic shock syndrome typically have GAS detectable from a normally sterile site such as blood or tissue. They also have shock, thrombocytopenia, and multiorgan failure, such as liver failure, kidney failure, and acute respiratory distress syndrome (ARDS).

M protein and the pyrogenic exotoxins are thought to play key roles in the pathogenesis of the streptococcal toxic shock syndrome. M protein has been shown to be important in the organism's ability to evade the immune system. In particular, two M types, M1 and M3, have been associated with the recent resurgence in invasive GAS disease. In addition, a variety of pyrogenic exotoxins produced by GAS have been reported to act as superantigens. SpeA in particular has been associated with this syndrome.

To understand what a superantigen is, it is important to understand the difference between the interaction of an antigen and that of a superantigen with the immune system. It is estimated that an antigen stimulates approximately 1 in 10,000 T cells after it is processed ("presented") by an antigen-presenting cell (macrophage or macrophage equivalent). By contrast, superantigens do not require processing by macrophages and nonspecifically stimulate up to 20% of T cells. This stimulation of large numbers of T cells leads to massive release of cytokines, including tumor necrosis factor (TNF), interleukin-1, and interferon. It is this massive release of cytokines that is believed to be responsible, in part, for the shock and organ system failure that are seen in patients with the streptococcal toxic shock syndrome.

7. Given the frequency and the potential seriousness of GAS infections, they would seem a logical candidate for the development of a vaccine. Vaccine development strategies for GAS are targeting the M protein and a variety of other virulence factors, including the C5 peptidase (important in the organism evading phagocytes), cysteine protease, and hyaluronic acid capsule. The molecule that has been the most attractive target for the development of a GAS vaccine is the M protein. This protein is known to play an important role in evasion of the immune system; it is located on the cell surface, and with modern biochemical techniques, it is fairly easy to purify. However, epitopes of M protein have been shown to share antigenic properties with several human tissue components, including myosin and sarcolemmal membrane proteins. Therefore, vaccines against M proteins have the potential to induce antibodies that could bind and damage a variety of tissues.

The challenge of making a vaccine against the M protein component of GAS is to identify epitopes that will induce the production of protective antibodies against as many different M types as possible while at the same time ensuring that the antibodies raised against these epitopes will not react with human tissues. Currently, there are no ongoing large-scale GAS vaccine clinical trials.

REFERENCES
1. **Bisno, A. L.** 2001. Acute pharyngitis. *N. Engl. J. Med.* **344:**205–211.
2. **Bisno, A. L., M. A. Gerber, J. M. Gwaltney, Jr., E. L. Kaplan, and R. H. Schwartz.** 1997. Diagnosis and management of Group A streptococcal pharyngitis: a practice guideline. *Clin. Infect. Dis.* **25:**574–583.
3. **Cunningham, M. W.** 2000. Pathogenesis of group A streptococcal infections. *Clin. Microbiol. Rev.* **13:**470–511.
4. **Ebell, M. H., M. A. Smith, H. C. Barry, K. Ives, and M. Cary.** 2000. Does this patient have strep throat? *JAMA* **284:**2912–2918.

5. **Kiska, D. L., B. Thiede, J. Caracciolo, M. Jordan, D. Johnson, E. L. Kaplan, R. P. Gruninger, J. Lohr, P. H. Gilligan, and F. W. Denny.** 1997. Invasive group A streptococcal infections in North Carolina: epidemiology, clinical features, and genetic and serologic analysis of causative agents. *J. Infect. Dis.* **176:**992–1000.

6. **Martin, J. M., M. Green, K. A. Barbadora, and E. R. Wald.** 2002. Erythromycin-resistant group A streptococci in schoolchildren in Pittsburgh. *N. Engl. J. Med.* **346:**1200–1206.

7. **Seppala, H., T. Klaukka, J. Vupio-Varkila, A. Muotiala, H. Helenius, K. Lager, P. Huovinen, and the Finnish Study Group for Antimicrobial Resistance.** 1997. The effect of changes in the consumption of macrolide antibiotics on erythromycin resistance in group A streptococci in Finland. *N. Engl. J. Med.* **337:**441–446.

CASE 7 The patient was a 64-year-old retired postal worker with a medical history of extensive facial reconstruction for squamous cell carcinoma of the head and neck. He had a 30-year history of smoking. The patient presented with progressive shortness of breath; a persistent, productive cough; purulent sputum; and fever to 39.0°C 2 days prior to admission.

On physical examination he had a temperature of 37.3°C, respiratory rate of 18/min, pulse rate of 103 beats/min, blood pressure of 154/107 mm Hg, and pO_2 of 92 mm Hg. Chest auscultation revealed coarse breath sounds at the left lower base with bibasilar fine crackles. He was found to have a left lower lobe infiltrate on chest radiograph. His admission white blood cell count was 10,600 with 70% neutrophils, and his hemoglobin was 9.4. Sputum Gram stain at admission revealed >25 polymorphonuclear cells and >25 squamous epithelial cells. Because of the high numbers of squamous epithelial cells, the specimen was not processed further. Two blood cultures obtained at admission revealed the organism seen in Fig. 1. The Gram stain from the blood culture bottle is shown in Fig. 2. Of note: this was the patient's third episode of this illness in the past month. Isolates from all three episodes belonged to the same serotype, type 23.

1. What is the organism causing this individual's infection? What are his risk factors for becoming infected with this organism?

2. What other patient populations are at risk for infection with this organism?

3. The organism infecting this patient produces two different virulence factors. What are they and what role do they have in the pathogenicity of this organism?

Figure 1

4. What strategies are available to prevent infections with this organism? Why are preventive strategies becoming of greater importance with this organism?

5. How do you explain the patient's having repeated episodes of infection with the same serotype of this organism? There are at least two and possibly more explanations.

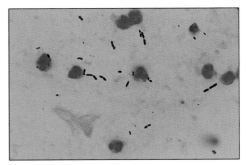

Figure 2

CASE DISCUSSION

1. Based on his physical and radiographic findings and the results of his blood cultures, this patient had bacterial pneumonia due to *Streptococcus pneumoniae* (also known as pneumococcus). *S. pneumoniae* is a catalase-negative, gram-positive diplococcus (Fig. 2). It is alpha-hemolytic on sheep blood agar and is susceptible to the copper-containing compound optochin. *S. pneumoniae* is the leading cause of community-acquired pneumonia, being responsible for approximately two-thirds of cases from which an etiologic agent is recovered. Approximately one-third of patients with pneumococcal pneumonia will have a positive blood culture, so the finding in this patient was consistent with this diagnosis. Frequently, pneumococcal pneumonia can be diagnosed by a characteristic Gram stain of sputum showing numerous polymorphonuclear cells and the presence of many lancet-shaped gram-positive diplococci. Such a diagnosis was not possible in this patient because he produced sputum of poor quality for analysis. Good-quality sputum specimens, i.e., those from the lower respiratory tract, are typically inflammatory with a high ratio of polymorphonuclear cells to squamous epithelial cells. Poor-quality specimens, on the other hand, typically have high numbers of squamous epithelial cells because of contamination of the specimen with oropharyngeal secretions. Oropharyngeal secretions contain high numbers of squamous epithelial cells. Because the pneumococcus can be part of the resident microflora of the oropharynx, the finding of this organism in a poor-quality sputum specimen cannot be reliably associated with the diagnosis of pneumococcal pneumonia. Such a finding may be a false positive.

The patient had several risk factors that put him at increased risk for pneumococcal pneumonia. These include his age (>60 years), immunosuppressed state due to his carcinoma, and long-standing smoking history.

2. Many different patient populations are at increased risk for invasive pneumococcal disease—pneumonia, bacteremia, and meningitis. Children less than 2 years of age have the highest risk for invasive disease with *S. pneumoniae*. Other patient populations in whom rates of pneumococcal invasive disease are increased include AIDS patients; patients who are anatomically or functionally asplenic (including patients with sickle-cell disease); patients with cardiovascular, liver, or kidney diseases; individuals with diabetes or malignancies; and individuals who are receiving immunosuppressive agents because of connective tissue disease or organ transplantation. Prevention strategies that target these populations are discussed in the answer to question 4.

3. The polysaccharide capsule is the major virulence factor of *S. pneumoniae*. Over 90 antigenically different capsular polysaccharides have been recognized, with seven types—4, 6B, 9V, 14, 18C, 19F, and 23F—being responsible for 80 to 90% of cases of invasive pneumococcal disease. Animal experiments done in the first part of the 20th century established the importance of capsule in the organism's ability to cause disease. It is well recognized that the capsular polysaccharide allows the pneumococcus to evade phagocytosis.

The second virulence factor is the cholesterol-dependent cytolysin, pneumolysin. Pneumolysin acts on both alveolar epithelial cells and pulmonary endothelial cells. Pneumolysin may contribute to the fluid accumulation and hemorrhage seen by directly damaging these two cell types. Animal studies of pneumococcal pneumonia indicate that pneumolysin plays a primary role in the inflammation, fluid accumulation, and hemorrhage that occurs in the alveoli during lobar pneumococcal pneumonia. The inflammatory response is due at least in part to pneumolysin up-regulating the synthesis of interleukin-6 in the airways.

4. Currently, there are two vaccines licensed for prevention of pneumococcal disease, a 23-valent polysaccharide vaccine and a 7-valent conjugate vaccine. The 23-valent vaccine is used in adults while the 7-valent conjugated vaccine was developed for use in children less than 2 years of age. These children are not able to reliably mount a T-cell-independent immune response, the type of immune response necessary to produce antibodies against polysaccharide antigens. However, they are able to mount a T-cell-dependent immune response.

A conjugate vaccine is one in which a polysaccharide antigen is coupled to a carrier protein. The coupling of a polysaccharide antigen to a protein creates a "new" antigen. This new antigen stimulates a T-cell-dependent immune response (see case 40 for further details). Therefore, the conjugated pneumococcal vaccine results in a protective immune response to capsular types present in the vaccine and perhaps to other related serotypes in children less than 2 years old. It has been shown to be highly efficacious (greater than 95%) in preventing invasive pneumococcal disease in this age group. It has been less effective in preventing a common pneumococcal infection in this age group, otitis media. The conjugated pneumococcal vaccine is now recommended for use in all children less than 2 years of age.

In adults, the 23-valent polysaccharide vaccine has been used successfully for many years. The efficacy of the 23-valent vaccine in adults is not as high (efficacy ranges from 50 to 90% in different populations) as the 7-valent conjugate vaccine is in children. Currently, there are no comparative data for the 23-valent polysaccharide vaccine versus the 7-valent conjugate vaccine in adults. In the United States, the goal is to vaccinate 90% of the population over 65 years of age with the 23-valent vaccine. This goal has not yet been met.

Alternatively, prophylactic antimicrobials have been used in selected populations, such as sickle-cell patients with a history of recurrent invasive pneumococcal infection.

Given the problem of emerging drug resistance in the pneumococci (see below), this is probably a preventive strategy which is becoming less viable.

The intense interest in pneumococcal vaccine is being driven to a significant degree by an alarming increase in the numbers of multidrug-resistant pneumococcal isolates being recovered from patients with invasive disease. Prior to 1990, pneumococcal isolates that were resistant to penicillin were quite unusual in the United States, as was the recovery of isolates that were resistant to other classes of antimicrobials. Beginning in the 1990s, pneumococcal isolates resistant to multiple antibiotics, including penicillin, macrolides, and trimethoprim-sulfamethoxazole, became increasingly common. Rates of resistance accelerated in the late 1990s. Some of this increase was due to the dissemination of selected clones of multidrug-resistant pneumococci, including the international dissemination of a multidrug-resistant type 23 strain. However, a common theme in the increasing drug resistance in this organism is the inappropriate use of antimicrobial agents. Several studies have been able to link increased use of specific antimicrobials, such as the macrolides and fluoroquinolones, with increased resistance. Because multidrug-resistant organisms are being seen with increasing frequency in invasive pneumococcal disease, it is clear that these multidrug-resistant strains have maintained their virulence, unlike some drug-resistant strains of other organisms which appear to be less virulent than nonresistant ones. Prevention of invasive infection with multidrug-resistant organisms by the two vaccines may be possible since greater than 90% of multidrug-resistant pneumococcal serotypes are either present in the vaccines or likely to cross-react with antibodies to the vaccine serotypes. It should be noted that in the pre-antibiotic era, mortality from invasive pneumococcal disease was 80%. It now stands at between 10 and 20%. With increasing resistance limiting the efficacy of antimicrobials, will mortality due to invasive pneumococcal disease begin to increase?

5. There are four potential explanations for why patients can have repeated episodes of infection with the same serotype. The first three fall under the category of inadequate treatment; the fourth involves reinfection.

In terms of inadequate treatment, the patient may have been treated with an antimicrobial to which the infecting organism was not susceptible. Given the increasing trend of multidrug resistance in pneumococci, this is a reasonable explanation. Susceptibility testing of this organism revealed it to be "pan-sensitive," meaning it was susceptible to all antimicrobials against which it was tested, including the antimicrobial with which he was treated. The second explanation is that the patient did not receive antimicrobials for a sufficient period of time to eliminate the organism. If hospitalized, it is likely that the patient would receive appropriate antimicrobial therapy during his stay. However, in the managed care era, hospitalizations are becoming shorter and shorter. He may have received 3 or 4 days of intravenous antimicrobials in the hospital and then oral antibiotics prescribed for use after discharge. If he failed to take his oral antibiotics, i.e., was noncompliant, his infection may have been inadequately treated, contributing to a relapse. A third possibility is that he had an undrained focus of infection which the antimicrobials did not adequately penetrate. In pneumococcal pneumonia, highly viscous pleural exudates may form that antimicrobials cannot penetrate. Removal of these exudates by drainage may be required for treatment of severe infections. Occasionally, drainage of exudates is not possible percutaneously. In these cases, a surgical procedure may be necessary to remove this focus of infection.

The fourth possible explanation is reinfection with the same serotype. Serotype 23 is one of the most common serotypes of *S. pneumoniae*, being responsible for 7% of invasive pneumococcal infections in a recent U.S. survey. It is not unreasonable to expect that the patient could become reinfected, especially if he is exposed to young children, who have a high nasopharyngeal carriage rate of pneumococci, especially in wintertime. It also is possible that he was carrying the organism in his nasopharynx and became reinfected in that manner, since it has been shown that antimicrobial therapy does not reliably eliminate nasopharyngeal colonization of pneumococci. What is more difficult to understand is why his original infection did not result in his mounting a protective immune response to this organism. A possible explanation is that his immunosuppressed state due to the carcinoma blunted his immune response. It is uncertain if vaccination would be an effective preventive strategy in this patient given the observation that he had three infections in a month with *S. pneumoniae* serotype 23, which is present in the vaccine.

REFERENCES

1. **Doern, G. V.** 2001. Antimicrobial use and the emergence of antimicrobial resistance with *Streptococcus pneumoniae* in the United States. *Clin. Infect. Dis.* **33**(Suppl. 3):S187–S192.

2. **Giebink, G. S.** 2001. The prevention of pneumococcal disease in children. *N. Engl. J. Med.* **345**:1177–1183.

3. **Jedrzejas, M. J.** 2001. Pneumococcal virulence factors: structure and function. *Microbiol. Mol. Biol. Rev.* **65**:187–207.

4. **Richter, S. S., K. P. Heilmann, S. L. Coffman, H. K. Huynh, A. B. Brueggemann, M. A. Pfaller, and G. V. Doern.** 2002. The molecular epidemiology of penicillin-resistant *Streptococcus pneumoniae* in the United States, 1994–2000. *Clin. Infect. Dis.* **34**:330–339.

5. **Schrag, S. J., B. Beall, and S. F. Dowell.** 2000. Limiting the spread of the resistant pneumococci: biological and epidemiologic evidence for the effectiveness of alternative interventions. *Clin. Microbiol. Rev.* **13**:588–601.

should be collected when the patient first awakens in the morning, a so-called "first morning sputum." It is believed that organism numbers in sputum are highest at that time.

This patient was begun on isoniazid for treatment of latent infection with *M. tuberculosis* (formerly called "prophylaxis") when she was a child. This was done because studies indicate that the risk of progression to clinical TB is decreased by more than half when patients complete a full course of daily oral isoniazid for a positive skin test. Lack of adherence to this regimen, as occurred in this patient, can often be prevented by the use of directly observed therapy (see below). Although there is some risk of hepatotoxicity due to this medication, the decrease in the risk of TB in individuals who complete a course of isoniazid results in the recommendation to use it in patients who have a positive tuberculin skin test and are under 35 years of age. Above the age of 35 years, there is a greater risk of hepatoxicity due to isoniazid than at the lower ages, but a recent reanalysis of available data suggests that the benefit may outweigh the risks of its use in patients over 35 years of age with a positive skin test.

5. *M. tuberculosis* is spread by respiratory droplets from infected individuals. Infection control measures to prevent the spread of this organism are based on this knowledge. Important components of respiratory isolation include housing patients suspected or known to be ill with TB in a "negative-pressure" room with the door closed, and the use of respirators by individuals entering the infected patient's room. A negative-pressure room is one in which the flow of air is into instead of out of the room. Patients must remain in the room until it is proven that they are either not infected or are no longer infectious. If the patient must leave the room, he or she must wear a face mask. For a patient to be considered no longer infectious, he or she must have negative acid-fast smears from three different sputum specimens obtained on different days.

One public health issue of importance is the screening of contacts of patients with active pulmonary TB. In fact, this patient's infant and four members of a family with whom the patient had lived were found to be PPD positive. They required evaluation to rule out the presence of active disease, including chest X rays, and were placed on isoniazid. The patient's mother did not cooperate with the public health nurse, frequently missed appointments, and subsequently developed a right upper lobe infiltrate with a cavity, consistent with active TB. Three additional contacts with clinical TB were subsequently identified as having acquired the same strain of *M. tuberculosis* on the basis of molecular typing of *M. tuberculosis* isolates by restriction fragment length polymorphism (RFLP) testing at the state public health laboratory. Thus, this patient was responsible for the infection of many other people with *M. tuberculosis*.

Besides infection control considerations, the other major factor in optimizing medical management is patient compliance with taking antituberculous therapy. Effective antituberculous therapy requires months to complete. Because of its length, patients are rarely hospitalized for its duration. That means that most antituberculous therapy is done in the outpatient setting, typically after the patient is no longer infectious. When patients are infected with *M. tuberculosis* that is resistant to multiple antimicrobial agents, the treatment regimen is not only extended but is also complex. As a result, an important factor in antituberculous therapy is directly observed therapy (DOT). In DOT, a reliable individual—anyone from a health care worker to a prison guard to a worker in a homeless shelter—watches the infected individual take his or her medicine to ensure adherence to the appropriate treatment regimen. Public health officials have the option of incarcerating infected individuals who refuse to follow their treatment regimen, to ensure compliance.

REFERENCES

1. **American Thoracic Society.** 2000. Targeted tuberculin testing and treatment of latent tuberculosis infection. *Am. J. Respir. Crit. Care Med.* **161**(4 Pt 2):S221–S247.

2. **Salpeter, S. R., G. D. Sanders, E. E. Salpeter, and D. K. Owens.** 1997. Monitored isoniazid prophylaxis for low-risk tuberculin reactors older than 35 years of age: a risk-benefit and cost-effectiveness analysis. *Ann. Intern. Med.* **127**:1051–1061.

3. **Selwyn, P. A., D. Hartel, V. A. Lewis, E. E. Schoenbaum, S. H. Vermund, R. S. Klein, A. T. Walker, and G. H. Friedland.** 1989. A prospective study of the risk of tuberculosis among intravenous drug users with human immunodeficiency virus infection. *N. Engl. J. Med.* **320**:545–550.

4. **Slovis, B. S., J. D. Plitman, and D. W. Haas.** 2000. The case against anergy testing as a routine adjunct to tuberculin skin testing. *JAMA* **283**:2003–2007.

CASE 9

The patient was a 5½-week-old male who was transferred to our institution with a 10-day history of choking spells. The child's spells began with repetitive coughing and progressed to his turning red and gasping for breath. In the prior 2 days, he also had three episodes of vomiting in association with his choking spells. His physical examination was significant for a pulse rate of 160 beats/min and a respiratory rate of 72/min (both highly elevated). The child's chest radiograph was clear. There was no evidence of tracheal abnormalities. His white cell count was 15,500/μl with 70% lymphocytes. The culture from the nasopharyngeal swab is seen in Fig. 1.

1. What was the organism infecting this child?

2. Were this child's clinical course and chest radiograph consistent with his infection? Explain your answer.

3. Why are specimens from the nasopharynx the specimens of choice in the diagnosis of this infection? Other than culture, what other methodology can be used to identify the presence of this pathogen in a specimen?

4. Why did this patient have a predominance of lymphocytes?

5. Vaccination is important in protecting children from infection with the organism infecting this child. How has the vaccine used to prevent this infection changed? What events led to this change?

6. The drug of choice to treat this infection is erythromycin. Clinically, the cough may persist for some time following therapy with erythromycin. Give possible reasons why a cough may persist in the face of erythromycin therapy.

Figure 1

CASE DISCUSSION

1. This child had a classic presentation for whooping cough, whose etiologic agent is *Bordetella pertussis*. The "mercury-like" colonies seen in Fig. 1 are typical of this organism. Special culture media, such as charcoal blood agar, are required for the isolation of *B. pertussis*.

2. Yes. This child's presentation is typical of whooping cough. This infection is usually limited to the upper airways, and pneumonia due to either *B. pertussis* or secondary bacterial agents is unusual. Therefore, normal chest radiographs are common.

Children with whooping cough often have paroxysms of coughing. The term "paroxysm" means a sudden recurrence or intensification. Children often cough repeatedly, and when they gasp for breath, the sound of this inspiration is the "whoop" of whooping cough. Because of repetitive coughing and resulting disruption of breathing, the children will have abnormal oxygen exchange and will often turn red and sometimes blue. The repetitive coughing may also result in vomiting or choking on respiratory secretions. All of these signs were seen in this child.

3. *B. pertussis* specifically binds to ciliated epithelial cells. This binding is mediated by filamentous hemagglutinin (FHA), an important virulence factor of this organism. Since the nasopharynx is lined with ciliated epithelial cells, culture of this site has a higher yield than culture from any other specimen source. However, culture is slow; it may take as long as 10 days to isolate this organism. In outbreak settings where this organism can be rapidly spread from person to person via inhalation, culture is too slow. For many years, direct fluorescent-antibody (DFA) testing for *B. pertussis* was done. This assay takes approximately 2 hours versus 7 to 10 days for culture, but it has a sensitivity of only 50 to 65%, and false-positive results may occur, especially when laboratorians are unaccustomed to reading these DFA smears. Where available, PCR (polymerase chain reaction) has become the method of choice for diagnosing whooping cough. This is especially important in settings where specimens must be transported long distances, because such transport compromises the sensitivity of culture but not the sensitivity of PCR. PCR is rapid, is much more sensitive than culture, and has a high negative predictive value. False-positive results do occur, but the overall performance accuracy of PCR is superior to that of culture. The Centers for Disease Control and Prevention (CDC) recommends that both culture and PCR be used diagnostically so that isolates of the organism will be available for susceptibility testing and molecular epidemiology studies.

4. *B. pertussis* produces a variety of virulence factors, including pertussis toxin. In earlier literature, pertussis toxin was described as many different entities, usually on the basis of a particular biological activity. One of the terms used to describe it was "lymphocytosis-promoting factor" because >50% of the peripheral white blood cells in mice injected with it were observed to be lymphocytes (the normal proportion is approximately 25%). Clinically, lymphocytosis, often as high as 70 to 80%, is routinely seen in patients with whooping cough and is a distinguishing characteristic of this infection.

5. Vaccination against pertussis using a whole-cell vaccine began in the 1940s. This vaccine was combined with diphtheria and tetanus toxoids to make the combination vaccine commonly known as DTP. With widespread immunization, a remarkable decline in disease incidence was achieved. This vaccine is far from perfect, however. There continue to be 3,000 to 7,000 cases of whooping cough per year in the United States. Most cases are in infants and young children, but some cases are in fully immunized children and adults. Cases among adults are particularly problematic. These cases often have an atypical presentation, with the patient having little more than a chronic cough. However, such an individual, especially if he or she is a health care provider or a day care center worker, can be the source of an outbreak of disease among immunologically more vulnerable infants and young children.

Another important problem with the whole-cell vaccine are side effects associated with its administration. Although almost all of the side effects are relatively mild and self-limited, rare cases of encephalopathy and death have been associated with administration of the whole-cell vaccine. A debate that was particularly intense during the late 1970s and early 1980s was whether this vaccine is more dangerous than the natural disease. Twice during the 1970s there were large-scale declines in the use of this vaccine in the industrialized world, once in Japan and once in Great Britain. Not surprisingly, large-scale outbreaks of disease soon followed, with a greater than 10-fold increase in the incidence of disease. Sadly, death rates from this disease also increased, from 5 deaths per year prior to the decline in vaccine use to 32 deaths per year during the height of the epidemic in Japan. Similar death rate increases were seen in Great Britain. With reacceptance of vaccination by a large segment of the population in those countries, declines in disease activity to pre-epidemic levels were achieved, emphasizing the importance of vaccination in disease prevention.

One of the benefits of the Japanese outbreak was that scientists in that country developed a prototype acellular vaccine. One of the problems with a whole-cell vaccine is that the antigen content is poorly standardized. In addition, with gram-negative organisms such as *B. pertussis*, the whole-cell vaccine often has significant lipopolysaccharide content, which is recognized to be reactogenic, especially with repeat administration. With increasing knowledge of the virulence factors involved in the pathogenesis of disease and also the protective immune response to the organism, protective antigens have been identified and tested. These antigens have been shown to elicit a protective immune response, with fewer side effects associated with vaccine administration. Currently, there are four DT acellular pertussis vaccines approved by the Food and Drug Administration (FDA) for immunization of children. These vaccines contain between one and four antigens (pertussis toxin, FHA, pertactin, and agglutinogens). The greater the number of antigens, the more protective the vaccines appear to be. Studies are under way to determine the safety and efficacy of these vaccines in adults. Pertussis vaccination is currently not recommended in adults because of the reactogenicity of the whole-cell vaccine in that population. However, because immunity wanes in adults, they can become infected with *B. pertussis* and spread it to vulnerable populations. Vaccination of adults at the same time they receive DT boosters may prevent infection in adults, further breaking the chain of transmission of a pathogen that only infects humans.

6. Although erythromycin is active against *B. pertussis*, the damage that the *B. pertussis* cytotoxin causes—ciliostasis and death of the tracheal epithelial cells—is not reversed by the administration of an antibiotic. Thus, the cough persists. Another possible reason why the cough may persist in the setting of erythromycin use is patient noncompliance, since erythromycin is often associated with gastrointestinal intolerance. An effective alternative therapy may be two new macrolide antimicrobials, clarithromycin and azithromycin. They are better tolerated than erythromycin and have similar in vitro activity. Data on the clinical efficacy of these agents against *B. pertussis* are limited. However, both appear to have similar if not superior efficacy to erythromycin in the treatment of whooping cough. Bacterial pneumonia, an occasional complication of this disease, must also be considered in a persistent cough, particularly if the patient worsens clinically.

Finally, the possibility that the organism is resistant to erythromycin must be considered. Resistant *B. pertussis* isolates have been identified in Arizona. However, susceptibility surveys suggest that *B. pertussis* resistance to macrolides is still rare in the United States.

REFERENCES

1. **Cherry, J. D., and P. Olin.** 1999. The science and fiction of pertussis vaccines. *Pediatrics* **104**:1381–1383.

2. **Guris D., P. M. Strebel, B. Bardenheier, M. Brennan, R. Tachdjian, E. Finch, M. Wharton, and J. R. Livengood.** 1999. Changing epidemiology of pertussis in the United States: increasing reported incidence among adolescents and adults, 1990–1996. *Clin. Infect. Dis.* **28**:1230–1237.

3. **Hill, B. C., C. N. Baker, and F. C. Tenover.** 2000. A simplified method for testing *Bordetella pertussis* for resistance to erythromycin and other antimicrobial agents. *J. Clin. Microbiol.* **38:**1151–1155.

4. **Loeffelholz, M. J., C. J. Thompson, K. S. Long, and M. J. R. Gilchrist.** 1999. Comparison of PCR, culture and direct fluorescent-antibody testing for detection of *Bordetella pertussis. J. Clin. Microbiol.* **37:**2872–2876.

CASE 10

The patient was a 3-year-old male who presented with a 4-day history of fevers. He became acutely ill and vomited during lunch. Over the next 4 days he developed fevers as high as 40°C that were controlled by Tylenol. He also developed cough, rhinorrhea, and conjunctivitis. He appeared to be fatigued, and his parents reported that he was "very sleepy." Over the past 2 days, his eyes had begun to itch and were painful. His parents noted that his eyes were puffy and he was sensitive to light. He had had no rashes. The patient's lips were dried and cracked, and he had a greatly reduced urinary output.

Other history pertinent to his illness is that he attended preschool twice per week, where he had multiple sick contacts (his illness occurred in late January). His 1-year old sibling had otitis media, some wheezing, vomiting, and a productive cough.

On physical examination he had a temperature of 38.6°C, pulse rate of 126 beats/min, and respiratory rate of 28/min with an oxygen saturation of 100% on room air. Significant findings included bilateral conjunctivitis with exudate in the left eye, bleeding, cracked lips, and rhinorrhea. He had shotty lymphadenopathy but no rash. His feet were slightly edematous. His respiratory examination was normal. Laboratory findings were all normal. A nasopharyngeal swab was sent for rapid antigen testing for respiratory syncytial virus and influenza A virus. The results of this test are shown in Fig. 1.

1. What is the agent causing his infection? What symptoms does he have which are consistent with his illness? What are the key virulence factors of this agent?

2. What is the usual outcome of this infection in this patient population? What patient populations are particular prone to infections with this agent?

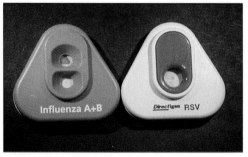

Figure 1 Influenza test: top well, influenza A; bottom well, influenza B.

3. The finding of conjunctivitis in this child made the pediatrician examining him attempt to elicit a history of rash. What was the clinical syndrome that this physician was concerned about in this child given his clinical presentation?

4. Why is it important to use acetaminophen or ibuprofen instead of aspirin to manage fever in children who have a presentation similar to this child?

5. How do you think this child became infected?

6. The infection this child has is vaccine preventable. Briefly describe the vaccine that is used to prevent this disease. How is this vaccine currently employed in the industrialized world? Are children generally vaccinated against this virus? Why is this policy followed?

7. What therapeutic options are available to treat this infection?

CASE DISCUSSION

CASE 10

1. During the winter months (Northern Hemisphere, late November to early April) in temperate areas, respiratory illnesses caused by both respiratory syncytial virus (RSV) and influenza virus are frequently epidemic in children 5 years of age and younger. Because the clinical presentation of these two illnesses in children can be similar (fever and upper respiratory tract symptoms), it is useful to distinguish between the two in the seriously ill child. The rapid enzyme immunoassay tests which can detect antigens of either influenza A and B or RSV directly in respiratory secretions were positive for influenza A virus. In addition to influenza A virus (the most common and most likely to cause severe disease), infection can occur with influenza B virus (also common) and with influenza C virus (less common, less severe, and does not occur in epidemics). Influenza viruses are single-stranded, segmented, negative-sense RNA viruses.

Influenza viruses have two well-characterized virulence factors on the surface of the virus, neuraminidase and hemagglutinin. Neuraminidase appears to have two functions in viral infections. First, it mediates virus penetration through the mucous layer overlaying the surface of the respiratory epithelium. Second, it plays a role in the release and spread of the virus from infected respiratory cells. Once the virus penetrates to the surface of the cell, binding to specific sialic acid-rich receptors is mediated by hemagglutinin. After the virus is endocytosed into the cell, hemagglutinin plays a role in the formation of channels through which viral RNA can enter the cytoplasm and initiate the viral replicative cycle.

2. Most cases of influenza in this age group are self-limiting. However, several observations have been made concerning influenza in nonimmunocompromised children. During "flu" season, the numbers of hospitalizations for respiratory illnesses, outpatient visits, cases of severe bacterial pneumonia, and the use of antibiotics all increase. It has been speculated that the number of cases of otitis media also increase. However, these data are confounded by the fact the RSV infections, which are more common than influenza infections in children, also peak at a similar time, and it is difficult to determine the contribution of each one of these viruses to these observations.

Mortality due to influenza in children, including deaths due to secondary bacterial pneumonia, is uncommon.

Influenza is a much greater threat to individuals over 50 years of age, especially those with underlying chronic diseases, particularly cardiopulmonary ones. There is significant excess mortality in that patient population due to influenza and accompanying bacterial pneumonia, especially pneumonia caused by *Streptococcus pneumoniae* and *Staphylococcus aureus*. Other patient populations recognized to be at increased risk for adverse outcomes with influenza infection include children with asthma and other chronic lung diseases, residents of long-term care facilities, and women in the second or third trimester of pregnancy.

3. The physician was concerned that the patient might have Kawasaki disease, a generalized vasculitis seen most frequently in children less than 5 years of age. This patient had many of the signs associated with this syndrome, including prolonged fever (usually at least 5 days), bilateral conjunctivitis, cracked and bleeding lips, swollen extremities, and lymphadenopathy. Patients with Kawasaki disease typically also have rash. Kawasaki disease is a diagnosis of exclusion; i.e., other possible explanations for the patient's symptoms are not found. The pathogenesis of this syndrome is not understood. It is believed to have an infectious etiology even though such an etiology has been sought for several decades without success. In this patient, the finding of influenza made Kawasaki disease much less likely. Cardiac manifestations play a significant role in the morbidity and mortality associated with Kawasaki disease.

4. Reye's syndrome occurs primarily in children who are treated with aspirin (salicylates) during a viral infection, particularly that caused by influenza viruses or varicella-zoster virus (chicken pox). This syndrome is characterized by fatty degeneration of the liver as well as acute encephalopathy with cerebral edema. Therefore, fever in children that is due to viral infection should be treated with antipyretic agents other than aspirin, such as acetaminophen or ibuprofen. Since the treatment of Kawasaki disease includes the use of high-dose aspirin, it was important to rule out viral etiologies as a cause of this child's syndrome before embarking on therapy for that disease. Laboratory findings showed that this patient's clinical syndrome was due to influenza A virus.

5. Day care centers have been shown to be an environment in which influenza A virus can spread rapidly. Since many of his classmates were sick during a time of the year (January) when both influenza A and RSV are likely to be endemic in the community, it would not be surprising that he became infected in his day care center. In addition, the observation that his 1-year-old brother also was sick with a syndrome consistent with influenza points out that a child in day care can bring home a variety of infectious agents, including influenza, resulting in illness among family members.

6. Influenza can be prevented by the use of a trivalent, inactivated vaccine. The strains present in the current vaccine include two subtypes of influenza A, H1N1 and H3N2, and influenza B. This vaccine must be given annually, as immunity to the virus is short-lived. The antigenic composition of the vaccine is determined by the types of viruses that circulated during the previous season. The efficacy of the vaccine is dependent on the level of change that may occur from year to year in the circulating virus. Changes in the antigenic structure of influenza A are classified in two ways, as "antigenic drift" and "antigenic shift." The antigenic variation occurs because of changes in the hemagglutinin (H) glycoprotein antigen and the neuraminidase (N) glycoprotein antigen. A total of three subtypes of hemagglutinin (H1, H2, and H3) and two subtypes of neuraminidase (N1 and N2) are typically associated with human influenza A virus epidemics. Small changes in the H and N antigens due to the accumulation of point mutations resulting in amino acid substitutions are responsible for antigenic drift. For influenza A virus, these changes will not necessarily result in the change of the classification of a viral strain (which is based upon the subtypes of the H and N antigens), but they may be sufficient to render patients with antibodies to the parent strain susceptible to the new mutant strain. This is the basis for the decision to reevaluate and potentially change the formulation of the influenza vaccine each year to include recent isolates, so that protective antibodies to the most recent isolates will be made in response to the vaccine.

The more dramatic antigenic shift is due to either the genetic reassortment of genes encoding the H and N antigens or, as occurred in a recent Hong Kong outbreak, direct introduction of new H and N antigens directly from avian sources. Since the influenza A virus contains a segmented RNA genome, coinfection of a cell with two different influenza A viruses can result in many different possible reassortments via the exchange of RNA segments. This may result in a new virus that differs dramatically from the parent strains in one or both of these antigens. Historically, antigenic shift (for example, the change from H2N2 to H3N2 in 1968), has been responsible for worldwide epidemics (pandemics) due to the susceptibility of much of the world's population to the new virus. This reassortment may occur in animals, such as pigs and birds. An H5N1 virus, believed to be of avian origin, was recently recovered from humans in Hong Kong. This unusual strain of influenza A was found to be highly virulent in people (killing 6 of the 18 patients with proven H5N1 infection), but through strong public health measures, including the killing of approximately 1 million chickens, ducks, and geese thought to be the reservoir for this virus, the outbreak was controlled. There are currently no data suggesting that this novel form of influenza virus is circulating in human populations. There is, however, a concern that at some time in the near future there will be a pandemic of influenza A due to the emergence in the human population of a novel strain as a result of either a significant antigenic shift or the introduction of an avian virus that is able to efficiently spread from person to person. Plans to ensure public health readiness for such an event are being made.

The vaccine is given to at-risk populations (see the answer to question 2 for a listing of at-risk populations) and health care providers who could transmit the virus to their at-risk patients. Numerous studies have proven the efficacy of this vaccine strategy. Recent studies suggest that immunocompetent children may also benefit from vaccination through reduction in hospitalizations, doctor office visits, antibiotic use, serious secondary bacterial infections, and spread to at-risk family members. Others would argue that vaccination in this patient population is not cost-effective. What is agreed on by both groups is that the current inactivated vaccine is not practical for widespread vaccination of children because it would add two additional vaccine injections annually to a list of soon-to-be 20 vaccine injections recommended in the first 2 years of life. The vaccine is not recommended for children less than 6 months of age, a population that at least some pediatricians feel would most likely benefit from influenza virus vaccination. A live, attenuated influenza vaccine that can be given intranasally has been shown to be protective in children older than 15 months. Adaptation of this vaccine for use in infants would reduce the barrier for widespread childhood vaccination against influenza. The risk-benefit ratio of this strategy has not been determined.

7. There are currently two classes of anti-influenza drugs. The first class of agents blocks formation of influenza-derived ion channels. The reason these virally derived ion channels are important is that they play an important role in the "uncoating" of the virus. This is a step in viral replication in which viral nucleic acid, in this case RNA, is released from the viral particle and enters the cytoplasm of the cell. The two drugs in this class are the oral agents amantadine and rimantadine. The drugs must be administered in the first 2 days of illness to be effective. They have been shown to reduce the disease course by 1 day. In addition, these agents prevent influenza illness in approximately 70 to 90% of individuals who take these agents prophylactically.

The second group of agents is the neuraminidase inhibitors. Two agents belong to this class of drugs—zanamivir, which is an inhaled agent, and oseltamivir, which is an oral agent. These agents are effective only if given in the first 2 days of illness and, like the ion channel-blocking agents, reduce the disease course by 1 day. They have not been approved for prophylactic use. The advantage of the neuraminidase inhibitors is that they are active against both influenza A and B viruses, while amantadine and rimantadine are active only against influenza A virus.

REFERENCES

1. **Couch, R. B.** 2000. Prevention and treatment of influenza. *N. Engl. J. Med.* **343:**1778–1787.

2. **Cox, N. J., and K. Subbarao.** 1999. Influenza. *Lancet* **354:**1277–1282.

3. **Dajani, A. S., K. A. Taubert, M. A. Gerber, S. T. Shulman, P. Ferrieri, M. Freed, M. Takahashi, F. Z. Bierman, A. W. Karchmer, W. Wilson, S. H. Rahimtoola,**

D. T. Durack, and G. Peter. 1993. Special report: diagnosis and therapy of Kawasaki disease in children. *Circulation* **87:**1776–1780.

4. **Horimoto, T., and Y. Kawaoka.** 2001. Pandemic threat posed by avian influenza A viruses. *Clin. Microbiol. Rev.* **14:**129–149.

5. **McIntosh, K., and T. Lieu.** 2000. Is it time to give influenza vaccine to healthy infants? *N. Engl. J. Med.* **342:** 275–276.

6. **Neuzil, K. M., B. G. Mellen, P. F. Wright, E. F. Mitchel, Jr., and M. R. Griffin.** 2000. The effect of influenza on hospitalizations, outpatient visits, and courses of antibiotics in children. *N. Engl. J. Med.* **342:**225–231.

CASE 11 The patient was a 4-month-old female who was admitted to the hospital in March with severe respiratory distress. Five days prior to admission she had developed a cough and rhinitis. Two days later she began wheezing and was noted to have a fever. She was brought to the emergency room when she became lethargic.

One sibling was reported to be coughing, and her father had a "cold." On examination she was agitated and coughing. She had a fever of 38.9°C, tachycardia with a pulse rate of 220 beats/min, tachypnea with a respiratory rate of 80/min, and blood pressure of 90/58 mm Hg. Her fontanelles were open, soft, and flat. Her throat was clear. She had subcostal retractions and nasal flaring. On auscultation of her lungs, there were rhonchi as well as inspiratory and expiratory wheezes.

A chest radiograph revealed interstitial infiltrates and hyperexpansion. Arterial blood gases on supplemental oxygen revealed a respiratory acidosis with relative hypoxemia. She was put in respiratory isolation in the pediatric intensive care unit and was subsequently intubated. Blood and nasopharyngeal cultures were obtained and sent to the bacteriology and virology laboratories. A rapid diagnostic test was positive (Fig. 1) and specific antiviral therapy was begun. She was also given the bronchodilator aminophylline to treat the bronchospasm that was resulting in her wheezing. She was extubated 5 days later and discharged home on day 8.

1. What is the differential diagnosis for this patient's pneumonia, and which of the possible viral agents is the most likely etiology?

2. Describe the epidemiology of the agent causing her infection.

3. Describe diagnostic strategies available for detection of the agent. Why is it important to establish this diagnosis quickly?

4. What treatment strategies are available for this infection?

5. What are the hospital's infection control issues related to this patient's diagnosis?

6. Besides infection control measures, what strategies are available for prevention of this infection?

7. Describe the pathophysiologic basis for wheezing.

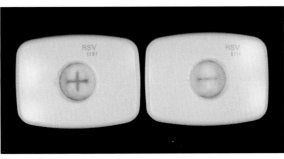

Figure 1

CASE DISCUSSION

1. The differential diagnosis for this patient's pneumonia includes respiratory viruses such as parainfluenza virus types 1, 2, and 3, adenovirus, influenza A and B viruses, and respiratory syncytial virus (RSV). *Mycoplasma pneumoniae* or *Bordetella pertussis* could also have caused her illness. This patient was found to be infected with RSV.

2. RSV is an enveloped, single-stranded RNA virus. It is the most important viral etiology of childhood respiratory illness in the industrialized world in terms of morbidity and mortality. It is particularly severe in preterm infants and in infants with chronic pulmonary or cardiovascular disease. Approximately 80% of RSV infections occur in children less than 5 years old, and disease with the greatest morbidity and mortality is seen in children less than 1 year old. Reinfections in older children and adults, as was probably the case here, may result in minimal respiratory tract symptoms. The elderly, especially those living in nursing home settings, and adults who are either immunocompromised or who have chronic cardiopulmonary disease may also develop severe RSV disease. It is estimated that from 2 to 9% of hospitalizations for pneumonia in adults more than 65 years old are for RSV.

Epidemics of RSV occur each winter in temperate climates. In the United States, peak disease prevalence is seen in January and February. The molecular epidemiology of RSV is becoming better understood. There are two antigenically distinct groups of RSV, designated A and B. Both circulate during epidemics, although type A tends to predominate in most but not all epidemics. Antigenic heterogeneity can occur in both group A and B viruses.

RSV is spread by large droplets and on fomites. In the hospital and day care center, it can be spread to the susceptible child on the hands of caregivers who do not use good hand-washing practices.

3. RSV can be detected within 1 to 2 hours by using either a direct fluorescent-antibody (DFA) technique or, as can be seen in Fig. 1, a membrane-enzyme immunoassay (EIA). Both methods have a sensitivity approaching that (or better than that) of culture. The advantage of the EIA method, and the reason why it is widely used, is that it is easy to perform. A disadvantage of this technique is that specimen quality cannot be determined. Specimen quality can be judged by the DFA technique. A new DFA reagent that detects both RSV and other respiratory viruses may result in DFAs becoming the method of choice for detection of RSV in respiratory specimens.

Culture detection can be performed relatively rapidly using the shell vial technique. (See case 18 for a description of this method.) This technique uses an antibody pool consisting of monoclonal antibodies specific for RSV, parainfluenza virus types 1 through 3, influenza A and B viruses, and adenovirus to stain the shell vial monolayer after 48 hours. If this stain is positive, then antibodies specific for individual viruses can be used to detect the specific agent causing the infection. This method would be used only if the rapid RSV assay was negative.

A rapid diagnostic technique is important for management decisions, including infection control and treatment, especially in children with congenital heart disease, bronchopulmonary dysplasia, cystic fibrosis, and immunodeficiency or immunosuppression.

4. Only one antiviral agent, ribavirin, is available for treatment of RSV in infants. It must be delivered by aerosol since oral (p.o.) administration may result in hepatic or bone marrow toxicity.

Because of its expense, difficulties in its administration, and debate about its clinical benefit, the use of ribavirin for RSV therapy remains controversial. The American Academy of Pediatrics recently changed its recommendations about the use of ribavirin from "should be used" to "may be considered" in RSV-infected children with congenital heart disease, bronchopulmonary dysplasia, cystic fibrosis, immunodeficiency or immunosuppression, or severe illness (defined as an arterial blood pO_2 of less than 65 mm Hg), or in children being mechanically ventilated. This change in recommendation was made because significant benefit from ribavirin therapy could not be demonstrated in several studies of RSV-infected patients. This child's illness occurred before the change in recommendations. She received ribavirin.

5. Because RSV can cause nosocomial infections, patients should be put in respiratory isolation. If patients are not isolated and stringent infection control practices are not followed, secondary infection rates of 20 to 50% can occur. Cohorting RSV-positive children and their health care providers, plus the use of gloves and gowns during contact with infected children and consistent hand washing before and after patient contact, has been shown to significantly lower RSV nosocomial infection rates. Though positive patients may be cohorted, many centers do not consider the sensitivity of the assay high enough to place a child with a negative RSV antigen assay in the same room as another child if there is a clinical suspicion of RSV. Nosocomial RSV infections are a hazard, particularly for hospitalized patients with congenital heart disease, lung disease, or immunodeficiency states who are at risk for life-threatening RSV infections.

6. Despite its importance as a cause of serious childhood respiratory infection and an increasing awareness of its importance as a cause of pneumonia in the elderly, no vaccine is currently available for RSV. Attempts to develop an inactivated RSV vaccine were disastrous: children receiving the candidate vaccine had more severe disease than the control populations.

Two approaches to RSV vaccines are under development. One approach has been to develop a live, attenuated vaccine. This approach thus far has been unsuccessful because candidate vaccines have either reverted to wild-type virus, or have not conferred immunity. A novel candidate vaccine that maintains its attenuated phenotype is currently being evaluated. The second approach is a subunit vaccine. Two transmembrane proteins, the F (fusion) protein, and the G (attachment) protein, have been used as target antigens because the presence of antibodies to these proteins has been associated with immunity. Two types of vaccines are currently under development. One is composed of purified F protein while the other is a chimeric FG glycoprotein. Only the F protein-containing vaccines have been tested in humans (older children and adults). They appeared to be safe, and approximately one-half to three-fourths of vaccinees produced increasing levels of neutralizing antibodies. However, large clinical trials in the major target population for RSV, children less than 1 year old, will be necessary before the potential usefulness of these vaccines can be judged.

Another approach has been the prophylactic use of human pooled RSV-immune globulin or a humanized mouse monoclonal antibody called palivizumab in high-risk populations. Palivizumab is directed against the RSV F protein. These products have been shown to have some benefit for children less than 2 years old with chronic lung disease but not heart disease and for premature infants in their first 6 to 12 months. Because of the very high expense of these products, efficacy needs to be demonstrated.

7. Wheezing and stridulous cough develop in infants with RSV infection because the virus has a tropism for the bronchial epithelium. As with other viral agents, this virus has been shown to be cytotoxic for ciliated epithelial cells. The resulting necrosis and edema can lead to collapse and blockage of the small-diameter bronchioles, with air trapping distally. The lung parenchyma can be involved with or without bronchiolitis. This patient apparently had involvement of both the bronchioles and the lung parenchyma.

REFERENCES

1. **American Academy of Pediatrics.** 1996. Reassessments of the indications for ribavirin therapy in respiratory syncytial virus infections. *Pediatrics* **97:**137–140.

2. **Domachowske, J. B., and H. F. Rosenberg.** 1999. Respiratory syncytial virus infection: immune response, immunopathogenesis, and treatment. *Clin. Microbiol. Rev.* **12:**298–309.

3. **Han, L. L., J. P. Alexander, and L. J. Anderson.** 1999. Respiratory syncytial virus pneumonia among the elderly: an assessment of disease burden. *J. Infect. Dis.* **189:**25–30.

4. **Landry, M. L., and D. Ferguson.** 2000. SimulFluor respiratory screen for rapid detection of multiple respiratory viruses in clinical specimens by immunofluorescence staining. *J. Clin. Microbiol.* **38:**708–711.

5. **Shay, D. K., R. C. Holman, R. D. Newman, L. L. Liu, J. W. Stout, and L. J. Anderson.** 1999. Bronchiolitis-associated hospitalizations among US children, 1990–1996. *JAMA* **282:**1440–1446.

6. **Sullender, W. M.** 2000. Respiratory syncytial virus genetic and antigenic diversity. *Clin. Microbiol. Rev.* **13:**1–15.

CASE 12

This 40-year-old male with multisystem failure secondary to bilateral pneumonia was transferred to our hospital via helicopter. He had presented to his local physician 3 days previously complaining of fevers, malaise, and vague respiratory symptoms. He was given amantadine for suspected influenza. His condition became progressively worse, with shortness of breath and a fever to 40.5°C, and he was admitted to an outside hospital 24 hours prior to transfer. A laboratory examination revealed abnormal liver and renal function. Therapy with Timentin (ticarcillin-clavulanic acid) and trimethoprim-sulfamethoxazole was begun. On admission, he underwent a bronchoscopic examination that revealed mildly inflamed airways containing thin, watery secretions. A Gram stain of bronchial washings obtained at bronchoscopy is shown in Fig. 1. Based on these findings, he was begun on appropriate antimicrobial therapy. Culture results are shown in Fig. 2.

Despite appropriate antimicrobial agents and supportive therapy, the patient never recovered adequate pulmonary function and died 9 months later in a long-term care facility.

1. Which organisms are common causes of community-acquired pneumonia?

2. What are bronchial washings and how are they obtained?

3. On the basis of the Gram stain of the bronchial washings and the patient's presentation, what is the most likely cause of this patient's catastrophic infection? Why must the laboratory be notified if this organism is considered in the differential diagnosis?

4. What techniques other than culture can be used to detect this organism within 24 hours of obtaining the culture?

5. What is the epidemiology of this organism? What infection control precautions are necessary with this patient?

6. What is the appropriate antimicrobial agent for treatment of this infection?

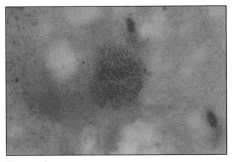

Figure 1

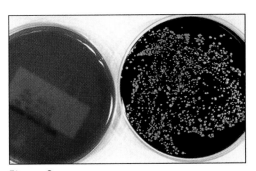

Figure 2

CASE DISCUSSION

1. The common bacterial causes of community-acquired pneumonia in immunocompetent patients are *Streptococcus pneumoniae; Haemophilus influenzae; Mycoplasma pneumoniae; Chlamydia pneumoniae; Staphylococcus aureus*, frequently following an initial infection with influenza virus; *Klebsiella pneumoniae*, especially in the elderly and alcoholics; and *Legionella pneumophila*. If there had been evidence of aspiration, mixtures of oral streptococci and anaerobic bacteria would also need to be considered. Viral causes include influenza A virus, for which this patient was treated, and other respiratory viruses, including adenovirus, parainfluenza virus, and respiratory syncytial virus. If the patient became ill during the winter months, influenza virus would be the most likely viral agent causing his illness.

2. Bronchial washings are obtained during bronchoscopic examination. The bronchoscope is introduced, and a small volume of saline is injected into the bronchi through a channel in the bronchoscope. The mixture of saline and bronchial secretions is then suctioned from the bronchi through the bronchoscope and sent to the laboratory for staining and culture. Bronchoalveolar lavage is another sampling method for detection of respiratory tract specimens. In this technique, the bronchoscope is wedged into a bronchus, and large volumes of saline (250 to 500 ml) are introduced into the airways in order to "lavage" the bronchus and the connecting bronchioles and alveoli. The saline is then quickly suctioned back out. This technique is actually much more sensitive than bronchial wash for detection of many infectious agents, including the organism causing his infection.

3. This patient has *L. pneumophila* pneumonia. *L. pneumophila* is an aerobic, poorly staining gram-negative rod. The Gram stain in Fig. 1 is what is known as an "enhanced" Gram stain, and thus the organisms appear more readily than they would with the Gram stain technique typically used. Legionellae are fastidious organisms and will not grow on media routinely used for cultivation of respiratory secretions. The laboratory must be informed that this organism is suspected so that a special medium, buffered charcoal yeast extract (BCYE) agar, which will support the growth of this organism, is used. This is the medium showing growth in Fig. 2. In addition, BCYE with antibiotics is used as a selective medium. Since *Legionella micdadei*, another species that may cause a clinically identical syndrome, may be inhibited by BCYE with antibiotics, BCYE without antibiotics must be used as well.

This case demonstrates the catastrophic nature of this illness in some previously healthy patients. Key findings in this case include hepatic and renal dysfunction, as well as the finding of thin, watery secretions, which are characteristic of pneumonia with this infection. Often patients with legionellosis have a "dry" cough. As a result, they do not cough up anything at all and require bronchoscopy for detection of this organism. Patients with bacterial pneumonia due to most other bacterial agents typically have thick, purulent secretions.

4. Culture for *L. pneumophila* typically takes up to 5 days before colonies are visible. Given this organism's potential for causing catastrophic illness, as we saw in this patient, more rapid diagnostic techniques are important in optimizing the management of this disease. *L. pneumophila* can be rapidly detected in two ways: by performing a direct fluorescent-antibody (DFA) test and by detecting a urinary antigen. Both tests have been used in clinical settings. The DFA test has a sensitivity of 60 to 70% compared with culture; i.e., 3 or 4 of 10 patients who are culture positive for *L. pneumophila* will not be detected by this test. However, the DFA test takes only 2 hours to become positive, compared with up to 5 days for culture, so it is useful. The urinary antigen test, an enzyme immunoassay, is more sensitive (80 to 90%) than the DFA test and is highly specific. It is positive primarily in patients infected with *L. pneumophila* serogroup 1, although cross-reactions with other serogroups and *Legionella* species have been reported. Epidemiologic studies have shown, and our experience has been, that *L. pneumophila* serogroup 1 causes approximately 80% of *Legionella* infections. Therefore, a percentage of cases will be missed by the urinary antigen test. For these reasons, culture continues to be essential.

PCR (polymerase chain reaction) detection of *Legionella* spp. appears to be a promising technique for detection of this organism in respiratory secretions. Because these specimens can contain substances that may inhibit the PCR reaction, purification of DNA from these samples appears to improve test performance. Unfortunately, it also adds time and expense. The technique has been shown to be at least as sensitive as culture, but false-positive results do occur. Because of these difficulties, PCR for detection of *Legionella* spp. is currently not widely available.

5. *L. pneumophila* was first recognized when an outbreak of pneumonia of unknown etiology called Legionnaires' disease occurred in 1976 during an American Legion convention in Philadelphia. The source of the organism was traced to the air-conditioning system at one of the convention hotels. Molecular epidemiologic studies of several outbreaks have shown that exposure to aerosols of water containing *L. pneumophila* is the major mode of transmission. *L. pneumophila* has been shown to cause sporadic nosocomial outbreaks of pneumonia. A common theme in all of these outbreaks was the aerosolization of water contaminated with *L. pneumophila*. The organism grows in air-conditioning systems, showerheads, tap water, and sinks. It is very difficult to eradicate from hospital water supplies, although superheating and hyperchlorination of water have both been used with various degrees of success.

Despite concern about nosocomial outbreaks of *L. pneumophila* infection, sporadic community-acquired cases are more common in the United States. It is estimated that *Legionella* spp. cause between 2 and 9% of cases of community-acquired pneumonia that require hospitalization. Chronic lung disease, immunosuppression, and advancing age all have been implicated as risk factors. *Legionella* pneumonia is of particular concern in solid-organ transplant recipients, in whom infection is associated with high mortality. Our patient had none of these. A possible risk factor in this patient may have been viral infection prior to his *Legionella* infection. Although prior viral infection is not recognized as a risk factor for *L. pneumophila* pneumonia, it is well recognized that bacterial superinfection can follow viral pneumonia, and this patient's original presentation may have been due to a virus such as influenza virus. Alternatively, *L. pneumophila* can cause a "flu-like" prodrome, and all his symptoms could be explained by this infection.

Unlike many respiratory tract infections, there is no evidence of person-to-person spread of this organism. Rather, infection is obtained by inhalation of aerosols of *Legionella*-contaminated water. As such, the finding of nosocomial pneumonia with *Legionella* spp. should lead to an investigation of various water sources in the patient's environment. Special respiratory precautions, including placing patients in negative-pressure rooms, are not necessary with *Legionella* infections. His respiratory illness, which was initially thought to be due to influenza A virus, which is highly transmissible by the respiratory route, resulted in his placement in respiratory isolation. In retrospect, given that the etiologic agent was *L. pneumophila*, this was unnecessary.

6. Erythromycin is the therapy of choice. This agent is usually considered to be active against two gram-negative rods that cause infection in the respiratory tract, *Legionella* spp. and *Bordetella pertussis*. A key characteristic of this antimicrobial agent is its ability to penetrate into white cells. This characteristic is probably important therapeutically since *Legionella* spp. survive and multiply within macrophages. Beta-lactam drugs have been proven ineffective against *L. pneumophila*, in part because of its ability to produce a β-lactamase. Although clinical experience is limited, the drugs clarithromycin and azithromycin are chemically closely related to erythromycin, and both appear to have good in vitro activity against *L. pneumophila*. Fluoroquinolones, including ciprofloxacin, levofloxacin, and others, have been used with some success in cases in which ototoxicity, a side effect of high doses of intravenous erythromycin, has necessitated a change from erythromycin to another drug. Finally, rifampin, which achieves excellent levels within eukaryotic cells, has been used in combination with erythromycin with some success in some severe *Legionella* infections.

REFERENCES

1. **Benson, R. F., P. W. Tang, and B. S. Fields.** 2000. Evaluation of Binax and BioTest urinary antigen kits for detection of Legionnaires' disease due to multiple serogroups and species of *Legionella. J. Clin. Microbiol.* **38:**2763–2765.

2. **Cloud, J. L., K. C. Carroll, P. Pixton, M. Erali, and D. R. Hillyard.** 2000. Detection of *Legionella* species in respiratory specimens using PCR with sequencing confirmation. *J. Clin. Microbiol.* **38:**1709–1712.

3. **Stout, J. E., and V. L. Yu.** 1997. Legionellosis. *N. Engl. J. Med.* **337:**682–687.

C A S E 13 The patient was a 34-year-old HIV-seropositive woman who presented with a fever, productive cough, diarrhea, headache, dizziness, anorexia, nausea, sore throat, and dysphagia. She also noted a 10-lb (ca. 4.5-kg) weight loss over the previous 2 months. She had not received any treatment for her HIV infection or prophylaxis for opportunistic infections since her diagnosis of HIV infection in 1988.

On physical examination, she was malnourished and in acute distress due to diarrhea. Her vital signs were normal. Her chest examination was abnormal with bibasilar crackles distributed up to the mid lung field. Her pO_2 was 77 mm Hg. Her chest radiograph showed diffuse interstitial infiltrates and a small pleural effusion in the right lung. Her white blood cell count was 3,700/μl with an absolute lymphocyte count of 600. Her HIV viral load was greater than 750,000 units/ml.

A bronchoscopy was performed. The organism detected in bronchoalveolar lavage (BAL) specimens from both lungs is seen in Fig. 1. A pathologic section from the lung of a patient infected with this organism is shown in Fig. 2.

1. What is the organism causing her pulmonary symptoms? Given her symptoms and her underlying disease, which other organisms would need to be included in your differential diagnosis?

2. A CD4 cell count was never done on this patient. What would you estimate her CD4 cell count to be, given the organism with which she was infected?

3. How might this infection have been prevented?

4. What technique is generally used in HIV patients to attempt to diagnose this infection prior to bronchoscopy? What type of bronchoscopically obtained specimen has the highest yield for detecting the organism causing her infection? How would the diagnostic strategy

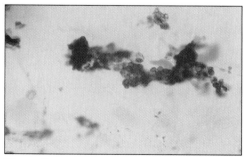

Figure 1

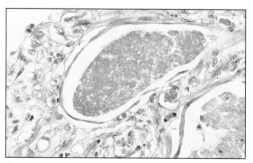

Figure 2

for detecting this infectious agent differ in non-HIV-infected, immunocompromised patients?

5. Describe the epidemiology of this agent in HIV-infected individuals versus non-HIV-infected patients. How is this organism typically acquired?

6. Patients with infection with this organism often have poor oxygen exchange, as evidenced by low pO_2 levels. Based on the findings in Fig. 2, how do you explain this?

7. Are the dysphagia and sore throat this patient described due to the organism causing her lung infection? What organism is most likely to cause these symptoms in HIV-infected patients (Fig. 3)? How would it be diagnosed?

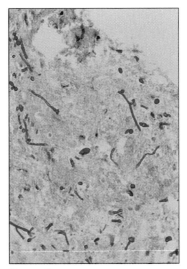

Figure 3

CASE DISCUSSION

CASE 13

1. Based on the silver stain in Fig. 1, which shows clusters of black cyst-like structures approximately 5 μm in diameter, this woman has *Pneumocystis carinii* pneumonia (PCP). PCP continues to be the leading cause of pneumonia in HIV-infected individuals even though the incidence of this disease in HIV-infected patients has declined dramatically in the industrialized world with the introduction of highly active antiretroviral therapy (HAART). PCP is now seen most commonly as an AIDS-presenting illness in an individual who was unaware that he or she was HIV-infected. Alternatively, it might be seen in patients who are noncompliant with their antiretroviral therapy or, as in this case, who have never received antiretroviral therapy. Compliance problems with antiretroviral therapy are a major concern in part because of the complexity of many of the antiretroviral regimens and in part because of the side effects associated with these agents.

Other organisms that might be infecting this woman include bacteria (such as *Streptococcus pneumoniae*, *Legionella pneumophila*, *Mycobacterium tuberculosis*, and *Haemophilus influenzae*); viruses (such as influenza A virus and respiratory syncytial virus in HIV-infected children); and fungi (including *Histoplasma capsulatum*, *Coccidioides immitis*, *Blastomyces dermatitidis*, and *Cryptococcus neoformans*). Noninfectious causes of pulmonary infiltrates in HIV-positive patients include Kaposi's sarcoma and lymphoma.

2. PCP is typically seen in HIV-infected patients with CD4 counts <200 cells/μl. Her absolute lymphocyte count is low, so it would not be surprising if her CD4 count was <200 cells/μl. The lower the CD4 count is below 200, the greater the risk of developing PCP.

3. Two approaches are important in preventing PCP in HIV-infected individuals. The first approach is the use of prophylactic antimicrobial agents in patients when their CD4 count falls to <200 cells/μl. Trimethoprim-sulfamethoxazole (TMP-SMX) has been the prophylactic drug of choice for preventing PCP because it is effective and inexpensive, and it has activity against other infections, including certain agents of bacterial pneumonia and *Toxoplasma gondii* infection. In patients who cannot tolerate TMP-SMX or who fail therapy, alternative prophylactic choices include dapsone, aerosolized pentamidine, and atovaquone.

The second approach is to effectively treat the HIV infection to either prevent damage to the immune system or "reconstitute" an HIV-damaged immune system. HAART has been shown to be effective both in delaying immune system damage by greatly slowing the decline in CD4 cell counts and in reconstituting the immune system. Recent studies have suggested that PCP prophylaxis can be stopped in patients receiving HAART whose CD4 cell counts increase to >200 cells/μl.

4. The least invasive and least expensive diagnostic approach is to examine an induced sputum sample for the presence of the typical *P. carinii* cysts (using silver stain) or cysts and trophozoites (using a direct fluorescent-antibody [DFA] stain). Conventional sputum examinations in AIDS patients give a low yield since sputum production in this disease is usually scanty. Sputum is induced by aerosolization of hypertonic (3%) saline into the airways, causing irritation, which results in coughing and expectoration of lower respiratory tract secretions. Because the number of organisms is so high in HIV-infected patients with PCP, this technique has a sensitivity of 60 to 80% if induction is performed properly.

BAL, as was used in this case, has a higher diagnostic yield than other types of bronchoscopic examinations and induced sputum. In BAL, large volumes of normal saline (100 to 200 ml) are introduced into a single lobe of the lung. This material lavages the bronchi and alveoli and is recovered by aspiration through the bronchoscope. This technique has a diagnostic yield of 90 to 95% for PCP in patients with AIDS. In selected patients with a negative BAL examination, open lung biopsies may be performed. This technique is considered definitive for detection of PCP. Biopsies are useful for detecting other agents of pneumonia as well, if appropriate cultures and stains are used.

In non-HIV-infected, immunocompromised patients, examination of induced sputum is not typically performed, in large part because there are no data on this test's reliability outside of HIV-infected patients. As a result, bronchoscopy with BAL would be the initial diagnostic step for detecting this organism in this patient population, followed by open lung biopsy if deemed necessary.

Interestingly, PCR (polymerase chain reaction) has not become a widely used technique for detection of PCP. This may be due in part to the rapidity and the ease of DFA and the expense and complexity of PCR. The performance of PCR to detect PCP appears to be quite good in HIV-infected individuals but relatively poor in individuals who are immunocompromised because of organ transplantation or malignancy.

5. Based on serologic surveys showing a high prevalence of *P. carinii* seropositivity early in childhood, it was long believed that these infections were due to reactivation of dormant organisms. Recent studies have shown clusters of PCP infections and the presence of *P. carinii* DNA as detected by PCR in the respiratory tract of individuals exposed to PCP-infected patients. These studies suggest that person-to-person spread of *P. carinii* does occur. Current thinking is that both reactivation of dormant organisms and new acquisition from infected patients can result in infection. Immunologically intact individuals do not develop respiratory infections due to this organism. Cell-mediated immunity (CMI) appears to be protective, and only when disruptions in CMI occur are patients at increased risk for PCP. Because HIV disrupts CMI by causing a decline in the number of CD4-positive T-helper cells, PCP is a common infection in this patient population. In fact, AIDS was first recognized when an unusual cluster of PCP cases was detected in homosexual men in southern California. In adults, the risk of becoming ill with PCP increases when the CD4 count drops below 200/μl. PCP incidence in HIV-infected patients has dropped dramatically in industrialized countries as a result of the introduction of HAART therapy, but it remains the leading serious opportunistic infection in this patient population. At the same time, studies have shown that PCP incidence is increasing in solid-organ and bone marrow transplant recipients and in patients with hematologic malignancies.

6. The organism binds to type 1 pneumocytes in the alveoli. This causes diffuse injury to the alveoli with leakage of exudate into the air space. The presence of this exudate, which is pink on the hematoxylin-and-eosin section in Fig. 2, prevents proper oxygen exchange between the alveoli and the bloodstream. This results in hypoxemia as measured by a low pO_2.

7. *P. carinii* caused this patient's pulmonary signs and symptoms but is an unlikely cause of dysphagia and sore throat. On physical examination, this patient had white, plaque-like lesions in her throat. A KOH wet mount of scrapings of one of these lesions was examined microscopically and revealed the organism seen in Fig. 3. In this figure, note the yeast forms and pseudohyphae. This is consistent with the presence of *Candida* spp., with *Candida albicans* the most likely organism. This is the typical way in which oral thrush is diagnosed. *C. albicans* is the leading cause of oral thrush, a common opportunistic infection in AIDS patients. This patient's dysphagia, sore throat, white, plaque-like lesions, and laboratory findings are consistent with oral thrush. She was treated with oral fluconazole, a triazole antifungal agent, and her symptoms resolved.

REFERENCES
1. **Hadley, W. E., and V. L. Ng.** 1999. *Pneumocystis*, p. 1200–1211. *In* P. R. Murray (ed.), *Manual of Clinical Microbiology*, 7th ed. ASM Press, Washington, D.C.

2. **Kaplan, J. E., D. Hanson, M. S. Dworkin, T. Frederick, J. Berolli, M. L. Lindgren, S. Holmberg, and J. L. Jones.** 2000. Epidemiology of human immunodeficiency virus-associated opportunistic infections in the United States in the era of highly active antiretroviral therapy. *Clin. Infect. Dis.* **30:**S5–S14.

3. **Masur, H., and J. Kaplan.** 1999. Does *Pneumocystis carinii* prophylaxis still need to be lifelong? *N. Engl. J. Med.* **340:**1356–1358.

4. **Nuesch, R., C. Bellini, and W. Zimmerli.** 1999. *Pneumocystis carinii* pneumonia in human immunodeficiency virus (HIV)-positive and HIV-negative immunocompromised patients. *Clin. Infect. Dis.* **29:**1513–1523.

5. **Sing, A., K. Trebesius, A. Roggenkamp, H. Russmann, K. Tybus, F. Pfaff, J. R. Bogner, C. Emminger, and J. Heesemann.** 2000. Evaluation of diagnostic value and epidemiological implications of PCR for *Pneumocystis carinii* in different immunosuppressed and immunocompetent patient groups. *J. Clin. Microbiol.* **38:**1461–1467.

6. **Vargas, S. L., C. A. Ponce, F. Gigliotti, A. V. Ulloa, S. Prieto, M. P. Munoz, and W. T. Hughes.** 2000. Transmission of *Pneumocystis carinii* DNA from a patient with *P. carinii* pneumonia to immunocompetent contact health care workers. *J. Clin. Microbiol.* **38:**1536–1538.

dysphagia = difficulty swallowing .

CASE 14

This 37-year-old man was admitted to the hospital with an increased white blood cell count and a peripheral smear consistent with acute leukemia. A bone marrow biopsy found 70 to 80% blast forms, diagnostic of acute myelomonocytic leukemia. The patient underwent induction chemotherapy. Following the chemotherapy, a repeat bone marrow biopsy again demonstrated blast forms. He therefore underwent a second round of induction chemotherapy, after which he became profoundly neutropenic (with fewer than 100 neutrophils per µl) and developed fevers without a clear source. Broad-spectrum antibiotic therapy was begun, but the fevers persisted. Empirical intravenous amphotericin B therapy was begun, and a subsequent chest radiograph revealed new bilateral fluffy pulmonary infiltrates. A bronchoscopy with biopsy was performed; the specimen demonstrated septate hyphae with acute-angle branching (Fig. 1). The organism recovered from the biopsy is seen in Fig. 2.

1. What is the differential diagnosis of pulmonary infiltrates in a leukemic patient?

2. Which fungus was seen on the specimen from bronchoscopy? Why was biopsy and not lavage necessary to make this diagnosis?

3. Would blood cultures have been useful in helping to make this diagnosis? Explain your answer.

4. Where in nature is this fungus found?

5. What predisposed this patient to this infection?

6. What other types of infections are caused by this fungus?

Figure 1

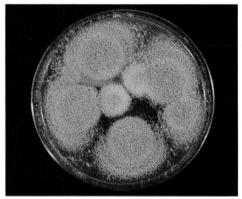

Figure 2

CASE DISCUSSION

1. Infectious causes of pulmonary infiltrates in leukemic patients include bacteria (especially gram-negative rods and *Staphylococcus aureus*), fungi (including *Aspergillus* species, zygomycetes, *Pneumocystis carinii*, *Cryptococcus neoformans*, and *Candida* species), and viruses (including cytomegalovirus). Noninfectious causes of pulmonary infiltrates in these patients include bleeding into the lung, leukemic infiltrates, and drug toxicity.

2. The presence of septate hyphae (3 to 4 μm in diameter) with acute-angle branching is consistent with the presence of an *Aspergillus* species. The species cannot be determined by the morphologic appearance of the fungus from a tissue specimen; culture is required. On culture (Fig. 2), the isolate was *Aspergillus flavus*. Other clinically significant species include *A. fumigatus* and *A. niger*. This patient had invasive aspergillosis, which produces a necrotizing pneumonia. In this disease, the organism actually grows into the tissue and often is not found superficially in the airway. Therefore lavage is not sufficient and a tissue sample is needed to make this diagnosis.

3. Because they are almost never positive in patients with invasive aspergillosis, blood cultures have little diagnostic value. In patients with invasive aspergillosis, the organism has a predilection for invading endothelial cells and, as a result, is rarely present in the bloodstream. In invasive disease, infection of the endothelial cells can lead to thrombosis and infarction of the infected vessels and the tissue supplied by those vessels. Hemorrhage may also occur if necrosis of large vessels occurs.

4. *Aspergillus* species can be isolated from grains, hay, decaying vegetable matter, soil, and plants. Aflatoxin, a potent carcinogen that has been linked to hepatocellular carcinoma, is produced by strains of *A. flavus* on improperly stored grains and nuts. *Aspergillus* spores are present in the air. Therefore, humans are constantly exposed to (and breathe) spores of these organisms. Because of the constant exposure to these spores, people may become colonized by *Aspergillus* species. Neutropenic patients who become colonized often will develop clinical disease. The finding of a positive respiratory tract culture in the neutropenic host should be managed aggressively (see answer 5 for more details). In contrast, the finding of this organism in the respiratory tract of an immunocompetent host is less likely to be clinically significant.

5. This patient had neutropenia, a dramatically decreased number of neutrophils in his peripheral blood. This condition predisposes to invasive infections not only by bacteria but also by fungi, including fungi of low virulence such as *Aspergillus* species. The risk of infection by fungi is related to both the severity and duration of the neu-

tropenia. Individuals undergoing bone marrow transplantation are at particular risk for infection with *Aspergillus* since they have to endure prolonged periods of neutropenia during the transplant process. This patient not only had very few neutrophils (sometimes none were detected) but also was neutropenic for a prolonged period. Despite treatment with intravenous amphotericin B, many leukemic patients succumb to this infection. In this case, the patient's neutrophil count began to rise and he survived the infection. The return of functioning neutrophils is central to the host's ability to resolve this infection.

6. In addition to invasive lung infection, *Aspergillus* species can cause pulmonary mycetoma (a "fungus ball" which often forms in a preexisting pulmonary cavity, such as in patients with prior cavitary tuberculosis), allergic bronchopulmonary aspergillosis (in patients with preexisting chronic lung disease), and infections of the external ear, nasal sinuses, eyes (following corneal trauma), and heart valves.

REFERENCES
1. **Denning, D. W.** 1998. Invasive aspergillosis. *Clin. Infect. Dis.* **26:**781–803; quiz 804–805.

2. **Stevens, D. A., V. L. Kan, M. A. Judson, V. A. Morrison, S. Dummer, D. W. Denning, J. E. Bennett, T. J. Walsh, T. F. Patterson, G. A. Pankey, and the Infectious Diseases Society of America.** 2000. Practice guidelines for diseases caused by *Aspergillus. Clin. Infect. Dis.* **30:**696–709.

C A S E 15
This 62-year-old man presented with a 4-day history of left eye swelling and left frontal headache. He also had noted the progression of left ptosis over the 4 days. He had an unremarkable medical history, although his family history was strongly positive for diabetes mellitus. On examination, the patient was febrile to 38.1°C and had complete left ptosis. Laboratory studies were notable for an elevated white blood cell count of 17,900/μl with 14,400 neutrophils/μl and an elevated blood glucose level of 484 mg/dl, indicating that he was a diabetic. A computed tomogram (CT) scan of the sinuses and orbits was notable for fluid in both ethmoid sinuses and inflammatory changes lateral to the left medial rectus muscle. The patient underwent surgery (a left external ethmoidectomy). A KOH preparation of the material from the left ethmoid sinus obtained at the time of surgery demonstrated broad, aseptate hyphae with right-angle branching (Fig. 1). The resulting culture is seen in Fig. 2.

1. Which organisms are consistent with these microscopic findings?
2. What is the natural habitat of this organism?
3. Which clinical conditions are associated with invasive infections with these organisms?
4. How is this infection managed? What is the prognosis for this patient?

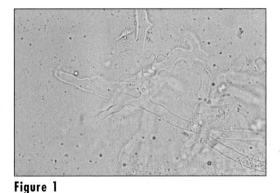

Figure 1

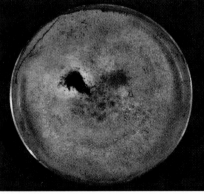

Figure 2

CASE	**CASE DISCUSSION**
15	**1.** The presence of broad, aseptate, or sparsely septate hyphae with right-angle branching is diagnostic of an agent of zygomycosis. This includes infections with molds from the genera *Mucor, Rhizopus, Rhizomucor,* and *Cunninghamella*. In

a section from clinical material these organisms are indistinguishable. They can often be differentiated from *Aspergillus* species in clinical materials since *Aspergillus* species have acute-angle branching (instead of right-angle branching as with agents of zygomycosis), frequent septations, and thin hyphae.

Zygomycotic infections are extremely aggressive and frequently fatal. They represent a true medical (and surgical) emergency, and differentiation from *Aspergillus* species can be crucial. Culture results showed that this patient was infected with a *Rhizopus* species. As can be seen in Fig. 2, there is abundant growth of this organism after only a few days of incubation, with the mycelial elements "filling" the plate.

2. The agents of zygomycosis are commonly found in the environment, on fruit and bread, and in soil. In fact, sterile bread devoid of preservatives (which may prevent the growth of zygomycetes) can be used as a sporulation medium for these organisms.

3. Zygomycosis is associated with several clinical conditions. Rhinocerebral zygomycosis is associated with diabetes mellitus (as with this patient), particularly in patients with ketoacidosis, and also occurs in patients with leukemia. Another clinical condition associated with zygomycosis is the use of iron chelation therapy with deferoxamine. When there is pulmonary involvement, leukemia and lymphoma (with neutropenia) are common underlying conditions, although pulmonary disease may occur with diabetes as well. The skin may be infected in burn patients and, rarely, in diabetic patients. The central nervous system may be involved via direct extension from the sinuses. Pathologically, the fungus directly invades blood vessels, causing thrombosis of the blood vessels and infarction of the area that normally receives its arterial blood supply via the invaded vessel. In some cases disseminated infection can occur.

Aspergillus — no vessel invasion/dissemination.
Zygomycosis — direct invasion of blood vessels.

4. Since the disease, when untreated, runs a progressive and fatal course, early recognition and a high degree of suspicion are necessary. It was the suspicion of zygomycosis that caused the surgeon in this case to obtain a frozen section intraoperatively, a procedure in which the pathologist examines the tissue as rapidly as possible (without the standard techniques used to fix tissue) while the patient is still in the operating room. Often, multiple frozen sections are examined during surgery. Debridement of the infected area will continue until a frozen section is obtained from the surgical site margin in which the organisms are no longer seen. Treatment includes aggressive surgical removal of infected and necrotic tissue, antifungal therapy (intravenous amphotericin B), and medical management, including the correction of the underlying condition, such as diabetic ketoacidosis.

REFERENCES

1. **Case Records of the Massachusetts General Hospital.** 2001. Weekly clinico-pathological exercises. Case 3-2001. A 59-year-old diabetic man with unilateral visual loss and oculomotor-nerve palsy. *N. Engl. J. Med.* **344:**286–293.

2. **Kontoyiannis, D. P., V. C. Wessel, G. P. Bodey, and K. V. Rolston.** 2000. Zygomycosis in the 1990s in a tertiary-care cancer center. *Clin. Infect. Dis.* **30:**851–856.

3. **Lee, F. Y., S. B. Mossad, and K. A. Adal.** 1999. Pulmonary mucormycosis: the last 30 years. *Arch. Intern. Med.* **159:**1301–1309.

4. **Ribes, J. A., C. L. Vanover-Sams, and D. J. Baker.** 2000. Zygomycetes in human disease. *Clin. Microbiol. Rev.* **13:**236–301.

CASE 16

This 38-year-old North Carolina man was in good health until 2 months prior to admission, when he developed a low-grade fever, myalgias, and a nonproductive cough. He was given oral erythromycin by his local physician. After 2 weeks of therapy, his condition had not improved. A chest radiograph demonstrated "right middle lobe air space disease," and therapy with oral ampicillin was begun. Over the next month, his condition worsened. He noted daily fevers, chills, night sweats, and a 15-lb (7-kg) weight loss. One month prior to admission, a chest radiograph demonstrated consolidation of the right middle lobe. A PPD (purified protein derivative) skin test was negative with positive controls, and an oral antibacterial agent was given. The patient's symptoms continued, and he was admitted to the hospital.

The patient had an unremarkable travel history and no animal exposure, was a nonsmoker, and had no HIV risk factors. He worked for the power company cutting tree limbs and tops. On physical examination he was febrile to 38.3°C. The skin examination was notable for a tender, raised, erythematous papule (1 by 1 cm) on the bridge of the nose (Fig. 1). A chest radiograph and subsequent computed tomogram (CT) scan were notable for a densely consolidated right middle lobe, a 3.5-cm subcarinal mass, and a small right hilar mass. Bronchoscopy was performed. KOH examination and acid-fast, modified acid-fast, and Gram stains gave negative results. Examination of the skin lesion using a silver stain demonstrated a large, round budding yeast with a broad base connecting the mother cell to the daughter cell (Fig. 2).

1. What is the differential diagnosis for this patient's pulmonary disease?

2. Which organism is causing his illness? What are its epidemiology and culture characteristics?

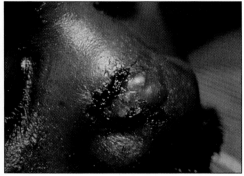

Figure 1

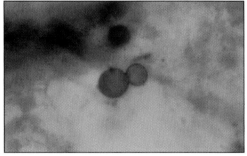

Figure 2

3. This patient's lungs and skin were involved with this infection. Which other sites are commonly involved?

4. What in this patient's history might alert a physician to think of this organism?

5. Which organisms may be detected by a KOH examination? An acid-fast stain? A modified acid-fast stain?

CASE DISCUSSION

CASE 16

1. The patient has a hilar mass and a densely consolidated right middle lobe. The differential diagnosis includes both noninfectious processes (such as malignant and benign tumors) and chronic infections with slowly growing organisms, such as fungi (including *Blastomyces dermatitidis*, *Coccidioides immitis*, and, less likely, *Histoplasma capsulatum*), mycobacteria (especially *Mycobacterium tuberculosis*), and other slowly growing bacteria such as *Actinomyces* and *Nocardia* spp.

2. The etiologic agent of this individual's illness is *B. dermatitidis*. The morphology of the organism seen in Fig. 2, a fairly large, broad-based budding yeast, is typical of this organism. It is a dimorphic fungus, so at room or ambient temperature it grows as a mold and at body temperature (37°C) it grows as a yeast. It is the etiologic agent of North American blastomycosis and should not be confused with *Paracoccidioides brasiliensis*, the agent of South American blastomycosis. *B. dermatitidis* is endemic in much of the southeastern United States. Other regions where it is endemic include areas within the Mississippi and Ohio River basins and parts of western New York State and bordering areas in Canada.

3. Other sites that are frequently infected are bone, joints, and the genitourinary tract. In fact, this patient later developed multiple bone lesions. He presented with pain in his shins, and a bone scan showed multiple lesions, especially in his long bones. An aspirate of a bone lesion subsequently grew *B. dermatitidis*. He also is at risk for infection of the prostate and epididymis, both common sites for disseminated infection.

4. The patient's symptoms were quite nonspecific. However, he failed to respond to three different regimens of antimicrobial therapy designed to treat common bacterial agents of community-acquired pneumonia, such as *Mycoplasma pneumoniae* and *Streptococcus pneumoniae*, and agents of bronchitis, such as *Haemophilus influenzae* and *Moraxella catarrhalis*. The weight loss, low-grade fevers, and indolent clinical course are all suggestive of *M. tuberculosis* infection. However, tuberculosis usually presents with upper lobe involvement, and a negative PPD skin test with positive skin test controls also argues against this infection. Patients with an indolent disease course and a nonproductive cough over extended periods may have pulmonary mycoses. The finding of the skin lesion on the face, a frequent occurrence in blastomycosis, further supports this diagnosis. The patient's occupation probably increased his risk for this infection. This organism can be recovered from decomposing wood. He probably was infected by inhaling spores while cutting down dead trees or branches. His skin infection was secondary to his primary pulmonary process.

5. On the basis of this patient's clinical presentation, a wide variety of microorganisms would be included in the differential diagnosis. Different techniques are required to best demonstrate the different organisms that need to be considered. An acid-fast stain was done to detect mycobacteria. Despite the negative skin test and atypical chest radiograph for tuberculosis, *M. tuberculosis* must still be considered, as must other mycobacteria. The modified acid-fast stain would be used to try to identify *Nocardia* spp., which could cause infections with a case presentation similar to this patient's. Finally, KOH examination is a commonly used technique to demonstrate fungi in clinical specimens. Fungi are fairly refractory to the activity of KOH while human tissues are dissolved, clearing the specimen and making the microscopic demonstration of the fungi much easier. Other special stains (such as methenamine-silver or Calcofluor white) may demonstrate the presence of fungal elements in histologic specimens.

REFERENCES

1. **Chapman, S. W., R. W. Bradsher, Jr., G. D. Campbell, Jr., P. G. Pappas, and C. A. Kauffman.** 2000. Practice guidelines for the management of patients with blastomycosis. Infectious Diseases Society of America. *Clin. Infect. Dis.* **30:**679–683.

2. **Klein, B. S., J. M. Vergeront, A. F. DiSalvo, L. Kaufman, and J. P. Davis.** 1987. Two outbreaks of blastomycosis along rivers in Wisconsin. Isolation of *Blastomyces dermatitidis* from riverbank soil and evidence of its transmission along waterways. *Am. Rev. Respir. Dis.* **136:**1333–1338.

3. **Pappas, P. G., R. W. Bradsher, C. A. Kauffman, G. A. Cloud, C. J. Thomas, G. D. Campbell, Jr., S. W. Chapman, C. Newman, W. E. Dismukes, and the National Institute of Allergy and Infectious Diseases Mycoses Study Group.** 1997. Treatment of blastomycosis with higher doses of fluconazole. *Clin. Infect. Dis.* **25:**200–205.

4. **Witorsch, P., and J. P. Utz.** 1968. North American blastomycosis: a study of 40 patients. *Medicine* **47:**169–200.

CASE 17

This 26-year-old man presented for evaluation of a neck mass and a right axillary mass. The patient, who came to the United States from Vietnam 6½ years ago, noted a right axillary mass 1 month prior to admission. The mass gradually increased in size. Approximately 3 weeks prior to admission, he noted a midline neck mass. These masses were incised and drained twice. Samples of purulent material were sent for routine bacterial culture at an outside hospital. The patient had lost 3 kg (ca. 7 lb) in the 2 months prior to admission but denied fever. Travel history was notable for his having lived in Arizona for 6 years prior to moving to Boston.

On examination, he was afebrile. An erythematous, fluctuant, nontender mass (8 cm by 6 cm) was present in the midline of the neck (Fig. 1; computed tomogram [CT] scan in Fig. 2). The right axilla demonstrated incision sites that were draining and were tender to palpation. He had no adenopathy elsewhere. His PPD (purified protein derivative) skin test was reactive, and his chest X ray was notable for apical scarring in the right lung.

1. Because of the positive PPD and apical scarring seen on the patient's chest X ray, the clinical suspicion of cervical tuberculosis (scrofula) was high, and the patient was begun on a four-drug antituberculous regimen. Other than tuberculosis, what is in the differential diagnosis of the neck and axillary mass?

2. Several days after the neck mass was drained of several milliliters of purulent material for culture, a mold was found to grow on the blood agar plates in the routine bacteriology section of the microbiology

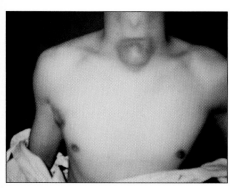

Figure 1

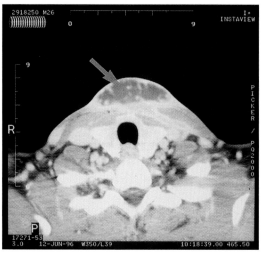

Figure 2

laboratory. The laboratory technologist did not notice the presence of the mold and opened the plates to examine them for bacteria. Why is this of concern?

3. The mold was white (Fig. 3) and initially did not have any identifying characteristics when a lactophenol cotton blue preparation was examined under a phase-contrast microscope. It was subcultured at both room temperature and body temperature, and after subculture began to demonstrate the presence of arthroconidia microscopically (Fig. 4). The identification was confirmed by using a commercially available genetic probe. What is this organism?

4. What in this patient's history makes him more likely than the U.S. population as a whole to have disseminated disease? What other body sites does this organism commonly involve?

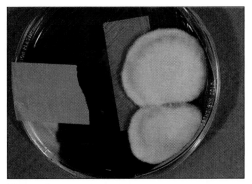

Figure 3

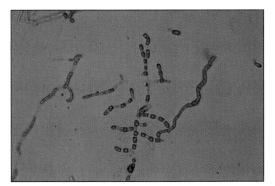

Figure 4

CASE DISCUSSION

1. The differential diagnosis includes several causes of subacute enlarging masses, both noninfectious and infectious. Of noninfectious causes, the most important are malignancies, such as lymphoma. Among the infectious causes, those etiologic agents that progress slowly typically include not only *Mycobacterium tuberculosis*, of which the reactive PPD and apical scarring on his chest X ray are supportive, but also other mycobacterial diseases and fungal and even some bacterial infections. Among the fungal causes, the patient's history of travel from Vietnam is suggestive of the dimorphic fungus *Penicillium marneffei*, which is found in Southeast Asia and is the third most common opportunistic infection among AIDS patients there (behind tuberculosis and cryptococcal disease). Another fungus that must be considered, given the patient's history of recently living in Arizona, is *Coccidioides immitis*. This dimorphic fungus is found in the soil in arid areas of the United States, Central America, and South America that correspond to the Lower Sonoran Life Zone. The patient does not have any relevant exposure history to the other dimorphic fungi (*Blastomyces dermatitidis*, *Paracoccidioides brasiliensis*, and *Histoplasma capsulatum*) that are found in rather well-defined geographic regions in the Western Hemisphere. Among bacteria, the rather slowly growing *Actinomyces* spp. and *Nocardia* spp. can certainly cause involvement of the neck, chest wall, and contiguous tissues. *Actinomyces* spp. must be sought by obtaining appropriate anaerobic cultures.

2. Laboratory-acquired infections with dimorphic fungi are a real risk to the clinical microbiologist. During the process of opening and examining a petri dish, it is possible for the arthroconidia of *C. immitis*, which are easily aerosolized, to become airborne and to infect a laboratory worker. This has been well documented with *C. immitis* and has actually resulted in the death of laboratory workers. As a result, clinical laboratories routinely require the use of biological safety cabinets for the isolation and identification of molds. In addition, all fungal cultures that are planted on petri dishes are routinely closed with either tape or a commercially available product such as Shrink Seal in order to prevent the plates from being inadvertently opened. This is not, however, the routine for bacterial cultures. Occasionally a mold grows on bacterial media. It is important that technologists be aware of this possibility and as a matter of good safety acquire the habit of looking at the plates before opening them, to minimize the risk of infection. The other, less important reason why fungal cultures should only be opened in a biological safety cabinet is to prevent cross-contamination of other cultures by the fungal conidial elements.

3. The presence of barrel-shaped arthroconidia is consistent with *C. immitis* (Fig. 4). Arthroconidia are formed by the fragmentation of hyphae during sporulation.

Although there are several other fungi that produce arthroconidia, they are not likely to produce this clinical syndrome. The patient, as noted above in the answer to question 1, lived in Arizona for 6 years, an area in which this agent is endemic. To definitively identify dimorphic fungi, it is necessary either to demonstrate that the fungus can convert from a mold form (at room temperature) to a yeast form (at body temperature), to demonstrate the presence of an antigen characteristic of the organism, or to identify the fungus with a genetic probe. This mold was sent to a reference laboratory that had the commercially available genetic probe. This helped to speed its identification.

4. Disseminated infection with *C. immitis* is more common in non-whites than it is in whites. In fact, studies have indicated that the risk of disseminated infection in Filipinos may range from 10 to 175 times the risk in Caucasians. Although there are no known data specifically on the risk of disseminated disease in persons of Vietnamese heritage, it is likely that there is an increased risk of dissemination among this population as well. The arthroconidia are inhaled, and the great majority of people develop either no symptomatic infection or disease limited to the lungs. In some patients, *C. immitis* may spread to bone, meninges (which may be life-threatening), or skin. Dissemination is more common among immunocompromised patients, including those with solid-organ transplants, patients with AIDS, and people who receive chemotherapy.

REFERENCES

1. **Duong, T. A.** 1996. Infection due to *Penicillium marneffei*, an emerging pathogen: review of 155 reported cases. *Clin. Infect. Dis.* **23:**125–130.

2. **Galgiani, J. N., N. M. Ampel, A. Catanzaro, R. H. Johnson, D. A. Stevens, and P. L. Williams.** 2000. Practice guidelines for the treatment of coccidioidomycosis. Infectious Diseases Society of America. *Clin. Infect. Dis.* **30:**658–661.

3. **Jones, J. L., P. L. Fleming, C. A. Ciesielski, D. J. Hu, J. E. Kaplan, and J. W. Ward.** 1995. Coccidioidomycosis among persons with AIDS in the United States. *J. Infect. Dis.* **171:**961–966.

4. **Rosenstein, N. E., K. W. Emery, S. B. Werner, A. Kao, R. Johnson, D. Rogers, D. Vugia, A. Reingold, R. Talbot, B. D. Plikaytis, B. A. Perkins, and R. A. Hajjeh.** 2001. Risk factors for severe pulmonary and disseminated coccidioidomycosis: Kern County, California, 1995–1996. *Clin. Infect. Dis.* **32:**708–715.

5. **Stevens, D. A.** 1995. Coccidioidomycosis. *N. Engl. J. Med.* **332:**1077–1082.

CASE 18

The patient was a 59-year-old female who underwent a cardiac transplant 6 months earlier for an idiopathic cardiomyopathy. At the time of transplant she was seropositive for cytomegalovirus (CMV) and seronegative for HIV, hepatitis B, and hepatitis C. Her heart donor was CMV seropositive and HIV, hepatitis B, and hepatitis C negative. Since the transplant she had done reasonably well, with the exception of two episodes of acute rejection which required increased doses of steroids to control rejection. One week prior to this admission, she complained of malaise, fatigue, a low-grade fever, and mild dyspnea on exertion. She was admitted to determine the etiology of her complaints. The physical examination was significant only for a temperature of 38.3°C and cushingoid body habitus (due to the steroids). Examination of her lungs revealed fine bibasilar rales. A stool specimen was guaiac positive. Her laboratory studies revealed a hematocrit of 24%, a white blood cell (WBC) count of 2,300/µl (leukopenia), and a normal platelet count. She was transfused with 3 units of blood and underwent upper gastrointestinal endoscopy, which revealed nodular gastric erosions. Biopsies and brushings were taken and submitted to the pathology and microbiology laboratories. A chest radiograph revealed diffuse infiltrates. A bronchoscopy was done, and transbronchial biopsy and bronchoalveolar lavage specimens were sent for histopathologic and cytologic examination and bacterial, fungal, viral, and mycobacterial culture. Gram stains were negative. The next day, hematoxylin-and-eosin stains of the gastric lesion brushings, as well as the lung tissue, revealed the cause of her infection (Fig. 1). Viral cultures were positive 2 weeks later, confirming the diagnosis.

1. What was the most likely etiology of the patient's infection? How did she become infected?

2. What is the typical clinical presentation of this organism in patients such as this woman who have received solid-organ transplants?

3. Which other two patient populations are subject to serious infections with this organism?

4. Which other opportunistic infections are seen with some degree of frequency in patients receiving cardiac transplants?

5. The agent infecting this patient grows slowly on routine viral culture, in her case requiring 2 weeks to detect. What other diagnostic approaches are available to directly detect this organism more rapidly?

6. What strategies are employed to attempt to prevent this infection in individuals receiving solid-organ transplants?

7. What agents are available for treatment of this infection? What drug resistance problems, if any, have been observed with these agents?

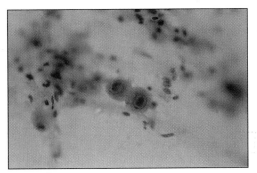

Figure 1

CASE DISCUSSION

1. The characteristic owl-eye cell seen in Fig. 1 is characteristic of CMV, one of the human herpesviruses. CMV is an enveloped, double-stranded DNA virus and is the most common infectious agent complicating transplantation. This virus was the cause of her gastritis and pneumonitis. CMV infections usually occur 1 to 4 months posttransplant. CMV, like other herpesviruses, can cause latent infection that can be lifelong. In patients who are CMV positive pretransplant, i.e., latently infected, infections posttransplant are usually the result of reactivation of latent viral infection. Because she received a CMV-positive organ, that too could have been the source of her infection. She would have received blood during her transplant. That too could have been a source, since 50 to 80% of the blood donor population is CMV positive. CMV infections are common in immunocompetent individuals, but clinical disease is rare. The immunosuppression of cell-mediated immunity necessary in transplantation greatly increases the likelihood of developing clinical disease with this virus, whether through reactivation of the patient's latent virus or through infection from the transplanted organ or blood products. This is an important point conceptually. Infection does not necessarily mean that an individual will be sick. One may become "infected" with an infectious agent, as measured by an immune response to that agent, without developing any clinical manifestations of that infection; i.e., they do not have "clinical disease." If the organism can cause latent infection, immunosuppression of a latently infected patient may result in reactivation and the development of clinical disease.

2. CMV infections following solid-organ transplantation are common, occurring in 8 to 50% of recipients depending upon the organ that was transplanted. The spectrum of CMV disease in these individuals ranges from asymptomatic infections to life-threatening disease. Most clinical CMV disease is classified as mild to moderate, with more severe disease being seen primarily in seronegative recipients of seropositive organs. Clinical manifestations of mild to moderate disease include fever and malaise. Leukopenia and thrombocytopenia may also be seen. More severe manifestations include infectious mononucleosis-like syndromes, hepatitis, esophagitis, gastritis, cholangitis, chorioretinitis, and pneumonia/pneumonitis. CMV infection has a predilection for some transplanted organs. For example, it causes hepatitis in liver transplant recipients, CMV pneumonitis in heart-lung transplant recipients, and pancreatitis in pancreas transplant recipients. CMV-associated myocarditis does occur in heart transplant recipients, but it is rare.

3. In addition to transplant recipients, AIDS patients and newborns can develop severe CMV infections. AIDS patients may develop CMV retinitis, which can result in blindness. The incidence of this is not high in patients receiving highly active anti-retroviral therapy (HAART). This complication is rarely seen in transplant patients. The CMV pneumonia that transplant patients develop usually is more severe than that seen in AIDS patients. The reasons for the different patterns of CMV disease in these high-risk populations are unclear.

Congenital CMV infection can result in deafness, psychomotor retardation, chorioretinitis, pneumonia, hepatitis, and rash.

4. Other opportunistic infections in cardiac transplant recipients include those caused by pathogens common in patients with impaired cell-mediated immunity. These infections include toxoplasmosis, *Pneumocystis carinii* pneumonia, varicella-zoster, cryptococcal pneumonia or meningitis, *Listeria monocytogenes* bacteremia or meningitis, and *Nocardia* spp. pneumonia. It is believed that CMV is itself immuno-suppressive and may increase the likelihood of these other opportunistic infections.

5. Three techniques—the shell vial assay, the direct antigenemia detection, and PCR (polymerase chain reaction)—all detect CMV more rapidly and with greater sensitivity than conventional tissue cultures. In the shell vial assay, fibroblast cells are grown as monolayers on a glass coverslip in a shell vial. Clinical specimens, including urine, respiratory secretions, biopsy tissue specimens, and/or WBCs, obtained from patients suspected of being infected with CMV are centrifuged onto the cell monolayer. After 1 to 2 days of incubation, the monolayer is stained with a fluorescent monoclonal anti-body specific for a CMV early antigen. This technique is more sensitive and much more rapid than conventional culture (1 to 2 days versus 2 weeks) but lacks sensitivity when compared with either antigenemia detection or PCR. It therefore is of limited value in monitoring solid-organ transplants for CMV infection.

In the direct antigenemia assay, CMV-infected leukocytes are directly stained with a monoclonal antibody which detects the CMV structural protein pp65. The number of infected cells is quantified. This method is preferred to culture for detection of CMV in peripheral blood, but it is very labor intensive. It is used almost exclusively in transplant recipients either to monitor their CMV infection status (a low number of infected cells in an asymptomatic patient may indicate a patient in the early stages of active infection) or to diagnose CMV disease in the acutely ill transplant recipient. The higher the number of infected cells the more likely the patient will have clinical disease although an exact cutoff of infected cells which predicts clinical disease, has not yet been established.

PCR is a third method that can be used to detect CMV in blood. One of the concerns of using PCR for detecting CMV infection is that in CMV-seropositive patients, it would be unable to differentiate latently infected patients from those with active infection. One approach to address this is to perform quantitative PCR on plasma or serum. The rationale for this approach is that virus would be "free" in plasma or serum only in patients with active infection, while the virus would be primarily cell associated in latently infected patients. The presence of higher amounts of CMV DNA in the plasma or serum should correlate with a higher likelihood of clinical disease. Studies comparing CMV antigenemia with quantitative PCR show them to have similar performance in the detection of both CMV infection and clinical disease.

6. The ideal approach for preventing CMV infection posttransplant is to transplant organs from seronegative donors into seronegative recipients. If these individuals require blood products, they should come from seronegative donors or should be leukocyte-depleted to remove as much CMV as possible. However, given the high rates of CMV seropositivity and a shortage of donor organs, this approach is the exception rather than the rule.

The patient population with the highest rate of developing CMV disease post-transplant is CMV negative recipients who receive organs from CMV-positive donors. Two approaches to prevent CMV infection can be used in this patient population: prophylactic or preemptive therapy. In prophylactic therapy, all transplant patients at risk for CMV disease are treated with a prophylactic regimen. The optimal anti-CMV prophylactic regimen is controversial and varies from center to center but may include hyperimmune CMV globulin, anti-CMV agents, or the combination of the two. The second approach is to use preemptive therapy. The strategy here is to screen patients for the presence of CMV with a highly sensitive technique such as antigenemia testing or quantitative PCR to detect an early stage of active infection. At that point, the patients would be treated with anti-CMV agents with or without hyperimmune CMV globulin to prevent them from developing clinical disease. Unlike prophylactic therapy, which is used on all at-risk patients, preemptive therapy is used only on those patients with evidence of CMV in their bloodstream. The rationale for preemptive therapy is that it is more cost-effective than prophylaxis and prevents at least some transplant patients from being exposed to potentially toxic antiviral agents.

7. Four drugs are currently available for treatment of CMV infections: ganciclovir, valganciclovir, foscarnet, and cidofovir. Valganciclovir is a prodrug of ganciclovir that is converted to the parent compound by intestinal and hepatic esterases. All four drugs inhibit viral replication by inhibiting the activity of CMV DNA polymerase. Phosphorylated forms of ganciclovir and cidofovir slow and then stop CMV DNA chain elongation. Foscarnet is a noncompetitive inhibitor of CMV DNA polymerase which acts by blocking the cleavage of pyrophosphate from the deoxynucleotide triphosphates preventing chain elongation. To have anti-CMV activity, ganciclovir but not cidofovir must be phosphorylated by CMV-derived protein kinase encoded by a sequence in the virus designated UL97. Mutants in the UL97 sequence can result in resistance to ganciclovir but not cidofovir. Certain mutations in the CMV DNA polymerase can result in resistance to both ganciclovir and cidofovir. Different mutations in the CMV DNA polymerase may result in resistance to foscarnet. Resistance to all four drugs has been reported in clinical isolates. Patients who fail or cannot tolerate ganciclovir or valganciclovir therapy are usually treated with foscarnet. Cidofovir is used primarily in CMV-infected AIDS patients, especially those with retinitis who have failed alternative therapies. CMV drug resistance is rare in solid-organ recipients, making ganciclovir and valganciclovir the drugs of choice in this patient population.

REFERENCES

1. **Erice, A.** 1999. Resistance of human cytomegalovirus to antiviral drugs. *Clin. Microbiol. Rev.* **12:**286–297.

2. **Patel, R., and C. V. Paya.** 1997. Infections in solid-organ transplant recipients. *Clin. Microbiol. Rev.* **10:**86–124.

3. **Sia, I. G., and R. Patel.** 2000. New strategies for the prevention and therapy of cytomegalovirus infection and disease in solid-organ transplant recipients. *Clin. Microbiol. Rev.* **13:**83–121.

CASE 19

This 2-year-old male child experienced an upper respiratory infection 2 weeks prior to hospital admission. Four days prior to admission, anorexia and lethargy were noted. The patient was seen in the emergency room 3 days prior to admission. At that time he had a fever of 39.9°C. Physical examination revealed a clear chest, exudative pharyngitis, and bilaterally enlarged cervical lymph nodes. A throat culture was taken, and a course of penicillin was begun. The child's course worsened, and he became increasingly lethargic; he developed respiratory distress on the day of admission. It was noted that the throat culture from 3 days prior to admission had not grown any group A streptococci. On examination, the patient was febrile to 38.9°C and had an exudate in the posterior pharynx that was described as a yellowish, thick membrane which bled when scraped and removed. The patient's medical history revealed that he had received no immunizations. The patient was admitted to the hospital and treatment was begun. Figure 1 shows the organism recovered from the patient's throat culture on special isolation medium. Figure 2 shows a Gram stain of this organism.

1. What was the pathogen? Why must the laboratory be notified if this organism is in the patient's differential diagnosis?

2. What special test is necessary to prove that this organism has the potential to cause disease?

3. This organism was recovered from a throat culture. When blood cultures are done for this organism they are always negative, yet the patient, even with the infecting organism confined to his throat, can have profound systemic pathologic effects, such as heart failure. Explain this observation.

4. How can this disease be prevented?

5. Where in the world have large outbreaks of infection due to the organism infecting this patient occurred? Why have they occurred? What is the importance of these outbreaks in the United States?

6. How is this infection treated?

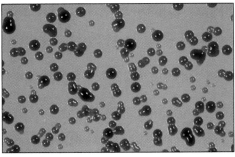

Figure 1

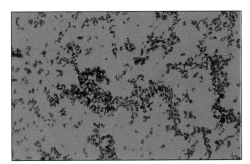

Figure 2

CASE DISCUSSION

CASE 19

1. The child appeared to have a strep throat that did not get better. On physical examination, the child had the classic pseudomembrane seen in cases of diphtheria. The pseudomembrane is composed of bacteria, fibrin, dead epithelial cells, and red and white blood cells. Aspiration of this pseudomembrane can cause death by suffocation. The etiologic agent of diphtheria is *Corynebacterium diphtheriae*, an aerobic, club-shaped, gram-positive rod.

It is important that the laboratory be notified when the diagnosis of diphtheria is being considered, as isolation of *C. diphtheriae* is problematic. Throat cultures are not routinely examined for *C. diphtheriae* because the disease is very rare and because saprophytic *Corynebacterium* spp., also called diphtheroids, are found in abundance in most throat cultures. These diphtheroids, by both colony and Gram stain morphology, are essentially indistinguishable from *C. diphtheriae*. However, a selective medium, cysteine tellurite agar, is useful in the isolation of *C. diphtheriae* (Fig. 1). The potassium tellurite salt in the selective medium also suppresses the growth of many organisms which normally inhabit the pharynx. On this medium, *C. diphtheriae* produces black colonies. It should be noted that other diphtheroids, i.e., short, club-shaped, gram-positive bacilli (Fig. 2), and *Staphylococcus aureus* may also turn this medium black. Black colonies should be Gram stained, and gram-positive rods should then be identified biochemically.

2. Only isolates of *C. diphtheriae* which are able to produce the exotoxin diphtheria toxin are considered pathogenic. Since nontoxigenic strains can be recovered from throat cultures, the isolation of *C. diphtheriae* does not prove that the patient has the disease. Rather, the pathogenic potential of any clinical isolate must be demonstrated by its ability to produce toxin. Three different approaches can be used to demonstrate toxin production. The classic approach is to do the Elek test. In this test, antitoxin against diphtheria toxin is reacted with culture supernatants of *C. diphtheriae* in a clear agar plate. The formation of a precipitin band that is due to a precipitation reaction between toxin present in the culture supernatant and antitoxin indicates that the organism is producing diphtheria toxin. This test is typically available at state public health laboratories in the United States. A second approach is to perform an enzyme immunoassay (EIA) on culture supernatants using specific antisera against diphtheria toxin. This is a highly accurate test but is not widely available. Third, PCR (polymerase chain reaction) can be done to look for diphtheria toxin-specific gene sequences. This test is highly sensitive but may give false-positive results since strains exist with the appropriate toxin gene sequences that do not produce toxin. Strains that are negative by PCR are reliably nontoxigenic, but some authors caution that positive PCR results should be confirmed by one of the other two toxin detection methods.

ABCDEFG

3. The pathogenesis of *C. diphtheriae* is one of the best understood among bacteria. The pathogenesis of disease is due to the organism's ability to produce a protein exotoxin called diphtheria toxin. In patients with diphtheria, the organism remains in the pharynx but the toxin can enter the circulation. Thus the organism can remain localized but cause systemic symptoms. Conceptually, this is known as "disease at a distance" and is seen with a number of toxigenic organisms. Diphtheria toxin is a classic A + B toxin. The B subunit of the toxin binds to a specific receptor on the surface of the mammalian cell. The toxin molecule is internalized, and the A subunit of the toxin is activated. The activated A subunit inhibits protein synthesis by inactivating elongation factor-2 (EF-2) by catalyzing its ADP-ribosylation. This prevents chain elongation in protein synthesis, which ultimately results in cell death. Diphtheria toxin has been shown to inhibit protein synthesis in a wide variety of mammalian cell types. The gene for toxin production is encoded on a lysogenic phage, and toxin synthesis is regulated, at least in vitro, by the concentration of iron in the environment of the organism. Protein synthesis is inhibited in a variety of tissues, with the heart, nerves, and kidneys being particularly targeted. Both myocarditis and neuropathy occur in patients with diphtheria.

4. Vaccination against diphtheria is mandatory in the United States, and children are not permitted to attend school without proof of vaccination. Because of these laws, it is estimated that >95% of school-age students are vaccinated against diphtheria. As the pathogenicity of this organism is due to its exotoxin, a toxoid vaccine has proved to be protective for this disease. Children receive a series of four vaccinations, with the first three doses given at approximately 2, 4, and 6 months of age and a booster dose given 6 to 12 months later. Children receive another booster dose just before entering school. After that, the individuals should receive booster vaccinations at 10-year intervals. Diphtheria vaccination is given in conjunction with vaccines for *Clostridium tetani* (tetanus) and *Bordetella pertussis* (pertussis) in a trivalent vaccine known as DTP. The 10-year booster doses are composed of only the diphtheria (D) and tetanus (T) components of this vaccine because of concerns about adverse reactions to the pertussis (P) component. Acellular pertussis (acP) vaccines are currently being tested in adults. They have been found to be much less reactogenic than the whole cell vaccine. A DTacP vaccine may eventually be recommended for use in adults. This vaccine is currently used in children.

The vaccination program for eradicating diphtheria in the United States has been highly effective. Usually fewer than 10 cases are reported per year. The cases that do occur are generally in adults, especially among the homeless, who do not receive routine medical care such as booster vaccinations. Cases are also seen in individuals who refuse vaccination for religious reasons.

5. Outbreaks of diphtheria were seen in several of the newly independent states of the former Soviet Union beginning in the late 1980s, with peak incidences of disease in 1993 to 1996. Large numbers of cases were seen in Russia, Ukraine, and the Central Asian Republics. Unlike epidemics in the pre-vaccine era, most cases occurred in adolescents and adults. These populations were at risk because they were not immune to diphtheria toxin. There are several reasons for this lack of immunity. First, many adults were born when diphtheria disease incidence was declining in the Soviet Union, and thus they had not been exposed and had not developed natural immunity. Second, many of these same individuals had not been vaccinated in childhood and had not acquired immunity. Third, for those who did receive childhood vaccination against diphtheria, adult booster doses were not recommended after the age of 14 to 16. This would result in waning immunity in a population not exposed to natural disease and would create a large population at risk for developing infection.

Other factors besides declining immunity or lack of immunity that contributed to the epidemic include poor living conditions in crowded urban and military settings, a decline in public health and sanitary conditions, and poor nutrition. In addition, availability of vaccine was not reliable in areas outside of Russia during the early 1990s, resulting in large populations of unvaccinated children who could contribute to the spread of this disease. The implementation of large-scale vaccination programs, which resulted in high vaccination rates in both children and adults, resulted in control of this epidemic.

There are some important lessons that can be learned from the epidemic that apply to the United States and other industrialized nations. First, although childhood vaccine rates are high, immunity wanes in U.S. populations. It is estimated that as many as 50% of adults are susceptible to diphtheria, with susceptibility increasing with age. Booster doses of diphtheria toxoid are recommended at 10-year intervals to maintain immunity. Second, these booster doses are particularly important for individuals who are traveling to areas of the world where this disease continues to be endemic. Diphtheria cases have been documented in Americans and western Europeans who have traveled to areas of Russia and other countries where diphtheria is endemic. Several of these individuals imported their infections into their home country. At least in theory, this importation of diphtheria puts susceptible populations in those countries at risk for infection, although documentation of secondary spread is rare.

6. The strategy for treating diphtheria is to use both antibiotics and diphtheria antitoxin. Antitoxin is given to attempt to neutralize circulating diphtheria toxin, since it is responsible for the symptoms caused by this organism. Antibiotics are given to eradicate the organism so that no more toxin can be produced.

The diphtheria antitoxin is prepared in horses, and, therefore, the development of serum sickness is a distinct possibility. Serum sickness is due to the patient making an antibody response to equine proteins, resulting in the formation of immune complexes. Clinical manifestations of serum sickness include fever, lymphadenopathy, rash, and joint pain. This child was unvaccinated. Because mortality rates of 10 to 20% have been reported in unvaccinated individuals, the risk of serum sickness was far outweighed in this case by the potential benefit of giving this antitoxin. Despite appropriate therapy, however, this child died of diphtheria.

REFERENCES
1. **Efstratiou, A., K. H. Engler, I. Z. Mazurova, T. Glushkevich, J. Vuopio-Varkila, and T. Popovic.** 2000. Current approaches to the laboratory diagnosis of diphtheria. *J. Infect. Dis.* **181**(Suppl. 1):S138–S145.

2. **Galazka, A.** 2000. The changing epidemiology of diphtheria in the vaccine era. *J. Infect. Dis.* **181**(Suppl. 1):S2–S9.

3. **Holmes, R. K.** 2000. Biology and molecular epidemiology of diphtheria toxin and the *tox* gene. *J. Infect. Dis.* **181**(Suppl. 1):S156–S167.

4. **Vitek, C. B., and M. Wharton.** 2000. Diphtheria in the former Soviet Union: reemergence of a pandemic disease. *Emerg. Infect. Dis.* **4**:539–549.

CASE 20

This 83-year-old man with metastatic adenocarcinoma of the prostate and with end-stage chronic obstructive pulmonary disease (COPD) was in his usual state of poor health (requiring home oxygen and corticosteroids) until he had an exacerbation of his COPD. He was seen by his home health nurse, who noted shortness of breath, and trimethoprim-sulfamethoxazole was prescribed by his physician, with subsequent improvement. Five days after discontinuing his antibiotic, he had another exacerbation of his COPD, this one requiring hospitalization, an increase in his dose of corticosteroids, and empiric intravenous antibiotics. After discharge from the hospital, the patient began to have nausea and vomiting, as well as shortness of breath and purulent sputum. A wet mount of his sputum is shown in Fig. 1. This organism was initially seen on Gram stain of his sputum.

1. List the nematodes that have a lung phase. Which one do you think is most likely in this patient? Does it help you decide if you learn on further history-taking that this patient was a World War II veteran who had been a prisoner of war helping to build the Thai-Burmese railroad?

2. Describe the life cycle of this parasite. How long can this parasite persist within the gut? How is it able to persist for this period of time?

3. In what way does corticosteroid therapy alter the host-parasite relationship in infections with this nematode? What may be done to try to prevent this infection in organ transplant candidates who will be immunosuppressed for life?

4. Which of the white blood cells is frequently increased in number in infections with this parasite? Would you expect them to be increased in this patient? Explain.

5. If this organism were to invade the bloodstream or the central nervous system, how might this be manifested clinically?

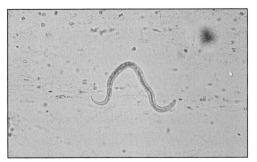

Figure 1

CASE 20

CASE DISCUSSION

1. Three common types of nematode larvae pass through the lung as part of their life cycle: *Ascaris lumbricoides*, hookworm (*Necator americanus* and *Ancylostoma duodenale*), and *Strongyloides stercoralis*. The findings of filariform larvae in sputum and the development of gastrointestinal symptoms after an increase in the patient's corticosteroid dose are consistent with hyperinfection with *S. stercoralis* (see answer to question 3 for further information). It is known that individuals who were prisoners of war in Southeast Asia during the Second World War have a very high rate of infection with *Strongyloides*. If this patient had given such a history (which he did not), this would be further strong evidence that he was infected with this nematode. Interestingly, strongyloidiasis has not been a significant problem in Vietnam War veterans, although the parasite is endemic in that country. Whether this is due to the younger age of these veterans, and thus a comparatively smaller pool of immunocompromised individuals at risk for hyperinfection, or to lower overall rates of infection in veterans of that war remains to be seen.

2. This parasite has a highly complex life cycle, with autoinfection being a prominent feature. Two larval forms are central to its life cycle: the filariform larvae (the infective form) and the rhabditiform larvae (the initial form of the worm, which develops into the filariform larvae). The life cycle begins with filariform larvae penetrating the skin from fecally contaminated soil. The larvae migrate via the bloodstream to the lung, where they break through the wall of the alveoli, crawl out of the bronchus and up the trachea, are swallowed, and reach the duodenum. There the parasite develops into an adult worm that invades and lives beneath the intestinal wall mucosa. Only female worms are present, and they reproduce by parthenogenesis. Eggs hatch as rhabditiform larvae. Most of these larvae are passed in feces. If feces are passed onto soil, the rhabditiform larvae either develop into filariform (infectious) larvae, and the cycle begins anew, or they develop into adult worms. These adult worms lay eggs in the soil that hatch into rhabditiform larvae and develop into filariform larvae. An important phase of the *Strongyloides* life cycle in hyperinfection states is the autoinfective stage. In this stage, rhabditiform larvae develop into filariform larvae in the intestinal tract. The filariform larvae then penetrate either the intestinal wall or the perianal skin, enter the bloodstream, migrate to the lung, and begin the infectious process again.

It is estimated that the parasite can live for 30 or 40 years in the human gastrointestinal tract. The ability of *S. stercoralis* to complete its entire life cycle within the human host is very unusual for a nematode, and it is this autoinfective cycle, plus the adult worm living within the intestinal mucosa rather than at the mucosal surface, which contributes to this parasite's unusual longevity in the human host.

3. Although the parasite may persist for many years, the parasite-host relationship appears to be kept in balance by the immune system. Infected, immunocompetent individuals frequently are asymptomatic, or they may have intermittent symptoms, which usually are gastrointestinal, including abdominal pain, diarrhea, nausea, or vomiting. A few patients might have intermittent "larva currens" (racing larvae), characterized by itching and skin rash. However, when patients receive immunosuppressive therapy, the balance between host and parasite is tilted in favor of the parasite. The worm burden increases dramatically, aided by autoinfection. This sharp increase in the number of parasites, with corresponding severe clinical disease due to tissue invasion by *S. sterocaralis* larvae, is called the hyperinfection syndrome. Patients with the hyperinfection syndrome frequently begin with worsening gastrointestinal symptoms similar to but more severe than those in symptomatic immunocompetent individuals. These patients also may have cough, shortness of breath, wheezing, and an abnormal chest radiograph, and they can rapidly progress to respiratory failure. In addition, the larvae can migrate to other organs, including the central nervous system.

Hyperinfection due to *Strongyloides* is of concern in patients who are organ transplant candidates because they will receive long-term immunosuppressive therapy. Therefore, in patients who have lived in endemic areas, multiple stool specimens should be examined for the presence of larval forms, since eggs are rarely if ever seen with this parasite. Even the examination of multiple stool specimens will not completely rule out the presence of the parasite since larval numbers are usually quite low in feces in all but hyperinfection states. The use of the agar plate culture technique has been shown to be more sensitive for the detection of *S. stercoralis* than are routine stool examination techniques. In this method, stool is placed on sterile agar plates, and the presence of characteristic furrows due to the migration of the parasites on the surface of the agar is sought.

4. As with many tissue-invasive parasites, eosinophil counts are increased in patients with strongyloidiasis. Although the eosinophil count was not available in this patient, the absence of eosinophilia in hyperinfection is not unusual. The reason is that these patients are frequently receiving immunosuppressive drugs that may reduce the numbers of eosinophils in the blood.

5. As mentioned previously, larvae can migrate to the central nervous system during hyperinfection. Individuals with invasion of the blood or of the central nervous system during the hyperinfection syndrome may have polymicrobial bacteremia or meningitis from which multiple species of enteric bacteria, such as *Escherichia coli*, *Klebsiella pneumoniae*, or *Enterococcus* spp., are simultaneously recovered. It is postulated that the migrating *Strongyloides* larvae are transporting enteric bacteria during their migration. Whenever enteric bacteria are recovered from cerebrospinal fluid (CSF) or blood of an immunocompromised host, especially if more than one species is found, hyperinfection syndrome due to *Strongyloides* should be considered.

REFERENCES

1. **Jongwutiwes, S., M. Charoenkorn, P. Sitthichareonchai, P. Akaraborvorn, and C. Putaporntip.** 1999. Increased sensitivity of routine laboratory detection of *Strongyloides stercoralis* and hookworm by agar-plate culture. *Trans. R. Soc. Trop. Med. Hyg.* **93:**398–400.

2. **Link, K., and R. Orenstein.** 1999. Bacterial complications of strongyloidiasis: *Streptococcus bovis* meningitis. *South. Med. J.* **92:**728–731.

3. **Mahmoud, A. A.** 1996. Strongyloidiasis. *Clin. Infect. Dis.* **23:**949–952.

THREE

Gastrointestinal Tract Infections

INTRODUCTION TO SECTION III

The major clinical manifestation of infections affecting the gastrointestinal tract is diarrhea. Diarrheal pathogens have two basic mechanisms by which they produce diarrhea. One is by the production of toxins called enterotoxins. Enterotoxins cause physiologic changes in the intestinal epithelium resulting in fluid and electrolyte secretion. *Vibrio cholerae*, which produces the enterotoxin cholera toxin, is a classic example of a diarrheal pathogen which produces a secretory diarrhea due to the action of an enterotoxin. Microscopically, the intestinal epithelium appears normal in patients with enterotoxin-induced diarrhea.

The other major mechanism of diarrheal disease is by damage to the intestinal epithelium. Organisms may also produce toxins that directly damage the intestinal epithelium. The protozoan *Entamoeba histolytica* produces such a cytotoxin. This cytotoxin is responsible for the characteristic ulcerative lesions that can be seen in individuals with amebic dysentery. Damage to intestinal epithelium can also occur as a result of direct invasion of the intestinal epithelium. A number of gastrointestinal pathogens, including *Salmonella* spp., *Shigella* spp., *Campylobacter* spp., and *Yersinia enterocolitica*, are capable of invading the intestinal epithelium. Inflammation frequently occurs in response to these pathogens. Patients with diarrhea due to organisms that damage the epithelium frequently will have white blood cells visible in their feces. However, these cells may also be present in feces of patients with noninfectious inflammatory bowel disease, so results of examination of feces for white blood cells should be interpreted cautiously.

Diarrheal diseases are almost always spread by the **fecal-oral route.** This means that individuals who become infected with diarrheal pathogens ingest either food or water that has been contaminated with human or animal feces. Improper handling or preparation of food or contamination of water due to poor sanitation are major means by which diarrheal pathogens are spread. In the industrialized world, the spread of diarrheal disease is particularly problematic in day care centers for children. In addition to spread by contaminated food and water, infected children can pass the organisms directly by placing contaminated hands in the mouths of other children or indirectly by handling toys with contaminated hands which are then mouthed by other children. The infectious dose of diarrheal pathogens varies greatly, from hundreds of thousands to millions in *Salmonella* spp. and *V. cholerae* to less than 100 organisms in *Shigella* spp.

Because the major pathophysiologic effect of diarrhea is dehydration due to fluid and electrolyte loss, the most important treatment is **rehydration.** In recent years, simple solutions of glucose, salts, and water given orally have been developed which are proven to be highly effective in treating patients with even the most severe forms of diarrhea. The widespread use of **oral rehydration** in the past 2 decades, especially in the developing world, has been credited with saving literally millions of lives, primarily young children in whom diarrheal disease takes the greatest toll.

In addition to diarrheal disease, hepatitis is an important infection in the gastrointestinal system. The epidemiology of hepatitis A and E viruses is the same as that of diarrheal pathogens. They are usually obtained by ingestion of raw shellfish taken from water contaminated by human sewage or from food handled by infected people with poor personal hygiene, i.e., individuals who fail to wash their hands after a bowel movement. Hepatitis B, C, and D are spread by contaminated blood. Contracting hepatitis used to be a major concern in individuals receiving blood transfusions. With the recognition of these agents and the development of screening tests for them, the epidemiology of hepatitis due to hepatitis B and C viruses has changed. Hepatitis B, C, and D infections (as well as HIV infections) are frequent in individuals who share needles while using illicit intravenous drugs. Hepatitis B is also spread sexually, especially in populations that practice anal intercourse. The frequency of spread of hepatitis C sexually is not as well understood. Unlike hepatitis A virus, which causes a relatively mild self-limited disease, hepatitis B can cause fulminant, sometimes fatal disease. Hepatitis B and C viruses can also cause chronic infections culminating in liver failure. Vaccines are available for hepatitis A and B viruses but not for hepatitis C.

Other important types of gastrointestinal infection are those in which the resident intestinal microflora or a pathogen escapes from the bowel and enters "sterile" tissues. One example is *E. histolytica* trophozoites that enter the liver and cause an amebic abscess. Another is when there is penetrating trauma to the intestines, as might occur with a gunshot wound to the abdomen or during bowel surgery. In either situation, microbes can escape from the intestines into the peritoneum where they can cause peritonitis or form an abscess. The organisms causing these infections are typically a mixture of both facultative and anaerobic bacteria that reside in the colon.

TABLE 3 SELECTED GASTROINTESTINAL TRACT PATHOGENS

ORGANISM	GENERAL CHARACTERISTICS	USUAL SOURCE OF INFECTION	DISEASE MANIFESTATION
Bacteria			
Bacteroides fragilis	Anaerobic, gram-negative bacillus	Endogenous	Abdominal abscess
Campylobacter spp.	Microaerophilic, curved, gram-negative bacilli	Poultry	Invasive diarrhea, sepsis in AIDS patients
Clostridium difficile	Anaerobic, toxin-producing, gram-positive bacillus	Endogenous; nosocomial	Antibiotic-associated diarrhea, pseudomembranous colitis
Clostridium perfringens	Anaerobic, gram-positive bacillus	Endogenous; high-protein foods	Gangrenous lesions of bowel or gall bladder; food poisoning
Enterohemorrhagic *Escherichia coli*	Sorbitol-nonfermenting (*E. coli* O157:H7) gram-negative bacillus	Improperly cooked ground beef	Enterohemorrhagic colitis, hemolytic uremic syndrome
Enterotoxigenic *Escherichia coli*	Lactose-fermenting, gram-negative bacillus	Fresh fruit and vegetables	Traveler's diarrhea, watery diarrhea
Salmonella spp.	Lactose-nonfermenting, gram-negative bacilli	Animal products; typhoid (human to human)	Invasive diarrhea, typhoid fever
Shigella spp.	Lactose-nonfermenting, gram-negative bacilli	Human to human; day care centers	Invasive diarrhea, dysentery
Staphylococcus aureus	Catalase-positive, gram-positive coccus	High-protein foods	Food poisoning
Vibrio spp.	Oxidase-positive, gram-negative bacilli	Raw fish and shellfish	Large-volume watery diarrhea
Yersinia enterocolitica	Lactose-nonfermenting, gram-negative bacillus	Meat and dairy products	Watery or invasive diarrhea
Parasites			
Ascaris lumbricoides	Roundworm	Food, soil	Diarrhea, abdominal discomfort, intestinal obstruction
Cryptosporidium parvum	Coccidian parasite	Fecally contaminated water; day care centers; farm animals	Malabsorptive diarrhea (chronic in AIDS)
Cyclospora spp.	Coccidian parasite	Water, fresh fruits and vegetables	Malabsorptive diarrhea

Organism	Description	Transmission/Source	Disease
Echinococcus spp.	Dog tapeworm	Ingestion of tapeworm eggs from infected dog	Hydatid cyst of liver
Entamoeba histolytica	Ameba	Water, fresh fruits and vegetables	Diarrhea, amebic dysentery, liver abscess
Giardia lamblia	Flagellated trophozoite, cyst	Fecally contaminated water; day care centers	Malabsorptive diarrhea (acute; chronic)
Necator americanus, Ancylostoma duodenale	Hookworm	Skin contact with larvae in soil	Anemia, gastrointestinal discomfort
Strongyloides stercoralis	Threadworm	Skin contact with larvae in soil, autoinfective cycle	Gastrointestinal discomfort, diarrhea, rash; larval invasion of lungs and other organs in immunosuppressed patients
Viruses			
Enteroviruses	Nonenveloped RNA viruses	Fecal-oral	Diarrhea, respiratory disease, aseptic meningitis, exanthems, myocarditis
Hepatitis A virus	Nonenveloped RNA virus	Shellfish, infected food handlers via fecal-oral	Acute, self-limited hepatitis
Hepatitis B virus	Enveloped DNA virus	Blood, secretions, direct sexual contact	Acute and chronic hepatitis, fulminant hepatitis, hepatic carcinoma
Hepatitis C virus	RNA virus	Blood	Acute and chronic hepatitis, fulminant hepatitis, hepatic carcinoma
Hepatitis D virus	RNA virus; requires co-infection with hepatitis B virus	Blood, secretions, direct sexual contact; can occur as superinfection of hepatitis B chronic carrier or coinfection with hepatitis B	Acute and chronic hepatitis, fulminant hepatitis, hepatic carcinoma; worse prognosis than in hepatitis B infection without hepatitis D
Hepatitis E virus	RNA virus	Fecal-oral, contaminated water, shellfish; possibly zoonotic (pigs, rats)	Acute, self-limited hepatitis; may be fulminant in pregnant women
Norwalk agent (calicivirus)	Nonenveloped RNA virus	Shellfish, common-source food outbreaks	Vomiting, diarrhea
Rotavirus	Wheel-like, nonenveloped RNA virus	Human to human (day care center)	Diarrhea, vomiting

CASE 21

This 18-year-old male presented to the outpatient medical clinic for evaluation of diarrhea and abdominal discomfort. The patient first noted mild abdominal discomfort and three loose bowel movements per day 1 week prior to evaluation. Two days prior to evaluation he noted intermittent, crampy periumbilical abdominal pain. He denied drinking well water, fever, blood in the stool, relation of the pain to meals, dysuria, or hematuria.

On examination, the patient was afebrile and had normal vital signs. The abdominal examination was notable for mild lower abdominal tenderness. The fecal examination demonstrated a greenish, watery stool that was negative for occult blood.

Laboratory evaluation included a normal white blood cell count, hematocrit, and platelet count. Examination of the feces microscopically was remarkable for the presence of white blood cells. The organism causing his illness is shown in Fig. 1 (Gram stain) and Fig. 2 (growth on special medium).

1. On the basis of the laboratory findings, what is the likely etiology of this patient's diarrhea? Is the finding of white cells in the feces consistent with the recovery of this organism? Explain your answer.

2. What special laboratory conditions are necessary to recover this organism?

3. What is the epidemiology of this organism? What simple precautions can be taken to prevent its spread?

4. How have modern means of food production contributed to an increasing incidence of infections with this organism?

5. What is the current status of drug resistance in this organism? What factors are believed to play an important role in this status?

6. Although the patient has evidence of local invasion in the intestinal tract with this organism, bacteremia due to this organism is unusual. Explain this observation.

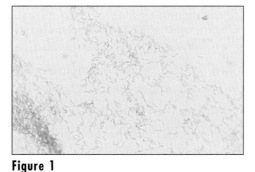

Figure 1

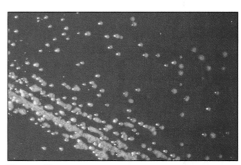

Figure 2

CASE DISCUSSION

1. Both *Vibrio* and *Campylobacter* spp. are slightly curved, gram-negative rods that cause diarrhea (Fig. 1). The pathogenesis of the most important *Vibrio* species, *V. cholerae*, is due primarily to the production of an exotoxin, cholera toxin, which causes a secretory diarrhea. The stools of patients with severe cases of cholera have a "rice water" appearance. Because of the secretory, noninflammatory nature of the diarrhea, white blood cells are rarely seen in the feces of patients with cholera. *Campylobacter* spp. cause an invasive diarrhea distinguished by the presence of white blood cells in the stool. As many as half of patients with *Campylobacter* diarrhea may have bloody stools due to the presence of red blood cells. The diarrhea seen in this patient is consistent with a *Campylobacter* infection, and *Campylobacter jejuni* was isolated from his stool (Fig. 2).

2. It is important to remember that the facultative aerobic fecal flora consists of approximately 10^7 to 10^9 CFU/g of feces and that finding an enteric pathogen, which may represent only a small fraction of this flora, is akin to trying to find a needle in a haystack. Selective media, such as Hektoen and MacConkey agar, used for the isolation of *Salmonella* and *Shigella* spp. from feces do not support the growth of *Campylobacter* spp. Therefore, several selective media have been developed for the isolation of *Campylobacter* spp. To further complicate matters, *Campylobacter* spp. are microaerophilic organisms, and so culture conditions that will support their growth must be used when attempting to isolate them. Finally, *C. jejuni*, the most frequently recovered *Campylobacter* species, grows optimally at 42°C, the body temperature of chickens, a natural host of this organism. Many laboratories inoculate fecal specimens onto campylobacter selective agar and incubate these plates at 42°C under microaerophilic conditions in an attempt to isolate *C. jejuni*. This approach is problematic since other *Campylobacter* spp. either fail to grow on certain types of campylobacter selective agar or cannot grow at 42°C. Alternative methods are available for the isolation of these species.

3. *C. jejuni* along with *Salmonella* spp. are the two most frequently recovered bacterial causes of gastroenteritis in the United States. The incidence of *Campylobacter* infections in the United States is 15 to 20 per 100,000. *C. jejuni*, like all enteric pathogens, is spread by the fecal-oral route. Improperly cooked poultry or cross-contamination of foods by raw poultry is postulated to be the most important source of infection. Studies have shown that as many as 98% of chicken carcasses in food markets are contaminated with *C. jejuni*. By contrast, 5% of ground beef packages have been shown to be contaminated with this organism. Outbreaks of *Campylobacter* infection have also followed the consumption of nonpasteurized milk. Contaminated water is an infrequent vehicle for this infection. There is no evidence of person-to-person spread of this organism. Adequate cooking of poultry and avoidance of cross-contamination of other foods will result in prevention of most *Campylobacter* cases. The infectious dose for this organism appears to be intermediate between those for *Shigella* spp. (low) and *Salmonella* spp. (high). Like *Salmonella* and *Shigella* spp., it is an organism that causes disease mainly during the warm-weather months, with a peak incidence in July. One of the interesting observations concerning this organism is that the highest incidence of infection is in infants (<1 year old) and adolescents and young adults (15 to 29 years old). It is the most frequent cause of bacterial gastroenteritis in college students in the United States, with isolation rates on certain campuses as high as 15% in individuals with diarrhea. In the adolescent and young adult age group, infections are more common in males. It is speculated that these individuals are preparing the bulk of their meals for the first time in their lives and may not practice the best food preparation hygiene.

4. Animals and animal products are the primary source of infection for *Campylobacter*. "Factory farming," in which large numbers of animals are grown in close quarters, results in high rates of colonization with this organism. For example, cattle "finished" on feedlots have a much higher rate of colonization than grazing animals. Essentially all chickens raised in commercial chicken barns, which can hold as many as 100,000 animals, are colonized with *Campylobacter* by the fourth week of life. Studies have shown that contamination of poultry carcasses can increase significantly during automated processing. Agricultural research is focusing on ways to reduce or prevent contamination during processing.

5. Recent studies have shown that resistance to fluoroquinolones, a class of antimicrobials frequently used to empirically treat suspected bacterial gastroenteritis, is increasing dramatically in *Campylobacter*. In one study in Minnesota, the rate of resistance to the fluoroquinolone ciprofloxacin between 1992 and 1998 increased from 1.2 to 10.3%. Ciprofloxacin resistance is even higher in Europe, with one study from Spain showing 75% of human isolates being ciprofloxacin resistant. Isolates resistant to ciprofloxacin are resistant to other fluoroquinolones as well. By contrast, rates of resistance continue to be low in the macrolides (erythromycin and azithromycin), the other class of antimicrobials recommended for treatment of *Campylobacter* gastroenteritis. A recent study done in Minnesota showed that 65% of patients with *Campylobacter* gastroenteritis received fluoroquinolones while only 25% received macrolides, even though there are higher rates of resistance to the fluoroquinolones. The reason for this difference may be that the macrolides are known to have gastrointestinal side effects which may cause physicians to avoid their use in patients with gastroenteritis.

Four factors seem to be important in the acquisition of drug-resistant *Campylobacter*. First, a single mutation in the DNA gyrase, the target protein for the fluoroquinolones, can result in high-level resistance. Second, one of the consequences of factory farming is the growth of large numbers of animals in close proximity. This creates ideal conditions for spread of illness that is frequently treated with nonhuman fluoroquinolones such as enrofloxacin. The use of fluoroquinolones in these animals results in the selection of mutant organisms, including *Campylobacter* spp., resistant not only to the "animal only" fluoroquinolone but also having cross-resistance to the "human" fluoroquinolones. The close proximity also results in the colonization of large percentages of animals with this resistant organism. Third, travel to foreign countries where hygiene during food preparation may not be ideal may result in infection with these drug-resistant organisms. For example, in Mexico, it is estimated that in 1997 each chicken was exposed to 0.5 liter of fluoroquinolone-medicated water, so the finding of a high rate of fluoroquinolone-resistant *Campylobacter* there would be expected. Not surprisingly, in a study done in Minnesota a major risk factor for acquisition of fluoroquinolone-resistant *Campylobacter* was travel to Mexico. Fourth, the empiric use of ciprofloxacin by travelers who develop diarrhea may further select for drug-resistant isolates.

6. *C. jejuni* was locally invasive in this patient, as evidenced by the presence of white blood cells in his feces. Like *Shigella* spp., this organism rarely causes bacteremia in the immunocompetent host. The most likely reason for this is that this organism, unlike *Salmonella* spp., does not survive within phagocytic cells. It is either locally ingested and killed by phagocytes in the intestinal wall or carried by lymphatic drainage to the Peyer's patches, where it is killed. Occasional cases of *C. jejuni* bacteremia occur, but most are transient because the reticuloendothelial system is able to eliminate this organism from the bloodstream.

Campylobacter-associated diarrhea and bacteremia are much more common in HIV-infected individuals who have CD4 counts of <200/µl. However, with the widespread use of highly active antiretroviral therapy (HAART) in the industrial world, *Campylobacter* bacteremia has declined dramatically in this patient population.

REFERENCES

1. **Altekruse, S. F., N. J. Stern, P. I. Field, and D. V. Swerdlow.** 1999. *Campylobacter jejuni*—an emerging foodborne pathogen. *Emerg. Infect. Dis.* **5:**28–35.

2. **Centers for Disease Control and Prevention.** 2000. Preliminary FoodNet data on the incidence of foodborne illnesses—selected sites, United States, 1999. *Morb. Mortal. Wkly. Rep.* **49:**201–205.

3. **Saenz, Y., M. Zarazaga, M. Lantera, M. J. Gastnares, F. Banquero, and C. Torres.** 2000. Antibiotic resistance in *Campylobacter* strains isolated from animals, foods, and humans in Spain in 1997–1998. *Antimicrob. Agents Chemother.* **44:**267–271.

4. **Smith, K. E., J. M. Besser, C. W. Hedberg, F. T. Leano, J. B. Bender, J. H. Wicklund, B. P. Johnson, K. A. Moore, M. T. Osterholm, and the Investigation Team.** 1999. Quinolone-resistant *Campylobacter jejuni* infections in Minnesota, 1992–1998. *N. Engl. J. Med.* **340:**1525–1532.

5. **Wegener, H. C.** 1999. The consequences for food safety of the use of fluoro-quinolones in food animals. *N. Engl. J. Med.* **340:**1581–1582.

CASE 22

The patient was a 4-year-old male who presented to the emergency room with a 2-hour history of vomiting, diarrhea, fever, irritability, and lethargy. The child had gone to sleep on the living room couch at 11 p.m. His grandmother found him on the floor at 3 a.m. covered with feces. When she picked him up to carry him to the bathtub, she noticed he was febrile. She bathed him and brought him to the emergency room. The patient's medical history was significant for his participation in group day care.

In the emergency room, he had two episodes of vomiting. He had a temperature of 38.9°C, pulse rate of 160 beats/min, and respiratory rate of 36/min, and he was noted to be dehydrated. His stool contained bloody streaks; a methylene blue stain of his feces is shown in Fig. 1. Other laboratory studies included a cerebrospinal fluid examination, which was within normal limits, done because of his lethargy; a peripheral white blood cell count of 13,200/μl with 85% neutrophils; a negative blood culture; and a negative stool examination for ova and parasites. Figure 2 shows a MacConkey agar plate culture of the organism recovered from the feces of this patient. Figure 3 shows the biochemical reactions obtained in a triple sugar iron (TSI) tube, a urea-motility-indole (UMI) tube, and a UMI tube with added Kovács reagent (to detect indole production).

1. Given his clinical picture, what bacterial pathogens are likely in this patient?

2. Based on the laboratory results seen in Fig. 2 and 3, what organism is likely causing his illness? Briefly describe the pathogenesis of this organism.

3. What factors contributed to his lethargy?

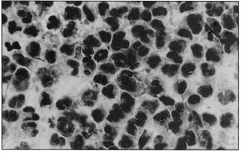

Figure 1

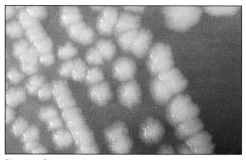

Figure 2

4. What would be the appropriate treatment strategy for this child?

5. Describe the epidemiology of this organism. What was the significance of his being in group day care? What special characteristics of this organism lead to its spread?

6. Was it surprising that this patient had a negative blood culture? Explain.

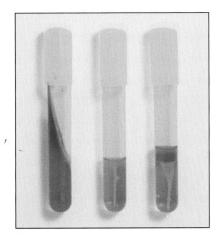

Figure 3

CASE DISCUSSION

1. The presence of white blood cells in feces (Fig. 1) indicates that the patient has an inflammatory type of diarrhea. Inflammatory diarrhea is frequently associated with infection with invasive bacteria and selected protozoans. Organisms that frequently cause invasive diarrheal disease include *Shigella* spp., *Salmonella* spp., *Campylobacter jejuni*, *Yersinia enterocolitica*, enteroinvasive *Escherichia coli*, and *Entamoeba histolytica*. *E. histolytica* was ruled out on the basis of a negative examination for parasites.

2. The colonial morphology and inability to ferment lactose on MacConkey agar (Fig. 2) are consistent with *Salmonella* and *Shigella* spp. The TSI agar slant shows an organism that does not ferment sucrose or lactose and does not produce H_2S as a product of metabolism. *Salmonella* spp. generally are H_2S positive, making this organism less likely. In the UMI tube, the isolate is nonmotile and urea and indole negative. All three results are typical for *Shigella* spp. In a clinical laboratory, further testing would be done to confirm the identity of this organism, but the results seen in Fig. 2 and 3 are consistent with *Shigella* spp. The particular species isolated from this patient was *Shigella sonnei*.

The pathogenesis of *Shigella* is due primarily to the organism's ability to invade intestinal epithelial cells. The virulence factors needed for invasion of intestinal epithelial cells are encoded on a large virulence plasmid. Over 20 different proteins encoded by this plasmid are involved in the invasion process. Many of these genes are necessary for the secretion of specific proteins that mediate invasion. Animal studies suggest *Shigella* enters the intestinal epithelium via M cells. From there, the organism invades the basolateral surface of the epithelial cells. *Shigella* secretes three invasin proteins called IpaB, IpaC, and IpaD, which mediate its invasion. Secretion of these proteins only occurs when the bacterium comes in contact with the host cells. This indicates that the resulting invasion is due to some type of signaling between the bacterium and host cell. These invasin proteins induce changes in the epithelial cell membrane that result in uptake of these bacterial cells. Once present in the host cells, these proteins are involved in the lateral spread of the bacterium throughout the intestinal epithelium. The combination of inflammation in response to the bacteria and the direct cellular damage done by the bacteria contributes to the nonspecific inflammation and ulceration which is characteristic of shigellosis.

In addition, *Shigella* can produce a toxin called Shiga toxin. This is an A + B toxin that inhibits protein synthesis. Although its exact role in the pathogenesis of shigellosis is not well understood, it is likely to contribute to the cellular damage seen in the intestinal epithelium.

3. Because of the patient's high fever and lethargy, a lumbar puncture was performed. Cerebrospinal fluid was normal, indicating that he did not have meningitis. Dehydration, which causes electrolyte imbalances, can cause lethargy and altered mental status. Alterations in consciousness are common in children with shigellosis. Although it has been suggested that Shiga toxin has a role in the pathogenesis of altered consciousness in cases of shigellosis, this is far from certain.

4. As with all patients with severe diarrhea and dehydration, rehydration is the key therapeutic modality. Usually oral rehydration is used because of its ease of administration and its low cost. Because this child was vomiting, intravenous rehydration therapy was given. Stool volume is relatively low in *Shigella* diarrhea, so rehydration can be achieved rapidly.

The use of antimicrobial agents in diarrheal disease is controversial. Most would agree that in the severely ill child, antimicrobial therapy is appropriate. In *Shigella* infection, antimicrobial therapy shortens the disease course and reduces the length of time organisms are excreted in feces. In the less severely ill patient, when antimicrobial therapy may not significantly shorten the duration of illness, some would argue that the infected individual should forgo antimicrobial therapy to limit the antimicrobial pressure on this organism and thus reduce antimicrobial resistance.

Antimicrobial resistance in *Shigella* has become a worldwide problem. For many years, the drugs of choice for treating *Shigella* were ampicillin and trimethoprim-sulfamethoxazole. Recent reports in the United States and elsewhere show that greater than 50% of *Shigella* isolates are resistant to these two antimicrobials, rendering both agents ineffective as treatment modalities for this organism. Nalixidic acid, which can be used safely in children, and the fluoroquinolones are currently the therapeutic agents of choice for treating severe *Shigella* infections. However, there are reports of resistance to these agents in both the developing world and the United States.

5. Unlike many other enteric pathogens which can infect a variety of domestic animals, e.g., *Salmonella* spp., *Shigella* infects only humans and nonhuman primates. Therefore, *Shigella* infections are a result of person-to-person spread typically, but not exclusively, due to contamination of water and food by human feces from *Shigella*-infected individuals. With the globalization of the food supply, outbreaks of *Shigella* infection in the United States and Canada have resulted from the importation of contaminated food from the developing world. Failure to use chlorinated water in food processing and packaging is recognized as one of the factors that may lead to contamination of imported foods, especially leafy vegetables such as parsley and lettuce.

Shigella is the third most common cause of bacterial gastroenteritis in the United States. It causes approximately four laboratory-confirmed cases per 100,000 annually. The actual incidence of disease is much higher. The disease is most common in children less than 5 years old. Infection rates have been found to be very high in Latino migrant workers who often live under poor hygienic conditions.

S. sonnei is the most common species of *Shigella* isolated in the United States, and *S. flexneri* is the second most common. Other species of this genus, *S. boydii* and *S. dysenteriae*, are infrequent causes of shigellosis in the United States.

Shigellosis is a common cause of outbreaks of diarrhea in the day care setting. The reason for this is that shigellae are much more resistant to the low pH found in the stomach than are other enteric pathogens, such as *Salmonella* spp. or *Vibrio cholerae*. As a result, only a small number of organisms (estimated at less than 100 viable bacteria) need be ingested for infection to occur. *Shigella* can spread from person to person by fecally contaminated hands. This is a common occurrence in the day care setting, whether it be from a day care worker who does not adequately wash his or her hands after changing the diaper of a *Shigella*-infected toddler or a *Shigella*-infected child who touches toys or places fingers which are fecally contaminated into the mouths of other children.

6. The finding that this child had a negative blood culture was not surprising. Although the organism is locally invasive, destroying cells in the intestinal mucosa, the organism rarely penetrates beyond the lamina propria because these organisms are susceptible to phagocytosis. Therefore, blood cultures are rarely positive with this organism.

REFERENCES

1. **Centers for Disease Control and Prevention.** 1999. Outbreaks of *Shigella sonnei* infections associated with eating fresh parsley—United States and Canada, July–August 1998. *Morb. Mortal. Wkly. Rep.* **48:**285–289.

2. **Centers for Disease Control and Prevention.** 2000. Preliminary FoodNet data on the incidence of foodborne illnesses—selected sites, United States, 1999. *Morb. Mortal. Wkly Rep.* **49:**201–205.

3. **Finlay, B. B., and P. Cossart.** 1997. Exploitation of mammalian host cell functions by bacterial pathogens. *Science* **276:**718–725.

4. **Jensen, V. B., J. T. Harty, and B. D. Jones.** 1998. Interactions of the invasive pathogens *Salmonella typhimurium, Listeria monocytogenes,* and *Shigella flexneri* with M cells and murine Peyer's patches. *Infect. Immun.* **66:**3758–3766.

5. **Mounier, J., F. K. Bahrani, and P. J. Sansonetti.** 1997. Secretion of *Shigella flexneri* Ipa invasins on contact with epithelial cells and subsequent entry of the bacterium into cells are growth stage dependent. *Infect. Immun.* **65:**774–782.

6. **Replogle, M. L., D. W. Fleming, and P. R. Cieslak.** 2000. Emergence of antimicrobial-resistant shigellosis in Oregon. *Clin. Infect. Dis.* **30:**515–519.

CASE 23

The patient was a 3-year-old female referred to the pediatric gastroenterology clinic with a 5-week history of diarrhea. Her diarrhea was characterized as green and often watery. Although potty-trained, she was occasionally incontinent of feces. She had no fevers, nausea, or vomiting. She was an only child who attended group day care, and her mother had had diarrhea for 3 days about 1 month previously. The family drank filtered water. Her physical examination was unremarkable.

The gastroenterologist asked the local physician to have the family send two stools from the child for parasitic examination. The organism seen in Fig. 1 was seen in both fecal specimens.

1. What is the organism causing this patient's diarrhea? What kind of diarrhea does this organism typically cause?

2. Briefly describe the pathogenesis of this organism, including any virulence factor that this organism possesses.

3. This organism is of particular concern in children in day care settings. Why?

4. Why can this organism cause outbreaks of infection that can affect thousands of individuals?

5. What populations have been found to be at increased risk for infections with this organism? Explain why they are at increased risk.

6. How are infections with this organism typically diagnosed?

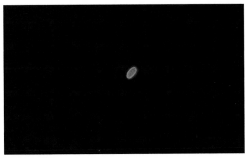

Figure 1

CASE DISCUSSION

1. The organism seen in Fig. 1 is the unicellular protozoan parasite *Giardia lamblia.* It is the most frequently detected parasitic cause of diarrheal disease in the United States. It is most commonly seen in children less than 5 years of age and in adults (frequently the parents of infected children) 31 to 40 years of age. The disease occurs most frequently in the summer and early fall.

This child's presentation was consistent with infection with this organism. Patients with clinical giardiasis typically are afebrile with constitutional symptoms of fatigue and malaise. Other symptoms are generally gastrointestinal in nature. *G. lamblia* is the most common infectious etiology of chronic diarrhea in childhood. It is probably the most common infectious diarrheal disease for which pediatric gastroenterologists are consulted; these children have chronic diarrhea and "failure to thrive." In children requiring hospitalization due to giardiasis, failure to thrive is documented in 20%. Most patients with chronic diarrhea due to *Giardia* have a malabsorptive type of diarrhea characterized by foul-smelling stools that frequently float because of fat malabsorption. These patients are frequently found to be deficient in fat-soluble vitamins such as vitamins A and B$_{12}$. Infected individuals frequently produce copious amounts of flatulence, although this was not noted in this case. Children with giardiasis often have lactose intolerance. The symptoms of chronic diarrhea may be attributed to "milk allergies" or other food allergies, delaying appropriate diagnosis and treatment. Lactose intolerance may persist for 2 to 4 months following appropriate treatment.

2. The pathogenesis of *G. lamblia* infections is fairly well understood. There are two stages in the life cycle of this protozoan, a cyst stage (infectious/environmental) and a trophozoite stage (reproductive/pathogenic). The infection begins with the ingestion of as few as 10 cysts, usually from contaminated water or via direct fecal-oral contact. The cyst passes through the stomach, where the change in pH is thought to signal the cyst to excyst. In the duodenum, pancreatic and *Giardia*-derived proteases act in concert to mediate the emergence of the trophozoite form from the cyst. The trophozoite form is flagellated and has a ventral sucking disk. It is postulated that the trophozoite "swims" to the brush border membrane. There, attachment is mediated by the sucking disk. Pathologic effects observed include blunting of the microvilli and reduction in disaccharidase enzyme production, both of which may explain in part the malabsorptive type of diarrhea that is such a common feature of infection with this protozoan. Trophozoites reproduce in the crypts of the duodenum and jejunum by binary fission. In the jejunum, trophozoites encyst when exposed to bile salts. Both trophozoites and cysts can be passed in feces, but only the cyst form survives outside the host.

3. The cyst form of *Giardia* is stable in the environment and somewhat resistant to chlorine-based disinfectants such as bleach. Over half of individuals infected with *Giardia* are asymptomatic and may pass cysts in stools for months. Toddlers often will have fecal material on their hands because of poor personal hygiene. Asymptomatic, infected children with fecally soiled hands may touch toys, eating utensils, or drinking cups. These soiled objects may find their way into the mouths of other children, infecting them. Infections with *Giardia* in day care centers are common. Children may become infected there and bring the infection home to parents and siblings. In the case discussed here, it is quite possible that the mother was infected by her daughter and had a mild *Giardia* infection herself.

4. Several large outbreaks of giardiasis have been described in the United States. One of the largest was in Pittsfield, Mass., in which an estimated 50,000 individuals were infected. In that outbreak there was a sanitation failure in the municipal water system. In most municipal systems, water is filtered and/or flocculated to remove particulate matter, including the parasites *Giardia* and *Cryptosporidium*. The water is also treated with chlorine. Since both *G. lamblia* and *Cryptosporidium parvum* are resistant to the chlorine levels used in water purification, failure of systems used to remove particulates may result in disease outbreaks with these parasites.

5. *Giardia* is common in several other populations. One population is hikers who drink untreated stream or surface water. Such water may be contaminated by infected mammals, including humans. With an increasing number of individuals involved in outdoor activities such as hiking, the U.S. Park Service has recognized drinking untreated stream and surface water as a health hazard and has posted signs in national parks advising against this practice. Many outdoor outfitters sell hand-held filtration devices plus iodine tablets which can be used effectively to treat stream and surface water.

A second population at risk are families with wells, especially those with relatively shallow ones whose water supply becomes contaminated either by animals or by leakage from cesspools or septic fields into the well water.

Travelers who visit areas of high endemicity for *Giardia* may also become infected. Typically, this occurs in individuals who travel to the developing world and ingest contaminated water or ice.

Another group of individuals who become infected with *Giardia* are people who swim in water that has been fecally contaminated by an individual who has *Giardia*. A typical scenario is an infected child wearing diapers who defecates into water at a swimming pool, water park, or recreational lake. Although water in swimming pools and water parks is typically treated with chlorine, the chlorine level is insufficient to kill the *Giardia* cysts. Parents should not allow children with diarrhea, especially those in diapers, to swim in community swimming areas. This will reduce but not eliminate the risk, since asymptomatic shedders exist.

6. *Giardia* can be diagnosed by detection of cysts or trophozoites in stool specimens. Two approaches are widely used in the United States. One is to use a direct fluorescent-antibody (DFA) assay to detect the organism microscopically. DFA reagents are available that will stain both *Giardia* and *Cryptosporidium* simultaneously. The other is to use an enzyme immunoassay (EIA) to detect *Giardia* antigens in stools. Single-test membrane EIAs have been developed that will simultaneously detect *Giardia* and *Cryptosporidium*. Both techniques are highly sensitive and specific. The major disadvantage of the use of DFA and EIA in the diagnosis of diarrheal disease due to parasitic agents is that, unlike light microscopy, the DFA and EIA do not detect any other parasites. The reality is that *Giardia* and *Cryptosporidium* are responsible for over 90% of the intestinal parasitic infections diagnosed in the United States. Therefore, the more labor-intensive and less accurate light microscopy technique should be reserved for patients who are likely to have other intestinal parasites, such as recent travelers to, or immigrants from, the developing world.

REFERENCES

1. **Adam, R. D.** 2001. Biology of *Giardia lamblia. Clin. Microbiol. Rev.* **14:**447–475.

2. **Furness, B. W., M. J. Beach, and J. M. Roberts.** 2000. Giardiasis surveillance—United States, 1992–1997. *Morb. Mortal. Wkly. Rep.* **49**(SS-7):1–13.

3. **Ortega, Y. R., and R. D. Adam.** 1997. *Giardia*: overview and update. *Clin. Infect. Dis.* **25:**545–550.

4. **Sharp, S. E., C. A. Suarez, Y. Duran, and R. J. Poppiti.** 2001. Evaluation of the Triage Micro Parasite Panel for the detection of *Giardia lamblia, Entamoeba histolytica/Entamoeba dispar,* and *Cryptosporidium parvum* in patient stool specimens. *J. Clin. Microbiol.* **39:**332–334.

CASE 24

The patient was a 1-year-old male admitted to the hospital in December because of fever and dehydration. His parents reported that he had a 1-day history of fever, diarrhea, emesis, and decreased urine output. On admission, his vital signs revealed a temperature of 39.5°C, slight tachycardia with a pulse rate of 126/min, and a respiratory rate of 32/min. He was very somnolent. His general physical examination was remarkable only for hyperactive bowel sounds. Laboratory tests showed a leukocytosis with a white blood cell (WBC) count of 14,200/µl with 80% polymorphonuclear leukocytes (PMNs). Urinalysis was significant for a high specific gravity and ketones (consistent with the patient's dehydration). Stool, blood, and urine samples were sent for culture. A stool sample was also checked for ova and parasites. There were no fecal leukocytes. The patient was given intravenous normal saline and had nothing by mouth. Over the next 48 hours his emesis abated. Once he was rehydrated and was tolerating oral feedings, he was discharged home. All cultures for bacterial pathogens gave negative results, but a rapid viral diagnostic test was positive (Fig. 1).

1. What is the differential diagnosis?

2. What is the most common cause of pediatric gastroenteritis? Briefly outline the pathophysiology seen with this organism.

3. Briefly describe the epidemiology of this agent.

4. What rapid diagnostic test was used? Why are rapid diagnostic tests useful in this setting?

5. What treatment is effective?

6. What special infection control precautions are necessary in the hospital setting when caring for a patient with gastroenteritis?

7. Briefly describe attempts to make a vaccine for this organism.

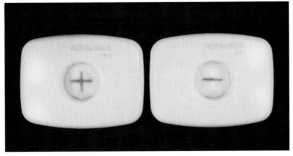

Figure 1

CASE DISCUSSION

CASE 24

1. The differential diagnosis for acute diarrhea includes bacterial, parasitic, and viral etiologies of gastroenteritis. Because of the absence of fecal leukocytes, agents of invasive diarrhea such as *Salmonella* spp., *Shigella* spp., *Campylobacter* spp., and *Entamoeba histolytica* are less likely, although certainly possible. The leading parasitic possibilities include *Giardia* and *Cryptosporidium* spp., especially if this child was in a day care center. The viruses that can cause gastroenteritis include rotavirus (most frequent), enteric coronaviruses and unclassified small round viruses, Norwalk and Norwalk-like viruses, enteric adenovirus, calicivirus, and astrovirus. Vomiting is seen frequently in viral gastroenteritis and less frequently in infections with the other agents listed, making a viral agent much more likely in this case.

2. Rotavirus, which is the most common cause of viral gastroenteritis, was the agent infecting this child. An enzyme immunoassay (EIA) performed on the patient's stool was positive for rotavirus antigen (Fig. 1). The clinical disease spectrum caused by this virus varies from asymptomatic infection to severe vomiting and dehydration. The disease is usually self-limited, lasting approximately 1 week. This duration of illness is much longer than that seen with most other viral agents of gastroenteritis, which usually resolve in 24 to 48 hours and rarely cause the type of severe symptoms that can be seen with rotavirus. Patients with rotavirus infections have watery diarrhea and frequent vomiting. These symptoms can be severe, resulting in dehydration which may require hospitalization. Rotavirus is the most common cause of diarrheal disease requiring hospitalization in the United States.

The pathophysiology of disease caused by this virus is very complex. It has long been known that the virus causes blunting and atrophy of the small intestinal villi and that this cellular damage is associated with the watery diarrhea typical of rotavirus. However, watery diarrhea has usually been associated with organisms that induce diarrhea through the action of an enterotoxin. A nonstructural rotavirus protein, NSP4, may act as an enterotoxin. It is speculated that this "toxin" may induce activation of the "enteric nervous system." This activation results in intestinal secretion of fluids and electrolytes. The evidence for a role of the enteric nervous system in the secretion of fluids and electrolytes seen in rotavirus disease is based on animal studies in which pharmacologic agents which block the activation of the enteric nervous system prevent fluid and electrolyte secretion in rotavirus-infected animals.

3. Rotavirus causes diarrheal disease primarily in children less than 5 years of age, with the most severe disease seen in children less than 2 years of age. As with all diarrheal diseases, it is spread primarily by the fecal-oral route. Common-source outbreaks outside of the day care setting are not well documented. Adults who become

infected usually are caregivers of an infected child. The disease is seen primarily during winter months in temperate zones. Because of this, it is frequently referred to as "winter vomiting disease." Disease seasonality is not as obvious in tropical areas. It is estimated that 30,000 to 50,000 children are hospitalized annually in the United States with rotaviral disease. Between 20 and 40 deaths are attributed to rotavirus annually in the United States. Fatalities are seen primarily in those who are malnourished and immunocompromised. In the developing world, rotavirus is a major cause of death in children less than 5 years of age, with an estimated 870,000 deaths occurring annually.

4. The EIA and the latex agglutination test are most commonly used to detect rotavirus. The virus was first discovered in the stools of children with vomiting and diarrhea by using electron microscopy. It was named for its characteristic wheel-like ("rota") morphologic appearance on electron microscopy. However, this technique is not routinely used because of the ease of EIA and latex agglutination.

Rapid testing for the detection of rotavirus is valuable for three reasons. First, children known to have rotavirus will not need any other expensive tests to determine the etiology of their disease. Second, appropriate rehydration therapy can be begun and the use of antibacterial agents can be avoided. Third, children who are infected with rotavirus can be cohorted. Because pediatric hospital beds are often at a premium during the winter months when rotavirus infection typically is seen, the ability to cohort children with the same illness allows the hospital to save isolation rooms for other children who need them.

5. Both oral and intravenous rehydration therapy are effective. Which therapy is used is based on the severity of the patient's vomiting. If the patient can tolerate it, oral rehydration is performed because of its low cost and ease of administration. There is no specific antiviral agent for rotavirus infections.

6. Because this pathogen can remain infectious on inanimate objects for days and on hands for as long as 4 hours, it is an important cause of both day care center and nosocomial diarrheal disease outbreaks. Strict hand washing and the use of gloves by health care workers delivering care to patients with gastroenteritis are necessary. Hospital outbreaks of rotavirus infection have occurred when health care workers have transmitted the virus from one patient to another.

7. With the recognition of rotavirus as the most clinically significant cause of infantile diarrheal disease, the development of a protective vaccine became an important public health goal. An attenuated, recombinant rhesus rotavirus vaccine was developed.

In clinical trials, the vaccine was found to prevent 50% of cases compared with the control group, and it reduced the severity of disease in those patients who developed disease postvaccination. Given its apparent efficacy, the vaccine was licensed in the United States and began to be used in September 1998. Over the next 10 months, there were several reports to the Vaccine Adverse Event Reporting System (VAERS) of intussusception (blockage of the intestines when the bowel folds over on itself) in the immediate postvaccine period. The majority of these individuals required surgical intervention. Two case control studies confirmed that intussusception was increased in the immediate postvaccine period. These findings resulted in a recommendation by the Advisory Committee on Immunization Practices that this vaccine be withdrawn from use. The obvious question was why this adverse event was not noted during clinical trials. Intussusception was observed in vaccinated children during the clinical trial. However, the rate was not statistically higher than that seen in the general population. Only when large numbers of patients were vaccinated was the association between vaccine and intussusception clearly established.

REFERENCES

1. **Centers for Disease Control and Prevention.** 1999. Intussusception among recipients of rotavirus vaccine—United States, 1998–1999. *Morb. Mortal. Wkly. Rep.* **48:**577–581.

2. **Centers for Disease Control and Prevention.** 1999. Withdrawal of rotavirus vaccine recommendation. *Morb. Mortal. Wkly. Rep.* **48:**1007.

3. **Glass, R. I., J. R. Gentsch, and B. Ivanoff.** 1996. New lessons for rotavirus vaccines. *Science* **272:**46–48.

4. **Keusch, G. T., and R. A. Cash.** 1997. A vaccine against rotavirus—when is too much too much? *N. Engl. J. Med.* **337:**1228–1229.

5. **Lundgren, O., A. T. Peregrin, K. Perrson, S. Kordasti, I. Uhnoo, and L. Svensson.** 2000. Role of the enteric nervous system in the fluid and electrolyte secretion of rotavirus diarrhea. *Science* **287:**491–495.

CASE 25 The patient was an 80-year-old female who 10 days previously had had a cystocele repair performed. At the time of that hospital admission, a urine culture was obtained that revealed >100,000 CFU/ml of an *Escherichia coli* strain that was susceptible to all antimicrobial agents against which it was tested. Postoperatively, she began a 7-day course of oral cephalexin. She was discharged after an uneventful postoperative course of 3 days. Ten days postoperatively, she presented with a 3-day history of diarrhea. The patient noted multiple watery, loose stools without blood, crampy abdominal pain, and vomiting. She presented with a temperature of 38.2°C, pulse rate of 90/min, respiratory rate of 20/min, and blood pressure of 116/53 mm Hg. Her white blood cell count was normal, but a large number (53%) of immature polymorphonuclear cells were seen. Physical examination, electrolytes, liver enzymes, and lipase were all within normal limits.

A methylene blue stain for fecal leukocytes is shown in Fig. 1. Cultures for *Salmonella*, *Shigella*, *Yersinia*, and *Campylobacter* spp. were all negative. An enzyme immunoassay (EIA) that was positive for the presence of a bacterial toxin in the stool established the patient's diagnosis.

1. What organism was causing this woman's diarrhea?

2. For most agents of bacterial diarrheal disease, culture is used to establish the etiology of disease. Why is this strategy not useful in this disease? Why is toxin detection employed to diagnose this disease? How well does it work?

3. What in her history was a predisposing factor for her development of this infection? How did it predispose her?

4. Why is this organism particularly problematic as a nosocomial pathogen?

5. Describe the disease spectrum seen with this organism.

6. What virulence factors does this organism produce, and what roles do these factors play in the pathogenesis of disease?

7. Discuss three different types of therapeutic strategies that can be used to treat this disease.

8. Recurrence of this disease occurs in 10 to 20% of patients. Give two possible explanations for why recurrences are common in this disease.

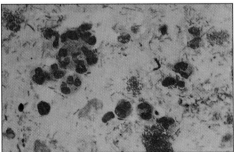

Figure 1

CASE DISCUSSION

1. This woman had *Clostridium difficile*-associated diarrhea. *C. difficile* is the most common infectious cause of diarrheal disease in hospitalized patients and in patients receiving antimicrobial agents. In the industrialized world, it is found in twice as many patients as the "traditional" enteric pathogens *Salmonella* spp., *Shigella* spp., *Campylobacter* spp., and *Yersinia enterocolitica*. *C. difficile* toxin A was detected in her feces by EIA.

2. There are two problems with culture as a means of diagnosis of *C. difficile*-associated diarrhea. First, there is a high (20%) asymptomatic carriage rate, so only specimens from symptomatic patients should be cultured. Second, this disease is toxin mediated. The organism produces two exotoxins, toxin A and toxin B. (See answer to question 6 for additional details.) The genes which encode the production of the two exotoxins are located in a region of the *C. difficile* chromosome called the pathogenicity locus, or PaLoc. Certain strains lack this series of genes, making them nontoxigenic and thus nonpathogenic. The number of patients who carry nontoxigenic strains varies from institution to institution. Although culture is considered highly sensitive for the diagnosis of *C. difficile*-associated diarrhea, it may lack specificity due to the inability to differentiate toxigenic from nontoxigenic strains. Therefore, toxin-based tests are believed to be most accurate for the diagnosis of this disease. Diagnosis is established by detecting one or both of the toxins produced by the organism in feces of infected individuals.

Several techniques have been developed for detection of *C. difficile* toxin. The reference method is a tissue culture cytotoxicity assay. This assay is very sensitive and highly specific but is also time-consuming (24 to 48 hours) and laborious. As an alternative, EIA is frequently used. This method is not as sensitive as tissue culture (80 to 90%), but it is highly specific (97 to 99%) and rapid (2 hours). EIAs that detect both toxin A and B are preferable to ones that detect only toxin A, since strains of *C. difficile* that produce only toxin B are being reported with increasing frequency.

3. Almost all patients who develop *C. difficile*-associated diarrhea have recently received or are currently receiving antimicrobial agents. This patient had just completed a course of oral cephalosporin when her diarrhea developed. Essentially all agents with antimicrobial activity, including agents not usually thought of as antimicrobials, such as the anticancer agent methotrexate, can induce this disease. Animal and in vitro studies indicate that organisms found in normal gut flora can suppress the growth of *C. difficile*. When antimicrobial therapy is given, it may alter the gut flora in such a way as to eliminate the *C. difficile*-suppressive elements. With their elimination, *C. difficile*, which either is an intrinsic component of bowel flora (a small percentage of patients) or is obtained from the environment (most patients), can grow and produce toxin, resulting in disease.

4. *C. difficile* is a spore-forming bacterium. Spores can remain viable for months in a hospital environment and are much more resistant to disinfectants than are vegetative cells. These spores are frequently found throughout the rooms of infected individuals. They may remain capable of infecting other patients weeks after the infected patient has left the room.

5. *C. difficile* causes a broad spectrum of disease. The most common manifestation of infection appears to be asymptomatic carriage of the organism. Infected patients can have mild diarrhea, often associated with concurrent antimicrobial therapy. They can have more severe diarrhea accompanied by nonspecific inflammatory changes in the intestinal tract. The most severe manifestation of *C. difficile*-associated diarrhea is pseudomembranous colitis. In this disease, the intestinal mucosa, which is severely damaged, is overlaid by a pseudomembrane composed of fibrin, bacteria, cellular debris, and white and red blood cells. This pathologic lesion has a characteristic appearance when observed during colonoscopy, and the diagnosis of pseudomembranous colitis is usually made in this way. Pseudomembranous colitis is a life-threatening condition that can be complicated by perforation and toxic megacolon. It must be aggressively treated.

6. Two exotoxins produced by *C. difficile* have been well characterized. These toxins are biochemically and immunologically distinct. One, an enterotoxin referred to as toxin A, is believed to be responsible for most of the pathologic events which occur with this disease. When purified toxin is given to laboratory animals, most of the pathologic effects observed in humans can be reproduced. The second toxin is a cytotoxin called toxin B, which appears to play a minor but apparently synergistic role in the *C. difficile*-induced disease process. Recent studies of patients infected with *C. difficile* strains that produce only toxin B suggest that toxin B plus other poorly defined factors can produce severe manifestations of *C. difficile*-associated diarrhea, including pseudomembranous colitis.

7. The therapeutic approach that is used is based on the disease course of the patient. The simplest treatment strategy is often to stop all antimicrobial agents the patient is receiving. Especially in patients with mild disease, this may be sufficient to allow resolution of the patient's diarrhea. However, many patients who develop *C. difficile*-associated diarrhea have serious primary infections which require continuing antimicrobial therapy. Stopping antimicrobial agents in those patients is not an option. Alternatively, many patients will develop the disease after completing a course of antimicrobial agents. Stopping antimicrobial therapy in those patients is not relevant. These patient populations typically receive either oral metronidazole or oral vancomycin. Both are equally effective, but oral metronidazole is preferred. This preference is due to the belief that increasing use of oral vancomycin to treat *C. difficile*-associated diarrhea may have been a major factor in the emergence of vancomycin-resistant enterococci as an important nosocomial pathogen.

A third strategy is to attempt to repopulate the gut with indigenous flora or organisms that have been shown to inhibit the growth of *C. difficile*. This therapeutic approach is used most commonly at the conclusion of antimicrobial therapy in patients who have recurrences of the disease.

8. Recurrences of *C. difficile*-associated diarrhea can be due to either relapse or reinfection. In relapse, spores which are resistant to the activity of antimicrobial agents remain dormant in the gut. When antimicrobial agents are discontinued, these spores can develop into vegetative cells. If a *C. difficile*-suppressive microbial flora is still present in the gut at the completion of *C. difficile*-directed therapy, it can prevent the spores from vegetating, thus preventing the cells from dividing, producing toxin, and causing relapse. If a suppressive flora is not present, the spore can vegetate and cells can divide, producing toxin and a relapse of diarrhea. Some patients actually have multiple relapses and require multiple antimicrobial courses before the infection finally resolves.

Alternatively, recurrences of *C. difficile*-associated diarrhea can be due to reinfection with another strain of *C. difficile*. Molecular fingerprinting studies have shown that relapse typically occurs sooner (mean 28 days) than reinfection (mean 38 days). Reinfection and relapse are responsible for similar numbers of episodes of recurrence.

It should be noted that recurrence is not due to the development of antimicrobial resistance by the vegetative cells of *C. difficile*. No *C. difficile* isolates that are resistant to either vancomycin or metronidazole have been reported.

REFERENCES

1. **Barbut, F., A. Richard, K. Hamadi, V. Chomette, B. Burghoffer, and J.-C. Petit.** 2000. Epidemiology of recurrences or reinfections of *Clostridium difficile*-associated diarrhea. *J. Clin. Microbiol.* **38:**2386–2388.

2. **Cohen, S. H., Y. J. Tang, and J. Silva, Jr.** 2000. Analysis of the pathogenicity locus in *Clostridium difficile* strains. *J. Infect. Dis.* **181:**659–663.

3. **Johnson, S., and D. N. Gerding.** 1998. *Clostridium difficile*-associated diarrhea. *Clin. Infect. Dis.* **26:**1027–1036.

4. **Karlstrom, O., B. Frykund, K. Tullus, L. G. Burman, and the Swedish *C. difficile* Study Group.** 1998. A prospective nationwide study of *Clostridium difficile*-associated diarrhea in Sweden. *Clin. Infect. Dis.* **16:**141–145.

5. **Moncrief, J. S., L. Zheng, L. M. Neville, and D. M. Lyerly.** 2000. Genetic characterization of toxin A-negative, toxin B-positive *Clostridium difficile* isolates by PCR. *J. Clin. Microbiol.* **38:**3072–3075.

CASE 26

The patient was a 3-year-old female. Her mother was an elementary school teacher, and on school days the patient attended a day care center. The patient presented with a 3-day history of increasingly severe diarrhea with three episodes of vomiting. In the previous 24 hours, she had 12 bowel movements. On physical examination, the child was lethargic and appeared dehydrated. She had a temperature of 38.3°C, blood pressure of 90/60 mm Hg, and a heart rate of 100 beats/min. The child was weighed and found to weigh 1 lb (ca. 0.4 kg) less than she had at a well-child visit 6 months previously. She had grown approximately 1.5 inches (4 cm) during that time. A stool specimen was obtained during her clinic visit. The stool specimen was reported as being watery, and no white or red blood cells were seen on microscopic examination. Cultures for enteric pathogens were negative. A modified acid-fast stain of a fecal specimen is shown in Fig. 1. The child's diarrhea persisted for approximately 1 more week before resolving.

1. What organism is infecting this patient? What other intestinal pathogens share similar staining characteristics with this organism?

2. What other techniques besides modified acid-fast stains are available for detection of this organism?

3. What is the typical disease course of this illness, and how should the disease be managed?

4. How do you think the child acquired this organism? How can the probable source of her infection be determined?

5. This organism has caused many large outbreaks of diarrheal disease. Briefly describe one of them. What four factors are thought to be important in this organism's ability to cause large-scale outbreaks?

6. When do you think she can return to group day care?

7. This organism is no longer an important cause of morbidity and mortality in HIV-infected patients in the industrialized world but continues to be an important cause of morbidity and mortality in the developing world. Explain this difference.

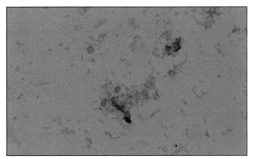

Figure 1

CASE DISCUSSION

1. The acid-fast organism with which this child was infected was *Cryptosporidium parvum*. The diagnostic forms seen in Fig. 1 are oocysts, which are generally round and 4 to 6 μm in diameter, similar in size to yeasts. On microscopic examination of feces when acid-fast staining is not used, *C. parvum* oocysts can be confused with yeasts.

Other parasites that are acid-fast and have been shown to cause diarrheal disease in humans include *Isospora belli* and *Cyclospora* spp. *Cyclospora* is slightly larger (8 to 9 μm) than *Cryptosporidium*. It also shows great variability in its uptake of the acid-fast stain, from taking up no stain to being light pink to dark red.

2. Both enzyme immunoassays (EIAs) and direct fluorescent-antibody (DFA) stains have been developed for detecting *Cryptosporidium*. As with *Giardia lamblia*, these tests are more sensitive and less labor-intensive than the acid-fast staining technique used for *Cryptosporidium* detection in feces. A DFA reagent containing fluorescent antibodies to both *Giardia* and *Cryptosporidium* is useful in screening for these two common parasites.

3. The patient's illness was typical for an immunocompetent child with *Cryptosporidium* infection. This child would be classified as having a "cholera-like" illness. This illness was characterized by multiple episodes (>10/day) of watery diarrhea, vomiting, low-grade fever, and dehydration. One characteristic finding of cryptosporidiosis, crampy abdominal pain, was not recorded for this patient. It should be remembered that only those individuals with the most severe manifestations of diarrheal disease seek medical care. This child, who had lost weight and had clear signs of dehydration (low blood pressure, lethargy, and rapid heart rate), was typical of such patients. It should also be noted that low blood pressure, lethargy, and rapid heart rate would also be seen in patients with sepsis.

As is the case with most enteric pathogens, the key to management of *Cryptosporidium* infection is adequate rehydration (either orally, if tolerated, or intravenously, if not) and maintenance of electrolyte balance. Close to 100 compounds have been evaluated for activity against *Cryptosporidium*. Only one, nitazoxanide, has shown promise in the treatment of this infection. In a small double-blind, placebo-controlled trial, it was shown to reduce both the duration of diarrhea and oocyst shedding. Further evaluation of this compound is needed before it becomes standard therapy.

4. The spread of diarrheal disease is very common in day care settings, especially when toddlers wearing diapers are present. Toys often become fecally soiled by infected toddlers who are only just learning the rudiments of good hygienic behavior. These fecally soiled toys are then mouthed by other children, and enteric pathogens including *Cryptosporidium* can be spread. Furthermore, day care workers who change diapers of ill children may not adequately wash their hands and will then contaminate food, thus spreading diarrheal disease in this setting.

There are two common sources of *Cryptosporidium* infections, humans and animals, especially cattle. Molecular genotyping studies have shown that genotype 1 organisms are of human origin and genotype 2 organisms are typically of bovine origin. During an outbreak, genotyping can be helpful in determining its source and in developing strategies to control it.

5. *Cryptosporidium* has been implicated in large-scale outbreaks of waterborne disease. The largest outbreak was in 1993 and affected an estimated 403,000 people in Milwaukee, Wis. One of the key factors in the outbreak was a sanitation failure of one of the two city water plants. This failure was documented by finding *Cryptosporidium* oocysts, the infective form of this parasite, in block ice made from water processed by that plant during the outbreak period. Molecular genotyping data have shown that genotype 1, the human genotype, was responsible for this outbreak. Contamination of the untreated water with human sewage, followed by the failure of the water treatment facility to remove this organism, is believed to be the cause of this outbreak. The *Morbidity and Mortality Weekly Report*, which is available electronically at http://www.cdc.gov, is an excellent source of information on the latest infectious disease outbreaks, including those caused by *Cryptosporidium*.

Four factors are important in understanding why *Cryptosporidium* is capable of causing large-scale outbreaks.

- *Cryptosporidium* is found in virtually all surface waters in the United States, including lakes, rivers, and reservoirs.
- Processing of drinking water in the industrialized world is generally accomplished by two steps, chlorination and flocculation-filtration. *Cryptosporidium* oocysts are resistant to chlorination.
- Flocculation-filtration is not always an efficient means of reducing oocyst number when processing drinking water.
- The infectious dose of *Cryptosporidium* is quite low, with a mean infectious dose of 130 oocysts. For some individuals the infectious dose may be as low as 1 oocyst.

These four factors when combined may result in contamination of municipal water supplies, especially when source water is heavily contaminated with the parasite. The large-scale outbreak in Milwaukee is an example of such a scenario.

6. The American Academy of Pediatrics recommends that children can return to day care once their diarrhea has resolved. It should be recognized that oocyst shedding has been documented as long as 2 months after resolution of symptoms. This finding further emphasizes the need to impress upon day care center workers the importance of hand washing after diaper changes or handling of objects such as toys, which potentially could be fecally contaminated.

7. In the early stages of the HIV epidemic in the United States, *Cryptosporidium* was an important cause of morbidity and mortality in HIV-infected patients who had CD4$^+$ cell counts of <200/µl. Unlike the self-limited nature of *Cryptosporidium*-induced diarrhea in immunocompetent hosts, diarrheal disease was not self-limited in HIV-infected patients with low CD4 counts. These patients could show four different disease courses with this organism. About one-third of patients developed a cholera-like illness with fluid losses measured in liters per day that required rehydration therapy. Another third had chronic diarrhea which did not require rehydration but was debilitating. Approximately one-sixth had intermittent diarrhea which varied in severity, and one-sixth had transient diarrhea. As immune function deteriorated, as measured by declining CD4 cell counts, diarrheal disease often worsened and continued until death. Approximately 10 to 15% of HIV patients chronically infected with this parasite developed cholecystitis or sclerosing cholangitis.

The incidence of severe, chronic *Cryptosporidium* disease in HIV-infected patients has declined significantly in the industrialized world. The reason for this decline has been the widespread use of highly active antiretroviral therapy (HAART). HAART has slowed immune function deterioration in most patients. In some patients with CD4 counts of <200/µl, HAART results in a significant rise in the CD4 count leading to "immune system reconstitution." HIV-infected patients with CD4 counts >200/µl typically have a self-limited *Cryptosporidium* disease course consistent with those seen in immunocompetent individuals.

The situation in the developing world is not as bright for HIV-infected patients. The cost of HAART is beyond the resources of all but a privileged few. *Cryptosporidium* has been reported in 30 to 50% of HIV-infected individuals in the developing world. It continues to be an important cause of diarrheal disease in these patients and is believed to play a significant role in "Slims disease," a wasting syndrome, which is a leading cause of morbidity and mortality in HIV-infected patients in the developing world.

REFERENCES

1. **Clark, D. P.** 1999. New insights into human cryptosporidiosis. *Clin. Microbiol. Rev.* **12:**554–563.

2. **MacKenzie, W. R., N. J. Hoxie, M. E. Proctor, M. S. Gradus, K. A. Blair, D. E. Peterson, J. J. Kazmierczak, D. G. Addiss, K. R. Fox, J. B. Rose, and J. P. Davis.** 1994. A massive outbreak in Milwaukee of cryptosporidium infection transmitted through the public water supply. *N. Engl. J. Med.* **331:**161–167.

3. **Manabe, Y. C., D. P. Clark, R. D. Moore, J. A. Lumadue, H. P. Dahlman, P. C. Belitsos, R. E. Chaisson, and C. L. Sears.** 1998. Cryptosporidiosis in patients with AIDS: correlates of disease and survival. *Clin. Infect. Dis.* **27:**536–542.

4. **Peng, M. M., L. Xiao, A. R. Freeman, M. J. Arrowood, A. A. Escalante, A. C. Weltman, C. S. L. Ong, W. R. MacKenzie, A. A. Lal, and C. B. Beard.** 1997. Genetic polymorphism among *Cryptosporidium parvum* isolates: evidence of two distinct human transmission cycles. *Emerg. Infect. Dis.* **3:**567–573.

5. **Rossignol, J.-F., A. Ayoub, and M. S. Ayers.** 2001. Treatment of diarrhea caused by *Cryptosporidium parvum:* a prospective randomized, double-blind, placebo-controlled study of nitazoxanide. *J. Infect. Dis.* **184:**103–106.

C A S E
27

This 12-year-old boy was brought in by his mother for evaluation. The patient had been having crampy abdominal pain and some diarrhea for the preceding 1 to 2 weeks. On the afternoon of his evaluation, the patient thought that he had had an "accident" in his pants. When he looked in his underpants, he saw "something move" that his mother "captured." She told his pediatrician that she thought she had recovered "an earthworm" (Fig. 1). He had no fever, cough, or blood-tinged sputum. His travel history was unremarkable. Physical examination was unremarkable as well, and no other worms or larvae were seen on anal examination.

1. What is a nematode? Which nematodes commonly cause gastrointestinal infections in humans?

2. Which parasite was probably found in this patient?

3. Describe the epidemiology of this infection, including the life cycle of this parasite.

4. What determines the severity of disease with this organism? What is the most severe manifestation of infection with this organism? How does it typically present clinically? How might you diagnose it? In what patient population in the United States or Canada is it most likely to occur?

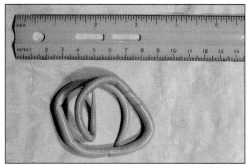

Figure 1

CASE DISCUSSION

CASE 27

1. A nematode is an unsegmented round worm. The most common, medically important intestinal nematodes include the etiologic agents of hookworm infection (*Necator americanus* and *Ancylostoma duodenale*), the pinworm (*Enterobius vermicularis*), the whipworm (*Trichuris trichiura*), the roundworm (*Ascaris lumbricoides*), and *Strongyloides stercoralis*.

2. The worm, when submitted to the parasitology laboratory, was identified as *A. lumbricoides*, the human roundworm. Superficially, *Ascaris* looks like an earthworm, especially when it is bile stained. This nematode, when mature, may measure up to 35 cm in length. Diagnosis of ascariasis is most commonly made by identifying eggs in a fecal specimen. Females may contain millions of eggs and may lay as many as 200,000 eggs per day. Fertilized eggs are oval and up to 75 µm long; unfertilized eggs are elongated and up to 90 µm long.

3. *Ascaris* is a human parasite. As such, infection is spread from person to person via the fecal-oral route. To understand the transmission of this parasite, it is important to understand its life cycle. Humans become infected by ingesting embryonated eggs after the eggs, excreted by humans in feces, have matured in the soil. This process usually takes 2 to 3 weeks. In soil, the eggs can remain viable and thus infective for up to 10 years. Once these infective eggs are swallowed, they hatch in the stomach and the small intestine. The larvae penetrate the intestinal wall and migrate via the portal circulation to the liver and then to the right side of the heart, the pulmonary vessels, and the lungs. The larvae develop in the lungs, migrate up the bronchi and trachea, and are swallowed. They mature in the intestines to the mature male and female and live in the lumen of the intestine, where females then lay eggs and begin the cycle again.

This infection is very common worldwide. The majority of infections occur in China and Southeast Asia because of this region's large population and the high prevalence of this disease, which approaches 90% in some populations living there. Contact with soil contaminated with human feces either directly, by ingesting eggs from soiled fingertips, or indirectly, by ingesting fresh vegetables which were fertilized with human feces, are the major modes of transmission. Children who eat dirt, so called "pica-eaters," are particularly susceptible to infections with this organism.

4. Worm burden is directly related to severity of illness. Since the worm does not replicate in the host, worm burden is determined by the number of eggs ingested. However, it should be understood that in highly endemic areas, worm burden is cumulative because the worms can live in the host for 12 to 18 months, and the host, exposed to an environment heavily contaminated with feces, may be frequently infected. Patients with a low *Ascaris* worm burden are frequently asymptomatic. As worm burden increases in the gastrointestinal tract, the patient may begin to experience nausea and abdominal discomfort such as intermittent cramping. Diarrhea is not a common symptom of *Ascaris* infection. In young children, malnutrition may be a manifestation of increasing worm burden. As with many other parasitic infections that migrate through various tissues, eosinophilia may occur in patients with ascariasis. Pulmonary symptoms may occur during the migration of the larvae through the lungs and may simulate asthma or pneumonia. Adult worms may migrate into open orifices such as the appendix and biliary tree. Biliary obstruction is relatively common and is usually treated conservatively with anthelmintic therapy. Worms may also migrate into other locations (such as the lacrimal duct or fallopian tubes), but this is uncommon.

The most severe complication of *Ascaris* infections is intestinal obstruction. These individuals have a worm burden large enough to mechanically block the intestinal tract. Fatal gastrointestinal perforation and peritonitis may occur. Patients with *Ascaris*-induced intestinal obstruction present with vomiting, abdominal pain and distension, and constipation. The diagnosis can be made by seeing the worms in vomitus or in the intestinal tract on plane films of the abdomen. This disease is most likely to be seen in individuals who have migrated from areas where the prevalence of *Ascaris* infection is high, such as Southeast Asia or China.

REFERENCES

1. **Garcia, L. S., and D. A. Bruckner.** 1997. Intestinal nematodes, p. 219–247. *In Diagnostic Medical Parasitology*, 3rd ed. ASM Press, Washington, D.C.

2. **Khuroo, M. S.** 1996. Ascariasis. *Gastroenterol. Clin. North Am.* **25:**553–577.

CASE 28

This 19-year-old Mexican man presented to the walk-in clinic complaining of a 5-day history of severe right-sided flank and right upper quadrant pain. He stated that the pain came on suddenly and got much worse over the few days prior to presentation. The pain was most severe on inspiration. He had some chills but denied fevers. He was anorexic, not having eaten anything over the preceding 24 hours. He denied vomiting, diarrhea, nausea, blood in stool, or shortness of breath.

On physical examination, he was flushed and mildly diaphoretic. His temperature was 37.5°C, his blood pressure was 153/75 mm Hg, and his heart rate was 102/min. Physical examination was significant for hepatomegaly of 20 cm with diffuse tenderness and guarding, greatest in the right upper quadrant and flank. His white blood cell count was 16,900/μl with 13,100 polymorphonuclear leukocytes (PMNs) per μl. Otherwise, his laboratory findings were within normal limits. Computed tomography (CT) of his abdomen showed a ring-enhancing lesion in the liver that on admission was 2.9 by 2.5 cm. Two days later the lesion had grown to 4 by 3.4 cm. A drain was placed. Three days later the lesion measured 4.3 by 4.1 cm, and the drain was found to be posterior to the lesion. An aspirate from the lesion had 1+ PMNs on Gram stain, but no organisms were seen. Bacterial and fungal cultures of his aspirate were negative.

He was treated with antibiotics, had progressively less fluid drained, had decreasing right upper quadrant pain, and was discharged on the eighth hospital day. He was never febrile throughout his hospital course. A serologic test revealed the etiology of his infection. Figure 1 shows what the organism causing this patient's infection looks like in stool specimens.

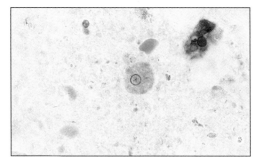

Figure 1

1. Explain why it is likely that this patient's disease process is infectious.

2. What two parasitic agents are likely to cause the type of infection seen in this patient? Briefly describe the epidemiology of each. Which is more likely in this particular patient? Why? What risks would aspiration of the liver lesion caused by these agents pose to the patient?

3. This patient denied having any nausea, vomiting, or diarrhea. Is this typical for the type of infection this patient has? What is the likelihood of finding the organism causing his liver process in this patient's feces?

4. Briefly describe the virulence factors produced by this organism that are responsible for the tissue damage it causes.

5. One of the observations made about intestinal tract infections with this agent is that up to 90% are asymptomatic. What new information concerning this genus may explain that finding?

6. Why is serologic testing the typical way in which this infection is diagnosed? What other techniques have been shown to be of value in diagnosing this infection?

CASE DISCUSSION

1. A circumscribed liver lesion is most likely due to one of two pathologic processes, either an abscess or a tumor. Given that the lesion grew rapidly in size (over 1 cm in 48 hours) an infectious process is more likely since even the most aggressive malignancies would grow much more slowly.

2. The most common cause of liver abscess in the United States is bacterial infection most frequently due to *Streptococcus intermedius*. In the developing world, amebic liver abscesses due to *Entamoeba histolytica* are common. Less common but still an important cause of liver abscess in the developing world are infections with the larval form of the dog tapeworm, *Echinococcus granulosus*.

E. histolytica infections are acquired by ingestion of water or food that has been contaminated with human feces. Alternatively, they can be acquired by direct oral-anal contact. Transmission of the disease is via the cyst form (Fig. 1).

The most common species of *Echinococcus* that infects humans is *E. granulosus*. Normally, the life cycle includes sheep, cattle, or other herbivores as "hosts" that ingest grass or feed that is contaminated with feces from dogs infected with the adult tapeworm. The herbivore acts as an intermediate host. An intermediate host is one in which the life cycle cannot be completed, and cysts containing the larval form of the worm are formed in either viscera or muscle. In the case of *E. granulosus* infections, eating the viscera of an infected herbivore infects the definitive host, the dog. Humans are an accidental host of this canine parasite and become infected by ingesting tapeworm eggs excreted by dogs or other canines. Infections are seen most frequently in individuals involved in animal husbandry with herbivores, such as sheepherders.

There are three reasons why this patient most likely has an *E. histolytica* liver abscess. First, *E. histolytica* is a much more common cause of liver abscess than *Echinococcus*. There is no information in his history that he had the appropriate animal exposure for acquiring echinococcal infection. Second, the acuity of his infection is consistent with *E. histolytica*. Patients with *E. histolytica* liver abscess tend to have an acute onset of disease. Signs and symptoms seen in this patient consistent with *E. histolytica* liver abscess include severe right upper quadrant pain that is more severe on inspiration, anorexia, chills, and diaphoresis. Third, the rapidity with which the lesion increased in size was much more consistent with *E. histolytica*. Echinococcal lesions grow much more slowly. Patients with *E. histolytica* liver abscess do, however, typically have fevers, which were not reported in this patient.

Aspiration of this patient's liver abscess involved risk. As has already been discussed, infection with *Echinococcus* was unlikely. However, aspiration of liver abscesses in patients with histories and clinical presentation consistent with this parasite should be avoided. If there is spillage of the abscess material that contains larval forms of the parasite, exposure to these echinococcal antigens may result in an anaphylactic reaction that may be fatal. Rather, the entire cyst formed by this parasite should be surgically removed.

3. Patients with amebic liver abscess infrequently have intestinal symptoms such as diarrhea, vomiting, or nausea. The parasite is rarely found (less than 5% in one study) in the stools of individuals with amebic liver abscesses. This is despite the fact that the cyst form is ingested and develops into the trophozoite phase in the intestinal tract before being carried hematogenously to the liver.

4. As the name of this parasite implies, this organism is histotoxic and is capable of producing ulcerative-type lesions in the intestinal tract as well as liver abscesses. It has been well documented that for this organism to cause its cytotoxic effect, it must directly adhere to its target cell. Three virulence factors are believed to play a role in the pathogenesis of *E. histolytica*. The Gal-GalNAc adhesin is important in the organism's ability to penetrate the thick mucous layer which overlays and protects the colonic epithelium, the target site for the parasite. This adhesin also plays a role in the parasite's direct adherence to the colonic epithelium. It is also speculated to be involved in stimulating colonic epithelial cells to produce the cytokine interleukin-8 (IL-8). This cytokine recruits and activates neutrophils, resulting in the inflammatory response which is so characteristic of this disease.

The second virulence factor is a polypeptide called the amebapore. It is well recognized that this polypeptide can form channels in the lipid bilayers of bacterial and cultured eukaryotic cells, causing them to lyse. It is less clear what role this virulence factor has in cytolysis of colonic epithelial cells, which is a central feature of the disease process. It is tempting to speculate that this molecule plays a central role in the cytolytic activity of this parasite. However, such a role has not been proved.

The third virulence factor is the cysteine proteinases. *E. histolytica* produces several different types of cysteine proteinases. These enzymes appear to have two roles in the pathogenesis of *E. histolytica* infection. First, they play a role in the invasiveness of the parasite by degrading the extracellular matrix of the colonic mucosa. Second, they can degrade a variety of proteins that may contribute to the cytolytic process produced by this protozoan.

5. Recent studies have shown that there are two closely related species of *Entamoeba*, *E. histolytica* and *E. dispar*. These studies have shown that *E. histolytica* is pathogenic while *E. dispar* is not. Although *E. dispar* produces several molecules that are similar to the virulence factors described for *E. histolytica* in the answer to question 4, these molecules do not have the same level of biologic activity or, in some cases, have different biologic activity than their *E. histolytica* counterparts. The diagnosis of *E. histolytica* infections typically has been made by examining feces microscopically for distinct morphologic features characteristic of the parasite. Since these two parasites are morphologically indistinguishable, microscopic examination is nonspecific. As a result, many of the clinical and epidemiologic data that are published about *E. histolytica* are misleading. Until recently, asymptomatic infection with *E. histolytica* was believed to be the most common clinical presentation of infection with this organism. It is now recognized that asymptomatic infections with *E. histolytica* are actually infrequent. Many of the patients who were previously reported to be asymptomatically infected with *E. histolytica* were most likely infected with *E. dispar*.

These parasites can be distinguished from each other by growing the organisms in culture and doing a complex analysis that determines the isolate's isoenzyme profile. This is something that is done in a few specialized research laboratories and is not practical for diagnostic laboratories. Molecular and immunologic techniques (described below) are being developed that will allow the direct, specific detection of *E. histolytica* in feces and liver aspirates in diagnostic laboratories.

6. Because most patients with *E. histolytica* can be treated without surgical intervention, noninvasive diagnostic tests are of value. As mentioned previously, the parasite is infrequently found in the feces, so alternative diagnostic approaches are necessary to diagnose liver abscesses due to *E. histolytica*. Detection of antibodies against whole-cell antigens using an indirect hemagglutinin assay (IHA) is highly sensitive in the diagnosis of amebic liver abscess, being positive in over 90% of patients. In a patient such as the one described in this case, who comes from an area of *E. histolytica* endemicity, positive IHA results can represent a previous rather than current infection. However, given his clinical presentation and radiographic and microbiologic findings, his positive serologic results are strong evidence of his having an *E. histolytica* liver abscess.

Direct microscopic examination of the aspirated material for the protozoan is of little value since the parasite is adherent to the wall of the abscess. Microscopic visualization of the parasite requires a biopsy of the abscess wall, something that is rarely done.

The organism can be grown in culture from the aspirated material. However, culture is usually limited to the research setting.

Three new approaches are available for the diagnosis of *E. histolytica* both in liver aspirates and from feces. Adherence lectin of *E. histolytica* can be detected both in serum and in aspirated material from patients with *E. histolytica* liver abscess using an antigen detection test. This test is highly sensitive (95%) in detecting the adherence lectin in the serum of patients with *E. histolytica* liver abscess provided they have not received antimicrobial therapy. This test is very insensitive in patients who have received antimicrobial therapy. The antigen can also be detected in material aspirated from the liver abscess, but the sensitivity does not appear to be as high as that found in serum.

A second approach is an assay that detects antibodies to the lectin. In patients with liver abscess, this test is sensitive and specific in those who have received antimicrobial agents but is less sensitive than the lectin antigen test in those who have not received antimicrobial agents.

It would seem that the most effective strategy for diagnosing amebic liver abscesses would be to use the antigen assay in patients who have not received antimicrobial therapy and the antibody assay in those who have.

A third approach is to use PCR (polymerase chain reaction) for detection of this parasite. This test has been shown to have sensitivity similar to that of antigen tests when compared with culture on stool specimens. The relatively poor sensitivity of PCR may be due to the presence of substances in stools that inhibit PCR.

REFERENCES

1. **Espinosa-Cantellano, M., and A. Martinez-Palomo.** 2000. Pathogenesis of intestinal amebiasis: from molecules to disease. *Clin. Microbiol. Rev.* **13:**318–331.

2. **Haque, R., I. K. M. Ali, S. Akther, and W. A. Petri, Jr.** 1998. Comparison of PCR, isoenzyme analysis, and antigen detection for diagnosis of *Entamoeba histolytica* infection. *J. Clin. Microbiol.* **36:**449–452.

3. **Haque, R., N. U. Mollah, I. K. M. Ali, K. Alam, A. Eubanks, D. Lyely, and W. A. Petri, Jr.** 2000. Diagnosis of amebic liver abscess and intestinal infection with the TechLab *Entamoeba histolytica* II antigen detection and antibody tests. *J. Clin. Microbiol.* **38:**3235–3239.

4. **Pillai, D. L., J. S. Keystone, D. C. Sheppard, J. D. MacLean, D. W. MacPherson, and K. C. Kain.** 1999. *Entamoeba histolytica* and *Entamoeba dispar:* epidemiology and comparison of diagnostic methods in a setting of nonendemicity. *Clin. Infect. Dis.* **29:**1315–1318.

CASE 29 The patient was a 61-year-old female textile worker who was in her usual state of good health until she was awakened at 3 a.m. on the day of her hospital admission with severe left upper extremity pain. She was seen in the emergency department, where she denied shortness of breath and chest pain. An electrocardiogram (EKG) was normal. Her physical examination was significant for cellulitis of the left upper arm. Her arm was very tender to palpation and had decreased range of motion. There was no history or evidence of trauma to her arm, and no scratches or other lesions were apparent. She did not have axillary or cervical lymphadenopathy. Her cellulitis progressed over the next 2 hours with a new finding of crepitance in the upper left arm. A radiograph of her left arm revealed gas in the soft tissues. Significant laboratory findings included a white blood cell count of 21,000 with 90% neutrophils and creatine kinase (CK) of 1,554 with a normal heart muscle fraction. She was begun on penicillin and gentamicin and taken to the operating room, where necrotic muscle was found. Gram stain of a biopsy from the left upper arm is shown in Fig. 1. Blood cultures obtained prior to the initiation of antimicrobial agents and the biopsy specimen revealed an organism that grew only under anaerobic incubation conditions. Growth on a blood agar plate is seen in Fig. 2.

1. What condition does this woman have? Explain her CK findings in light of the infection she has.

2. What is the usual therapy for this type of infection?

3. What is the organism causing this patient's infection? Hint: There is a structure visible within the bacteria seen on tissue section that is an important clue to which organism is likely causing her infection. The colonial morphology in Fig. 2 is typical of this organism as well.

4. Few white cells are seen in the tissue section, and often few are seen on Gram stain of tissue biopsies in patients infected with this group of organisms. Why?

5. When the surgeons knew the identity of her organism, they immediately examined her bowel. For what were they looking and why?

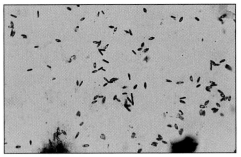

Figure 1

Figure 2

CASE DISCUSSION

1. This woman had nontraumatic, spontaneous clostridial myonecrosis, which is commonly referred to as gas gangrene. Extreme pain in the upper left arm is frequently associated with myocardial infarction, so cardiac enzymes and an EKG were obtained on admission. An elevated CK level, as seen in this case, indicates the presence of muscle damage. CK levels are usually elevated in patients with myocardial infarctions. However, isoenzyme analysis showed normal levels of cardiac muscle-derived enzyme, suggesting that damage was occurring in skeletal muscle. Her lack of other symptoms associated with myocardial infarction, such as shortness of breath or chest pain, and her normal EKG make the diagnosis of myocardial infarction unlikely. The findings of rapid progression of infection, cellulitis, severe pain at the site of the infection, the presence of crepitance, the radiographic observation of gas in tissue, and the finding of necrotic tissue at surgery are all consistent with clostridial myonecrosis.

2. One of the favorite expressions of surgeons is, "To heal, use steel." Probably in no other disease process is the use of surgical debridement more important in patient outcome than it is in clostridial myonecrosis. It literally can be a life-saving operation; delay in surgery for as brief a period as a few hours can result in an inoperable, fatal infection. The need for surgical intervention is due in large measure to the fact that the gas gangrene-producing clostridia (*Clostridium perfringens, C. septicum, C. noyvi, C. sordelli, C. histolyticum, C. bifermentans,* and *C. fallax*) grow rapidly and produce a variety of histotoxic factors that destroy tissue and allow the rapid spread of the organism. In addition, the use of penicillin G, to which the gas gangrene-producing clostridia continue to be susceptible, is an important adjunct therapy to surgery. Gentamicin was used to cover the patient for the presence of facultative, enteric gram-negative rods. They were not present in this infection. Gentamicin is not active against anaerobic bacteria such as clostridia.

3. The most common cause of gas gangrene is *C. perfringens.* However, most cases of gas gangrene due to *C. perfringens* typically follow trauma in which the spores of this organism are introduced into devitalized tissue. Devitalized tissues create an ideal anaerobic environment for the vegetation of these spores, which grow rapidly and, through the action of the various extracellular enzymes they produce, spread through tissue. However, this patient has an unusual form of gas gangrene described in the surgical literature as nontraumatic, spontaneous gas gangrene. This patient's disease course is characteristic, with her having no history of trauma, extreme pain in an extremity with rapidly progressing cellulitis, crepitance, and gas in tissue radiographically. The organism typically associated with this disease process is *C. septicum,* the second most common cause of gas gangrene after *C. perfringens*. Spontaneous myonecrosis due to *C. septicum* often progresses rapidly and, because of the absence of trauma, may not be recognized. As a result, mortality with this infection as high as 65% has been reported. A possible explanation for why *C. septicum* can cause spontaneous gas gangrene is that it is somewhat aerotolerant compared with *C. perfringens* and may be able to initiate infection in normal rather than devitalized (i.e., anaerobic) tissue.

Microbiology laboratory findings in this case are consistent with *C. septicum*. Organisms growing in tissue containing subterminal spores are typical for *C. septicum*, while spores are generally not observed in *C. perfringens* infections. In addition, *C. perfringens* characteristically produces clear zones of hemolysis on sheep blood agar while *C. septicum*, which is motile, is one of the few human pathogens (*Proteus mirabilis* and *Proteus vulgaris* are others) to show a swarming form of growth, as seen in Fig. 2.

4. One of the characteristics of infection with both *C. perfringens* and *C. septicum* is the absence of white blood cells in clinical specimens obtained from infected tissue. The reason for this is that both organisms produce virulence factors that destroy a wide variety of cell types, including white blood cells. *C. perfringens* produces a phospholipase C hemolysin while *C. septicum* produces a toxin called alpha-toxin. Alpha-toxin is an extracellular toxin that is produced as a protoxin and must be activated, typically by a host cell protease, before it has biologic activity. It binds to a specific receptor on a wide variety of cells, including neutrophils, forming a pore in the membrane that results in lysis. Both *C. perfringens* and *C. septicum* produce other extracellular enzymes such as DNase and hyaluronidase, that allow them to spread through damaged tissue.

5. One of the important questions in nontraumatic, spontaneous gas gangrene is how the organism gets to the site of infection. It is clear that *C. septicum* sepsis and myonecrosis is associated with malignancy, most frequently colonic, but also with hematologic malignancy, such as leukemia and lymphoma. The incidence of the disease is also increased in patients with cyclic neutropenia and diabetes. It is thought that the malignancies cause a breach in the integrity of the gut mucosa, allowing the organism to enter the bloodstream from the gut. One of the unresolved questions this presents is why this infection is almost always due to a single organism rather than multiple gut organisms.

The surgeons examined her colon for a malignancy because of the strong association (80%) between the finding of *C. septicum* in the bloodstream and the presence of a gastrointestinal tract malignancy. This woman had a colon carcinoma identified on endoscopic examination.

REFERENCES

1. **Gordon, V. M., K. L. Nelson, J. T. Buckley, V. L. Stevens, R. K. Tweten, P. C. Elwood, and S. H. Leppla.** 1999. *Clostridium septicum* alpha toxin uses glycosylphosphatidylinositol-anchored protein receptors. *J. Biol. Chem.* **274:**27274–27280.

2. **Larson, C. M., M. P. Bubrick, D. M. Jacobs, and M. A. West.** 1995. Malignancy, mortality, medicosurgical management of *Clostridium septicum* infection. *Surgery* **118:**592–598.

3. **MacFarlane, S., M. L. Hopkins, and G. T. MacFarlane.** 2001. Toxin synthesis and mucin breakdown are related to swarming phenomenon in *Clostridium septicum. Infect. Immun.* **69:**1120–1126.

4. **Stevens, D. L., D. M. Musher, D. A. Watson, H. Eddy, R. J. Hamill, R. Gyorkey, H. Rosen, and J. Mader.** 1990. Spontaneous, nontraumatic gangrene due to *Clostridium septicum. Rev. Infect. Dis.* **12:**286–296.

CASE 30 The patient, a 39-year-old Asian-American man originally from China, was known to be a chronic carrier of hepatitis B virus (HBV). For the past 2 years, he had received treatment with the antiviral drug lamivudine. He presented to his care provider for a follow-up visit for the management of his chronic infection. He appeared well and was anicteric.

Laboratory studies obtained included normal levels of aspartate aminotransferase (AST), alanine aminotransferase (ALT), γ-glutamyltransferase (GGT), alkaline phosphatase, INR (International Normalized Ratio, a coagulation test that depends on the liver's biosynthetic ability), and direct and total bilirubin. His HBV serologic studies demonstrated a positive hepatitis B surface antigen (HBsAg) that was confirmed by neutralization, a negative antibody to HBsAg, a positive hepatitis B core antibody, a positive hepatitis Be antigen (HBeAg), and a negative assay for antibody to HBeAg.

1. How is HBV infection transmitted from person to person? Does the epidemiology of the infection differ between Asian and Western patients?

2. How can HBV infection be prevented?

3. In the setting of the serologic results seen here, if you did not have a history of chronic infection with HBV, what additional serologic test would help to determine if the patient has an acute HBV infection or is a chronic carrier?

4. An additional study demonstrated >200,000 copies of HBV DNA per ml. Given that this patient was being treated with an antiviral agent and early in the course of his therapy had a low level of HBV DNA, how would you explain the current high level of viral DNA in this patient?

5. What is the natural course of untreated HBV infection? What viral agent can, in the setting of coinfection, be associated with severe disease? What complications of this infection occur?

6. What are the goals of therapy in patients with chronic HBV infection? What other therapy can be used in the treatment of chronic HBV infection?

CASE 30

CASE DISCUSSION

1. HBV can be found in the body fluids of infected individuals, including, most notably, blood and semen. Transmission of HBV occurs from sexual exposure, perinatally from an infected mother to the infant, and via blood either directly (such as by transfusions of blood products prior to effective donor screening) or indirectly (such as by sharing contaminated needles). Needle-stick injuries in health care workers are of particular concern, as HBV is transmitted more efficiently than HIV.

There are an estimated 350 million carriers of HBV worldwide, many of whom are found in Asia. Asian patients, unlike those in the United States and western Europe who are usually found to be infected as adults, are usually infected perinatally. Perinatal infection is not typically associated with an acute hepatitis clinical syndrome. These perinatally infected patients almost always become chronically infected and may develop the complications of chronic HBV infection that are discussed in the answer to question 5.

2. The most important method used to prevent HBV infection is an immunization given as a series of three intramuscular injections. Effective vaccines against HBV have been available since 1982. The initial vaccine was derived from individuals who were chronic carriers of HBV. The vaccine that is now in use is a recombinant vaccine. A subset of people do not respond to the vaccine (i.e., do not produce detectable antibodies to HBsAg). With repeat vaccinations, some of these individuals will respond, as demonstrated by the production of an adequate antibody level to HBsAg.

The neutralizing antibodies that are induced by vaccination are directed to a hydrophilic region of the HBsAg. One concern that has been voiced in recent years is that HBV mutants with amino acid substitutions within this region can allow the replication of HBV in people who have received vaccination. There is the theoretical possibility that an increase in the frequency of these mutations will result in a decrease in the efficacy of the vaccine. This will have to be watched closely and, if it occurs, may result in the need for a modification of the current vaccine.

An additional method of protection, which is used in the setting of a known exposure to HBV in a nonimmune individual, is the use of hepatitis B immunoglobulin (HBIg). This is given, for example, following a needle-stick injury in a nonimmune individual.

Both the vaccination and HBIg have been shown to decrease the rate of chronic infection due to HBV in infants born to mothers who are infected with HBV.

3. The interpretation of the serologic testing for HBV infection is of great importance clinically. Knowledge of the time course of the typical serologic patterns seen in the acute infection (Fig. 1) and in patients who become chronic carriers of HBV after the initial infection (Fig. 2) is needed to interpret the serology.

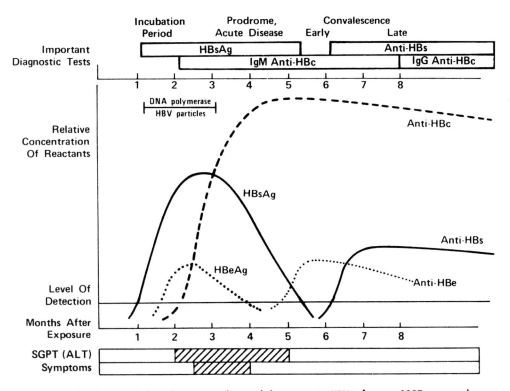

Figure 1   Serologic and clinical patterns observed during acute HBV infection. SGPT, serum glutamic pyruvic transaminase. (Reprinted from *Manual of Clinical Microbiology*, 7th ed., ©1999 ASM Press, with permission.)

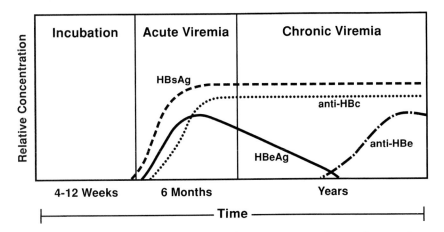

Figure 2   Typical sequence of serologic markers in patients with acute hepatitis B who develop persistent infection after exposure to HBV. (Reprinted from *Manual of Clinical Microbiology*, 7th ed., ©1999 ASM Press, with permission.)

The presence of HBsAg in this patient's blood is consistent with either an acute infection or a chronic infection with HBV. He has hepatitis B core antibody, which has not been further characterized. If these antibodies can be demonstrated to be of the IgM class, it would be consistent with an acute infection; if the hepatitis B core antibody is composed only of IgG, the infection is consistent with a chronic infection (Table). Thus, the IgM hepatitis B core antibody assay can be helpful in differentiating acute from chronic HBV infection.

4. HBV infections often respond clinically to the oral nucleoside analogue lamivudine. This can be seen in both an improvement in the severity of liver disease in many patients and a decrease in the serum levels of HBV DNA in nearly all patients. In fact, this patient had an initial decrease in the serum level of HBV DNA. Unfortunately,

TABLE INTERPRETATION OF HEPATITIS B SEROLOGIC STUDIES

TEST	RESULT	INTERPRETATION
HBsAg	Negative	
Anti-HBc	Negative	Susceptible
Anti-HBs	Negative	
HBsAg	Negative	
Anti-HBc	Positive	Immune due to natural infection
Anti-HBs	Positive	
HBsAg	Negative	
Anti-HBc	Negative	Immune due to hepatitis B vaccination
Anti-HBs	Positive	
HBsAg	Positive	
Anti-HBc	Positive	
IgM anti-HBc	Positive	Acutely infected
Anti-HBs	Negative	
HBsAg	Positive	
Anti-HBc	Positive	
IgM anti-HBc	Negative	Chronically infected
Anti-HBs	Negative	
HBsAg	Negative	
Anti-HBc	Positive	Four interpretations possible[a]
Anti-HBs	Negative	

[a]1. May be recovering from acute HBV infection. 2. May be distantly immune, and the test is not sensitive enough to detect very low levels of anti-HBs in serum. 3. May be susceptible with a false-positive anti-HBc. 4. May be an undetectable level of HBsAg present in the serum, and the person is actually a carrier.

during therapy with lamivudine, in some patients there is the emergence of HBV sub-populations with a particular mutation in the YMDD (tyrosine, methionine, aspartate, and aspartate) motif of the HBV polymerase. This mutation has been associated with HBV resistance to lamivudine. This is the most likely reason for the patient's high serum level of HBV DNA. Another possibility is that the patient has stopped taking his lamivudine.

5. The great majority of adult patients who are infected with HBV have an acute illness characterized by jaundice due to elevated bilirubin, fatigue, and abnormal liver function tests. Only 3 to 5% of adults become chronically infected. This contrasts with infected neonates, 95% of whom will become chronic asymptomatic carriers and children infected after the neonatal period but prior to age 6, 30% of whom will become chronically infected.

Clinically, liver function tests indicative of hepatocellular injury (AST and ALT) are elevated in patients with acute hepatitis. Other tests of hepatic function, such as coagulation times (INR) are typically abnormal because of a decrease in the hepatic synthesis of clotting factors. The great majority of infected patients will have a self-limiting illness. A small subset of acutely infected patients will develop fulminant hepatitis, which can be fatal. This is more likely to occur in those patients who are coinfected with hepatitis D virus (previously called delta hepatitis), an RNA-containing passenger virus that requires the presence of HBV to replicate and cause an infection.

Chronic infection, associated with a significant rate of cirrhosis, may lead to end-stage liver disease and to the development of hepatocellular carcinoma.

Other complications, due to immune complex formation, may result in extrahepatic manifestations, such as in polyarteritis nodosa and other vasculitides. These can result in injury to the kidney, lung, and other organs.

6. The goal of therapy is to stop or delay the progression of hepatic injury due to HBV infection. If the carrier state can be eliminated, as can be demonstrated by the loss of HBsAg from the serum (which occurs in the minority of treated patients) the therapy is successful. Histologic improvement, as can be demonstrated by following sequential liver biopsies while the patient is on therapy, often occurs even in those patients who do not have elimination of the carrier state. This is thought to be due to reductions in viral replication.

In addition to specific antiviral therapy with lamivudine, chronic infection with HBV can be treated with interferon. Interferon therapy is not known to be associated with the emergence of HBV mutations, which limits the long-term efficacy of lamivudine. Unfortunately, interferon therapy has a high rate of side effects and is more difficult to administer than lamivudine, which is taken orally. HBeAg is a peptide that is a marker of active viral replication. In patients who are treated with interferon, sustained reductions in HBV DNA occur primarily in those patients who lose serum HBeAg and produce antibodies to HBeAg (demonstrate seroconversion). In patients treated with lamivudine, the reduction occurs in HBV DNA even in those patients who do not seroconvert to HBeAg.

REFERENCES

1. **Dienstag, J. L., E. R. Schiff, T. L. Wright, R. P. Perrillo, H. W. Hann, Z. Goodman, L. Crowther, L. D. Condreay, M. Woessner, M. Rubin, and N. A. Brown.** 1999. Lamivudine as initial treatment for chronic hepatitis B in the United States. *N. Engl. J. Med.* **341:**1256–1263.

2. **Lee, W. M.** 1997. Hepatitis B virus infection. *N. Engl. J. Med.* **337:**1733–1745.

3. **Zuckerman, A. J.** 2000. Effect of hepatitis B virus mutants on efficacy of vaccination. *Lancet* **355:**1382–1384.

CASE 31

The patient was a 49-year-old woman who presented to the emergency room with a 3- to 4-week history of nausea, vomiting, and periodic fevers. She had recently returned from a visit to rural areas of India. She admitted to swimming in the Ganges River, eating in restaurants, drinking water without boiling or filtration, and having "raw" milk in her tea. She was exposed to rats, cattle, and mosquitoes. She did not take malaria or gamma globulin prophylaxis. Two days prior to admission, she began feeling tired and "started sleeping all day." She complained of vomiting after meals and 2 to 3 days of watery diarrhea at the onset of the illness that has since resolved. She had fevers every 5 days with associated arthralgias and myalgias. Her fevers were documented to 38°C.

On physical examination, she was a thin woman in no apparent distress. She had a temperature of 36.9°C, pulse rate of 70/min, and blood pressure of 134/80 mm Hg reclining and 104/76 mm Hg standing. Her physical examination was significant for scleral icterus and a liver palpable at the costal margin. Laboratory tests were significant for the following values: aspartate aminotransferase (AST), 4,872 U/liter; alanine aminotransferase (ALT), 3,682 U/liter; γ-glutamyltransferase (GGT), 174 U/liter; and bilirubin, 6.0 mg/dl. The results of her hepatitis serologic tests were as follows: hepatitis B virus (HB) surface antigen, negative; HB surface antibody, positive; anti-HB core antibody, positive; anti-HB core immunoglobulin M (IgM) antibody, negative; anti-hepatitis A virus (HAV) antibody, positive; anti-HAV IgM antibody, positive.

1. Given her travel history and her exposures, name three infectious agents with which she has an increased likelihood of being infected. What were her risk factors for each of these agents?

2. Minimally, what organisms should have been ruled out in this patient, and how would that have been done?

3. What is the agent causing her present illness? Explain the results of her serologic tests and how they helped you come to the conclusions that you did. What results from her physical examination and laboratory tests other than serologic tests are consistent with this illness?

4. How do you think she obtained her infection? What feature of this agent allows it to be spread in this way?

5. What is the usual outcome of infection with this organism?

6. What prophylactic strategies are available for this agent? (Note: on previous trips to India, she employed one of them.)

CASE DISCUSSION

1. Her travel history puts her at increased risk for a number of infectious diseases. In particular, this woman appeared to have taken none of the precautions that would lessen the likelihood of her getting the most common infectious diseases that travelers to India encounter. These include a variety of diarrheal diseases, such as typhoid fever, nontyphoidal salmonellosis, shigellosis, amebic and shigella dysentery, cholera, and traveler's diarrhea due to toxigenic *Escherichia coli* (risk factors: bathing in Ganges, eating in restaurants, drinking nonboiled or nonfiltered water); hepatitis A (same risk factors as the diarrheal pathogens plus failure to be vaccinated against HAV or to take gamma globulin); hepatitis E (risk factor: ingestion of contaminated water); malaria (risk factors: failure to take malarial prophylaxis, failure to use protection against mosquito bites, such as insect repellent, netting, and protective clothing); brucellosis (risk factors: drinking "raw," i.e., unpasteurized milk); tuberculosis (risk factor: high rate of endemic infection in India); plague (risk factor: exposure to rats); and leptospirosis (risk factor: cutaneous exposure to water that was potentially contaminated with infectious animal urine). For further information on agents for which travelers are at increased risk in the Indian subcontinent and how they can be avoided, visit the website http://www.cdc.gov/travel/indianrg.

2. Given her history of fever, two things should have been done for this woman. Malaria smears should have been done to rule out *Plasmodium* infection. Patients who visit malaria-endemic regions and fail to take steps to prevent malaria and present with fever have malaria until proven otherwise. Onset of disease is usually within the first 4 weeks of returning from a malaria-endemic region, but it can be much longer with *Plasmodium ovale* and *Plasmodium vivax*, both of which can have persistent forms within the liver (hypnozoites). Symptoms of nausea, vomiting, diarrhea, and icterus can all occur in malaria. *Plasmodium falciparum* infections can be life threatening, so ruling out malaria infection is important. Second, this patient should have blood cultures done because of her risk of typhoid fever and enteric fever due to nontyphoidal serotypes of *Salmonella* spp. as well as brucellosis.

In addition, given her symptoms and her travel history, she is at increased risk for HAV infection. Appropriate serologic tests should be done to rule out this infection. Since all hepatitis infections have similar clinical presentations acutely, hepatitis B and C viruses should also be considered. Since both hepatitis B and C are typically spread parenterally or via sexual contact (hepatitis B), these latter two viruses are less likely because there is nothing in her history to suggest that she had exposures putting her at risk for these infections. A relatively recently described agent, hepatitis E virus, an enterically transmitted virus that can be acquired by drinking contaminated water, is found in developing countries. It is not typically severe, except in pregnant women, in whom it may be fatal. Diagnostic studies for hepatitis E are not routinely requested in the United States.

3. The patient has an acute illness consistent with hepatitis with positive serologic tests for antibodies to hepatitis B surface antigen, hepatitis B core antigen, and IgM and total antibodies to HAV. Clinical and laboratory findings consistent with hepatitis include markedly increased liver enzymes (ALT, AST, and GGT), markedly increased bilirubin levels, hepatomegaly, and scleral icterus. Symptoms of nausea, vomiting, malaise, fever, arthralgias, and myalgias are much less specific and could be associated with a wide array of agents to which she was potentially exposed.

Her serologic profile is consistent with an acute hepatitis A infection and prior infection with hepatitis B. In patients with acute hepatitis A infection, IgM antibody is almost always present at the time of acute illness. Anti-HAV IgM antibodies can persist for months after the resolution of the acute illness. In acute hepatitis B infections, patients typically have hepatitis B surface antigen in their blood, with an absence of antibodies to either the surface or core antigens. IgM antibodies to core antigens would be expected to be present in patients with acute hepatitis B infection. Her core IgM antibodies were negative. Was her serologic profile consistent with her having received hepatitis B virus vaccine? It is not, because the vaccine contains only the surface antigen so vaccine-derived immunity should consist of only antibodies to the surface antigen. Vaccinated individuals do not have antibodies to core antigens.

4. Hepatitis A is spread by the fecal-oral route, usually by ingestion of fecally contaminated water or food, especially shellfish. It is a disease controlled by good sanitation practices and a public health infrastructure. Where these do not exist, the disease is common. In many parts of the developing world, including India, essentially all adults have serologic evidence of prior hepatitis A infection. Infections are common in children, who probably serve as the source of much of the fecal contamination of water sources in those locales. This patient was likely exposed to the virus while bathing in the Ganges, which is known to be highly contaminated with human fecal material. Approximately 5% of hepatitis A infection in the United States occurs in individuals traveling to the developing world who are exposed to hepatitis A virus-contaminated food and water. Infections in the industrialized world are most common in intravenous drug users, in children who attend day care centers, and in adults who work there. In addition, shellfish taken from water contaminated with human feces has been shown to be a source of hepatitis A, both in the industrialized world and in the developing world. Shellfish are filter feeders; as part of the filter feeding process, they may ingest hepatitis A-contaminated human feces, and the virus persists in their flesh. In locales in the United States where sanitation is poor, the incidence of hepatitis A infection is high.

HAV belongs to the family *Picornaviridae*. They are nonenveloped, single-stranded RNA viruses and are similar to other groups of *Picornaviridae* viruses, the rhinoviruses ("common cold" viruses) and enteroviruses (polioviruses, among others), which infect humans and are known to be stable in the environment. The pathogenesis of hepatitis A infection begins by ingestion of viral particles. The virus can survive pH extremes found in the stomach. It is absorbed into the bloodstream either in the stomach or small intestine and specifically infects and replicates in the liver, where it has its pathologic effect. Viral particles are excreted in bile in large numbers (10^8 viral particles per ml) and can then be passed into the environment in feces. Once in the environment, the virus can survive for weeks both in fresh and salt water.

5. Hepatitis A is almost always an acute, self-limited disease. In children, the infection is frequently subclinical. Severe manifestations of the disease, although rare, are typically seen in adults. Fulminant hepatitis due to hepatitis A is also rare, with a mortality rate of less than 1.5% in patients hospitalized with hepatitis A infection. Unlike hepatitis B and C viruses, chronic infection with hepatitis A does not occur.

6. A formalin-inactivated hepatitis A vaccine was approved for use in all individuals over 2 years of age in 1995. The vaccine is safe and highly efficacious, with seroconversion occurring in more than 99% of vaccinated individuals and a low rate of minor side effects. The vaccine is recommended for adults who travel to areas outside the industrialized world. It is also recommended for children who live in communities with high rates of hepatitis infections. Universal childhood vaccination against hepatitis A has also been advocated.

Prior to the availability of the vaccine, immunoprophylaxis using human gamma globulin was used for individuals traveling to the developing world. (This patient had used it on previous visits to India.) The protection is relatively short-lived, estimated to be 2 to 2.5 months. In addition, gamma globulin has been shown to be protective for exposed individuals during HAV outbreaks. For an individual who will immediately embark on a trip to a developing country and who, therefore, will not seroconvert to the hepatitis vaccine during the trip, gamma globulin can be given to afford immediate, though relatively short-lived, protection.

REFERENCES

1. **Bell, B. P., C. N. Shapiro, M. J. Alter, L. A. Moyer, F. N. Judson, K. Mottram, M. Fleenor, P. L. Ryder, and H. S. Margolis.** 1998. The diverse pattern of hepatitis A epidemiology in the United States—implications for vaccination strategies. *J. Infect. Dis.* **178:**1579–1584.

2. **Blair, D. C.** 1997. A week in the life of a travel clinic. *Clin. Microbiol. Rev.* **10:**650–673.

3. **Cuthbert, J. A.** 2001. Hepatitis A: old and new. *Clin. Microbiol. Rev.* **14:**38–58.

FOUR

Skin and Soft Tissue Infections

INTRODUCTION TO SECTION IV

The resistance of skin to infection is due to the integrity of the keratinized skin, the presence of inhibitory fatty acids produced by sebaceous glands, the dryness of the skin, and the inhibitory effect of the resident normal skin flora. Skin and soft tissue infections can be caused by either direct penetration of a pathogen through the skin or hematogenous spread of the pathogen to the site. Normal skin flora includes organisms that, in the setting of a disruption in the integrity of the skin (such as the presence of a surgical suture or an insect bite), may cause infection. In the setting of severe damage to the skin, as occurs with burns, even normally innocuous organisms, including endogenous bacteria, can cause severe disease. Similarly, when the skin is no longer dry, as may occur in moist intertriginous spaces or when occlusive dressings are present, the patient is at increased risk of infection.

Cutaneous manifestations of systemic disease are common. Rocky Mountain spotted fever, meningococcemia, enteroviral infection, and toxic shock syndrome can all present with fever and a diffuse erythematous macular rash. Other systemic infections that can present with a diffuse rash include scarlet fever, measles, and German measles. The characteristic rash of Lyme disease, erythema migrans, is specific enough to establish the diagnosis. The nature of the lesion (macular, papular, vesicular, pustular, or bullous) may help to narrow the differential diagnosis. For example, varicella-zoster virus infection typically results in vesicular skin lesions. The rash of secondary syphilis, on the other hand, may present clinically as macular, papular, maculopapular, or pustular skin lesions but does not present as a vesicular rash.

Skin and soft tissue infections can be classified on the basis of the anatomic level at which infection occurs. The more superficial infections, such as folliculitis caused by *Staphylococcus aureus* or cellulitis caused by *Streptococcus pyogenes*, are important to treat at an early stage. Delay in treatment may result in invasion of the deeper structures, as in necrotizing fasciitis, which has a high mortality rate.

Damage to the skin and soft tissues, as occurs in traumatic injuries, may allow the entry into the wound of soil organisms such as *Clostridium perfringens*, an anaerobic gram-positive rod. Under favorable conditions, potentially fatal soft tissue infections (myositis, gas gangrene) may occur.

Important agents of skin and soft tissue infection are listed in Table 4. The presence of ectoparasites, such as lice, is not designated an infection but rather an infestation. Ectoparasites are, however, included for completeness.

TABLE 4   SELECTED SKIN AND SOFT TISSUE PATHOGENS

ORGANISM	GENERAL CHARACTERISTICS	SOURCE OF INFECTION	DISEASE MANIFESTATION
Bacteria			
Bacillus anthracis	Spore-forming, aerobic gram-positive bacillus	Exogenous; livestock or animal products; bioterrorism agent	Cutaneous, gastrointestinal, and inhalation anthrax, meningitis, bacteremia
Bartonella henselae	Fastidious gram-negative bacillus	Exogenous; cats appear to be primary host	Cat scratch disease, bacillary angiomatosis (in immunocompromised individuals)
Borrelia burgdorferi	Spirochete	Tick-borne	Lyme disease; rash, arthritis, nervous system and cardiac manifestations
Clostridium perfringens	Anaerobic gram-positive bacillus	Exogenous (wounds), endogenous (bowel flora)	Gas gangrene, bacteremia, food poisoning, emphysematous cholecystitis
Clostridium tetani	Anaerobic gram-positive bacillus	Exogenous (wounds)	Tetanus
Corynebacterium diphtheriae	Aerobic gram-positive bacillus	Exogenous	Diphtheria (pharyngeal) and wound diphtheria
Group A streptococci (*Streptococcus pyogenes*)	Catalase-negative, gram-positive cocci	Endogenous and exogenous	Cellulitis, bacteremia, scarlet fever, necrotizing fasciitis, pharyngitis, pneumonia, rheumatic fever, poststreptococcal glomerulonephritis
Group B streptococci (*Streptococcus agalactiae*)	Catalase-negative, gram-positive cocci	Endogenous	Cellulitis, sepsis, meningitis, UTI[a] (diabetics)
Neisseria gonorrhoeae	Oxidase-positive, gram-negative diplococcus	Direct sexual contact, vertical (mother to child)	Genital tract involvement, pharyngeal infection, ocular infection, bacteremia, arthritis with dermatitis
Neisseria meningitidis	Oxidase-positive, gram-negative diplococcus	Endogenous (from colonization)	Meningitis, bacteremia, pneumonia
Pasteurella multocida	Oxidase-positive, gram-negative bacillus	Zoonosis (often animal bite or scratch)	Cellulitis, bacteremia, osteomyelitis, meningitis, pneumonia
Pseudomonas aeruginosa	Lactose-nonfermenting, oxidase-positive, gram-negative bacillus	Exogenous	Skin infections in burn patients, community and nosocomial UTI, nosocomial pneumonia, nosocomial bacteremia, ecthyma gangrenosum

(continued next page)

203

TABLE 4 SELECTED SKIN AND SOFT TISSUE PATHOGENS (continued)

ORGANISM	GENERAL CHARACTERISTICS	SOURCE OF INFECTION	DISEASE MANIFESTATION
Staphylococcus aureus	Catalase-positive, coagulase-positive, gram-positive coccus	Endogenous	Cellulitis, bacteremia, endocarditis, septic arthritis, abscesses, pneumonia
Treponema pallidum	Spirochete (does not Gram stain)	Direct sexual contact, vertical (mother to child)	Primary (painless chancre), secondary (diffuse rash), latent, and late syphilis; can affect any organ
Fungi			
Blastomyces dermatitidis	Dimorphic mold	Exogenous	Cutaneous infection, pneumonia, meningitis, bone infection
Candida albicans	Yeast, often germ tube positive	Endogenous	Thrush, vaginal yeast infection, diaper rash, esophagitis, nosocomial UTI, nosocomial bloodstream infection
Candida spp., non-*albicans*	Yeasts, germ tube negative	Endogenous	Thrush, vaginal yeast infection, nosocomial UTI, nosocomial bloodstream infection
Cryptococcus neoformans	Encapsulated yeast	Exogenous (environmental, rarely zoonotic)	Meningitis, pneumonia, bloodstream infection, cellulitis
Epidermophyton floccosum	KOH-positive skin lesions; club-shaped macroconidia, absent microconidia	Anthropophilic	Dermatophyte infection of keratinized tissue (rarely nails)
Microsporum spp.	KOH-positive skin lesions; fluoresces yellow-green under Wood's light	May be zoophilic (e.g., *M. canis*), geophilic (e.g., *M. gypseum*), or anthropophilic (e.g., *M. audouinii*)	Dermatophyte infection of keratinized tissue (rarely nails)
Trichophyton spp.	KOH-positive skin lesions	May be zoophilic (e.g., *T. mentagrophytes*) or anthropophilic (e.g., *T. schoenleinii*)	Dermatophyte infection of keratinized tissue (including nails)

Parasites

Organism	Characteristics	Transmission	Manifestations
Ancylostoma braziliense	Hookworm of dog	Exogenous	Cutaneous larva migrans
Ancylostoma caninum	Hookworm of dog	Exogenous	Cutaneous larva migrans
Leishmania tropica	Protozoan	Exogenous (sand fly)	Ulcerative skin lesions
Pediculus spp.	Ectoparasite	Exogenous	Body lice
Phthirus pubis	Ectoparasite	Exogenous	Crab louse
Sarcoptes scabiei	Ectoparasite	Exogenous; zoonotic varieties less common than human varieties	Scabies infestation

Viruses

Organism	Characteristics	Transmission	Manifestations
Enteroviruses	Nonenveloped, ssRNA[b]	Usually fecal-oral	Aseptic meningitis, rash, myocarditis
Herpes simplex virus types 1 and 2	Enveloped, dsDNA[c]	Person to person; reactivation of latent infection; during passage of the neonate through the birth canal	Genital ulcers; oral, ocular infections; encephalitis; neonatal infection; esophagitis (immunocompromised individuals)
Human herpesvirus 6	Enveloped, dsDNA	Person to person	Exanthem subitum (roseola)
Human immunodeficiency virus (HIV)	Enveloped RNA retrovirus	Bloodborne and sexual transmission; mother to child	AIDS, mononucleosis-like syndrome with rash in primary infection
Rubella virus (German measles)	Enveloped, ssRNA	Vertical, mother to child	Inapparent or subclinical infection in adults; birth defects in infants
Rubeola virus (measles)	Enveloped, ssRNA	Respiratory spread	Measles; pneumonia, encephalomyelitis, subacute sclerosing panencephalitis
Papillomavirus	Nonenveloped, dsDNA	Person to person	Warts
Varicella-zoster virus	Enveloped, dsDNA	Respiratory spread	Chicken pox; zoster (may disseminate)
Variola virus	Enveloped dsDNA	Person to person, respiratory spread; bioterrorism agent	Smallpox; vesicular, pustular, hemorrhagic rash

[a]UTI, urinary tract infection.

[b]ssRNA, single-stranded RNA.

[c]dsDNA, double-stranded DNA.

CASE 32

The patient was a 45-year-old male who was in his usual state of good health when he awoke at 3 a.m. with pain in the lateral aspect of his left calf. He looked at his calf and thought that the pain was due to an ingrown hair and went back to sleep. At 10 a.m., he expressed a small amount of pus from the ingrown hair. Over the next 8 hours, the patient developed an area of cellulitis on the lateral aspect of the calf of approximately 5 by 10 cm. At that time, a small amount of pus was again expressed from the area of the ingrown hair. The next morning, the area of cellulitis extended from just below the knee to just above the ankle. The patient visited his physician. His vital signs at that visit, including pulse, respirations, blood pressure, and temperature, were all within normal limits. Physical exam was significant for an area of cellulitis as described that was red and warm to the touch but with no area of obvious fluctuance. No lymphadenopathy was observed. The central area of the cellulitis, near the area that the patient described as where the ingrown hair had been, was punctured three times with a 20-gauge needle but no pus was drained. The patient was referred to the surgery service. The surgeons examined the patient and said they would follow him. The patient was given 2 g of ceftriaxone intramuscularly and begun on oral cephalexin.

The patient returned to the surgical clinic 48 hours later with an obvious area of fluctuance in the center of the area of cellulitis. Over the preceding 48 hours, the patient reported low-grade fevers. Approximately 1 ml of pus was aspirated and was sent for Gram stain and culture (Fig. 1 and 2). When pus was aspirated from the lesion, the surgeon decided to excise and drain the lesion (Fig. 3).

1. Based on Gram stain and culture, what is the organism most likely causing this patient's infection? How do you think he became infected with this organism?

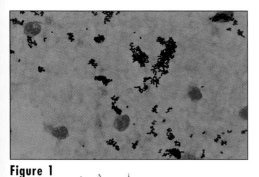

Figure 1

Staph = clumps
Strep = chains

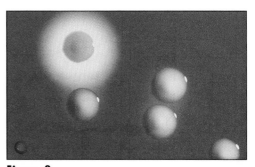

Figure 2

β - hemolytic

2. Why were incision and drainage necessary to treat this infection? Why would antimicrobial agents alone not be effective in treatment of this infection?

3. What other types of infections does this organism frequently cause?

4. Two days after the incision and drainage, the susceptibility results of this patient's isolate were available (Fig. 4). When these results were known, the patient's physician changed his antimicrobial therapy. Why was this done? Explain the mechanisms of resistance seen in this isolate. Hint: there are two different mechanisms at work here. How common is each one for this organism?

5. A new drug resistance problem is emerging with the organism infecting this patient. What is it? Is this patient at risk for becoming infected with a strain of this organism? Explain.

6. If this patient initially presented with fever, diffuse skin rash, low blood pressure, and diarrhea, what virulence factor might the organism causing his cellulitis be producing? Briefly explain how it causes its pathophysiologic effect.

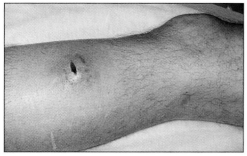

Figure 3

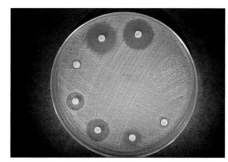

Figure 4 Disks (clockwise, starting at 4 o'clock): penicillin, oxacillin, gentamicin, vancomycin, erythromycin, clindamycin, trimethoprim/sulfamethoxazole.

CASE DISCUSSION

1. The finding of gram-positive cocci in clusters on Gram stain is consistent with staphylococci. The finding of a yellowish colony which is beta-hemolytic on 5% sheep blood agar is consistent with *Staphylococcus aureus*. The staphylococci are divided into two groups based on the biochemical test called the coagulase test; *S. aureus* is positive, while a group of over 30 staphylococcal species are negative. This group of organisms is referred to as the coagulase-negative staphylococci (CoNS). Two of the CoNS species are frequently encountered clinically. *Staphylococcus epidermidis* can infect implanted foreign bodies, such as pacemakers, intravascular catheters, and artificial joints. The other frequently encountered CoNS species is *Staphylococcus saprophyticus*, which causes urinary infections primarily in young, sexually active women. The isolate recovered from this patient was coagulase positive and was identified as *S. aureus*.

The patient's infection began as a folliculitis at the site of the ingrown hair and progressed to a cellulitis and, ultimately, evolved into an abscess. Approximately 20% of adults are chronic nasal carriers of *S. aureus*, while an additional 60% may carry the organism intermittently. From the nose, the skin can become colonized. Studies have shown intermittent skin carriage rates as high as 40%, although most studies target the skin carriage rate at 10 to 15%. In all likelihood this individual's initial folliculitis was a result of the infecting *S. aureus* coming from skin colonization. Manipulation of the skin resulted in the spread of the organism to the dermis, leading to cellulitis and abscess formation.

2. The standard of care for an abscess is twofold: incision and drainage (Fig. 3) and antimicrobial therapy. The reason why antibiotics alone would not be sufficient is that abscess formation results in a loss of blood flow to the center of the infected area (the abscess). As a result, antibiotic levels in the center of the abscess would be low or, in a large abscess, completely absent, allowing the survival of the infecting organisms present there. Incision and drainage removes a large number of organisms and reduces the infected area, making penetration of high levels of antimicrobial agents to the infected tissue and killing of the infecting organism more likely.

3. *S. aureus* is an important cause of both nosocomial and community-acquired infections. It is the most common cause of nosocomial wound infection, the second most common cause of nosocomial bacteremia, and an important cause of nosocomial pulmonary infection. It is either the leading or a major cause of each of the following community-acquired infections: skin and soft tissue, such as was seen in this case; pneumonia, especially following respiratory viral infection; osteomyelitis, endocarditis, and septic arthritis. Ingestion of enterotoxins produced by *S. aureus* in food is the leading cause of food poisoning in the United States.

4. The Kirby-Bauer susceptibility results indicate that this organism is resistant to penicillin and oxacillin, a penicillinase-stable penicillin ("OX" disk on the susceptibility plate [Fig. 4]). Resistance to penicillin is due to the organism's ability to produce a β-lactamase enzyme that degrades the beta-lactam ring of penicillin inactivating the antibiotic. Approximately 95% of *S. aureus* clinical isolates produce this enzyme. This resistance mechanism was seen with increasing frequency in the early 1950s and led to the development of the penicillinase-stable semisynthetic penicillins. Isolates of *S. aureus* which are resistant to penicillinase-stable penicillins are referred to as methicillin-resistant *S. aureus*, or MRSA. Resistance in MRSA strains is due to modifications in a cell wall protein termed penicillin-binding protein 2 (PBP 2). The modified PBP 2 in MRSA is designated PBP 2A. PBPs are enzymes involved in cell wall synthesis. PBP 2A has greatly reduced affinity for all beta-lactam antimicrobial agents, rendering these organisms resistant to all penicillins, cephalosporins, monobactams, and carbapenems. The gene that encodes this modified PBP is called *mecA*, and analysis of this gene indicates that it arose in an organism other than *S. aureus*. All MRSA clinical isolates carry this gene. When first recognized, MRSA strains were predominantly nosocomial pathogens. Recent studies have indicated that these organisms are no longer limited to nosocomial settings, but are now found in as many cases of community-acquired infections as in nosocomial ones. This suggests that these organisms have become widely disseminated in human populations instead of being limited to health care institutions. Recent studies show that approximately 35% of *S. aureus* isolates in the United States are methicillin resistant. Worldwide surveillance shows that MRSA prevalence varies widely, from as low as 2% in Switzerland to 70% in Japan and other western Pacific countries.

MRSA strains often are susceptible only to vancomycin and are therefore problematic to treat. In this particular patient, the isolate was susceptible to several other agents, so treatment options were not as limited. An oral agent, clindamycin, was chosen.

5. There have been several cases of glycopeptide-intermediate *S. aureus* (GISA) reported in the literature. These organisms have reduced susceptibility to vancomycin with MIC values of 8 to 16 µg/ml (susceptible = ≤4 µg/ml; resistant = >16 µg/ml). Patients infected with GISA typically were chronically infected with *S. aureus* and had received prolonged vancomycin therapy prior to the recovery of the GISA isolate. All patients had a foreign body that was the probable source of infection. Vancomycin is currently the choice for treating serious infections, such as endocarditis in patients with MRSA or staphylococcal infections in patients with allergies to beta-lactam antimicrobial agents. A recently published report has identified an *S. aureus* isolate that was fully resistant to vancomycin. The isolate was found to contain the *vanA* vancomycin-resistance gene seen in enterococci. If vancomycin resistance emerges in

S. aureus, then only bacteriostatic agents will be left in the antimicrobial armamentarium to treat these serious infections. The effectiveness of these agents is unproven in serious *S. aureus* infections, and, theoretically at least, they are not likely to be as effective as bactericidal ones. This patient was not at risk for infection with GISA because he was not chronically infected with *S. aureus*, did not have a foreign body in place, and had not received vancomycin.

6. The clinical syndrome that is briefly described is that of toxic shock syndrome. Toxic shock syndrome occurs when an *S. aureus* strain produces an exotoxin called toxic shock syndrome toxin 1, or TSST-1. *S. aureus* enterotoxins, specifically enterotoxin B (SEB) and enterotoxin C (SEC), have also been associated with this syndrome. *S. aureus* toxic shock syndrome first gained prominence in the early 1980s when a large number of menstruating women developed this syndrome in association with the use of superabsorbent tampons. Although the incidence of *S. aureus* toxic shock syndrome has declined in menstruating women as a result of withdrawal of these tampons from the market, this continues to be the patient population in whom this syndrome is most prevalent. Staphylococcal toxic shock syndrome occurs in other populations as well. Interestingly, in menstruating women all strains causing toxic shock syndrome produce TSST-1. In other patient populations, only about one-half of the strains produce TSST-1, while the other half are believed to produce SEB (47%) or SEC (3%). Approximately 20% of *S. aureus* strains causing bacteremia carry the *TSST-1* gene, but probably only a small fraction cause toxic shock syndrome in the clinical setting. TSST-1, SEB, and SEC can act as superantigens. Superantigens work by nonspecifically stimulating T cells. In a typical immune response, an antigen is processed by a macrophage and is "presented" to an antigen-specific T cell. Approximately 1 in 10,000 T cells is stimulated to release cytokines that activate the immune response to this antigen. Superantigens nonspecifically bind to T cells, stimulating 5 to 20% and resulting in massive release of cytokines such as tumor necrosis factor and interleukins. In addition, TSST-1 has been shown to potentiate the activity of endotoxin and to have a direct effect on endothelial cells, resulting in hypotension. All of these superantigen activities together are believed to be responsible for the diarrhea, skin rash, hypotension, and multiorgan failure characteristic of the *S. aureus* toxic shock syndrome.

REFERENCES

1. **Centers for Disease Control and Prevention.** 2002. *Staphylococcus aureus* resistant to vancomycin—United States, 2002. *Morb. Mortal. Wkly. Rep.* **51:**565–567.

2. **Diekema, D. J., M. A. Pfaller, F. J. Schmitz, J. Smayevsky, R. N. Jones, M. Beach, and the SENTRY Participants Group.** 2001. Survey of infections due to

Staphylococcus species: frequency of occurrence and antimicrobial susceptibility of isolates collected in the United States, Canada, Latin America, Europe, and Western Pacific region for the SENTRY antimicrobial surveillance program, 1997–1999. *Clin. Infect. Dis.* **32**(Suppl.):S114–S132.

3. **Dinges, M. M., P. M. Orwin, and P. M. Schlievert.** 2000. Exotoxins of *Staphylococcus aureus. Clin. Microbiol. Rev.* **13**:16–34.

4. **Llewelyn, M., and J. Cohen.** 2002. Superantigens: microbial agents that corrupt immunity. *Lancet Infect. Dis.* **2**:156–162.

5. **Lowy, F. D.** 1998. *Staphylococcus aureus* infections. *N. Engl. J. Med.* **339**:520–532.

6. **Moreno, F., C. Crisp, J. H. Jorgenson, and J. E. Patterson.** 1995. Methicillin-resistant *Staphylococcus aureus* as a community organism. *J. Infect. Dis.* **21**:1308–1312.

7. **Shapiro, M., K. J. Smith, W. D. James, W. J. Giblin, D. J. Margolis, A. N. Foglia, K. McGinley, and J. J. Leyden.** 2000. Cutaneous microenvironment of Human Immunodeficiency Virus (HIV)-seropositive and HIV-seronegative individuals, with special reference to *Staphylococcus aureus* colonization. *J. Clin. Microbiol.* **38**:3174–3178.

8. **Smith, T. L., M. L. Pearson, K. R. Wilcox, C. Cruz, M. V. Lancaster, B. Robinson-Dunn, F. C. Tenover, M. J. Zervos, J. D. Band, E. White, W. R. Jarvis, and the Glycopeptide-Intermediate *Staphylococcus aureus* Working Group.** 1999. Emergence of vancomycin resistance in *Staphylococcus aureus. N. Engl. J. Med.* **340**:493–501.

CASE 33

This 65-year-old woman was bitten by her cat on the dorsal aspect of the right middle finger at 8:00 a.m. She rinsed the bite with water, and at 4:30 p.m. she noted pain and swelling in the finger and the dorsum of the right hand. She then noted pain in the axilla, red streaking up the forearm, and chills. On examination, she had a temperature of 38°C and her right upper extremity was notable for swelling, erythema, warmth, and tenderness on the dorsum of the hand. Two small puncture wounds were seen on the proximal phalanx of the long finger, and erythema was visible over the extensor surface of the forearm. Axillary tenderness was also noted. Laboratory studies demonstrated an elevated white blood cell count of 12,000/μl with a left shift (the presence of immature neutrophils in the peripheral blood). Aspiration of an abscess on her finger was sent for culture, and the patient was taken to the operating room for incision and drainage of the abscess. Gram stain of the organism causing this woman's infection is seen in Fig. 1, and Fig. 2 shows cultures on sheep blood and chocolate agars. There was no growth on MacConkey agar.

1. Which organism was isolated on culture of the abscess?

2. What is the reservoir of this organism? How do humans most commonly become infected by this organism?

3. How can infection with this organism be prevented?

4. What other clinical syndromes can be caused by this organism?

Figure 1

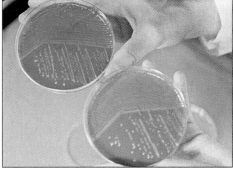

Figure 2

CASE DISCUSSION

1. The organism that was isolated from this patient's abscess was *Pasteurella multocida*. This organism is a gram-negative coccobacillus (Fig. 1) that will grow on most nonselective media, i.e., blood or chocolate agars (Fig. 2), but fails to grow on media selective for gram-negative bacilli such as MacConkey agar. Another gram-negative coccobacillus that grows on chocolate agar but not on MacConkey agar and is frequently encountered clinically is *Haemophilus influenzae*. *H. influenzae*, however, is unlikely to be associated with a wound infection and abscess following an animal bite. It also does not grow on sheep blood agar. Another feature of this case, which is typical of *P. multocida*, is the rapid onset of clinical signs of infection following the animal bite. One point which is worth emphasizing is that infections following cat and dog bites are commonly polymicrobial, often including both aerobic and anaerobic bacteria, with a median of five different bacterial isolates per culture when appropriate techniques are employed for the isolation of anaerobes.

2. *P. multocida* is widely distributed throughout nature and is part of the normal flora in the nasopharynx of many mammals (both wild and domestic) and birds. Human infection is most likely to be associated with cat bites or scratches and less likely (though still quite commonly) to be caused by dog bites. Infections following bites by other members of the cat family, including lions, have been reported to cause *P. multocida* wound infections. In a minority of human infections the patients have had no known animal exposure. Particular organisms are often associated with bites from specific animals. For example, *Capnocytophaga canimorsus* (*cani* = dog; *morsus* = bite) infection may be transmitted by dog bites, and both *Streptobacillus moniliformis* and *Spirillum minus* are transmitted by rat bites.

3. Infection can be prevented by limiting contact with cats and dogs. If a person is bitten or scratched by a cat or dog, the wound should be thoroughly cleaned as soon as possible.

4. In addition to soft tissue infection with rapid onset, other infections seen with this organism following animal bites include osteomyelitis, tenosynovitis, abscess formation, and arthritis. Serious infections are more frequent after cat bites than after dog bites. It is speculated that the cat tooth, which is long and thin, is more likely to cause puncture wounds that penetrate the tendon sheath (causing tenosynovitis) or periosteum (causing osteomyelitis). These infections are particularly problematic because they often occur on the hands and wrists. Because of the extraordinarily complex anatomy involved, infections of the hand and wrist, if neglected, can require complicated surgical debridement and loss of important motor function for the patient, either temporarily or permanently. Other unusual complications include bacteremia with septic shock, meningitis, brain abscess, and peritonitis.

REFERENCES

1. **Talan, D. A., D. M. Citron, F. M. Abrahamian, G. J. Moran, E. J. Goldstein, and the Emergency Medicine Animal Bite Infection Study Group.** 1999. Bacteriologic analysis of infected dog and cat bites. *N. Engl. J. Med.* **340:**85–92.

2. **Weber, D. J., J. S. Wolfson, M. N. Swartz, and D. C. Hooper.** 1984. *Pasteurella multocida* infections: report of 34 cases and review of the literature. *Medicine* **63:**133–154.

CASE 34 The patient was an 18-month-old female who presented to the emergency room with fever, a diffuse rash (onset 5 days before), and a swollen right hand. On examination she was irritable but alert. Her temperature was 39°C and her heart rate was increased at 180 beats/min. She had diffuse vesiculopustular lesions over her entire body (Fig. 1), with some areas showing older, crusted lesions. She had cellulitis of the right hand manifested by marked erythema, swelling, and tenderness. There were no mouth lesions, the lungs were clear, and the liver and spleen were not enlarged. Laboratory data were significant only for leukocytosis with a white blood cell count of 15,800/µl with 88% neutrophils. The chest radiograph was clear. A radiograph of the right hand showed only soft tissue swelling. The patient was treated with intravenous cefazolin. Improvement in the condition of her right hand was notable within 48 hours. This patient had a systemic viral infection with a complication of bacterial superinfection (cellulitis).

1. This patient had a characteristic rash (Fig. 1) at various stages of evolution. What was her underlying viral illness? What other causes of her skin rash should be considered in the differential diagnosis?

2. How is the diagnosis of infection with this pathogen made?

3. Describe the epidemiology of this viral infection.

4. What complications other than bacterial superinfection (as seen in this case) can occur as a result of this viral infection?

5. Which specific antiviral therapy has been shown to be efficacious?

6. After acute primary infection with this virus, latent infection develops. What illness may occur years later as a result of viral reactivation? How do the clinical manifestations of this reactivation infection differ from those of primary infection?

7. What are the infection control issues related to this patient's illness?

8. How can this disease be prevented?

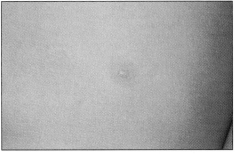

Figure 1

Figure 2

CASE
34

CASE DISCUSSION

1. The underlying viral illness was varicella (chicken pox). This illness is due to primary infection with varicella-zoster virus (VZV), which is a member of the herpesvirus group. These are enveloped, double-stranded DNA viruses. Varicella lesions develop in "crops" such that lesions can be seen in various stages of evolution, including vesicular, pustular, and crusted.

The differential diagnosis in this case includes impetigo (group A streptococcal infection), disseminated enteroviral infection, and disseminated herpes simplex virus infection in a child with underlying skin disease (e.g., eczema). This child had no history of a preexisting dermatologic disorder. Contact dermatitis, drug reactions, and insect bites are noninfectious causes of skin rashes that may be confused with varicella. Because of concerns about bioterrorism, the specter of smallpox must also be considered. In smallpox, unlike chicken pox, all the lesions are at the same stage of development whereas these patients' lesions simultaneously included vesicular, pustular, and crusted lesions.

2. In immunocompetent children, the diagnosis of chicken pox is often made on the basis of clinical findings alone. For adults and immunocompromised children, laboratory confirmation of VZV infection is frequently sought. A method that combines rapidity with sensitivity is direct fluorescent-antibody staining of scrapings taken from vesicular lesions. Polymerase chain reaction (PCR) assays have also been developed, including quantitative assays. They also are highly sensitive and rapid but as of yet are not widely available. Culture techniques for detection of VZV include shell vial techniques (Fig. 2), and standard tissue culture. Shell vial assays, which take 1 to 3 days, are both more rapid and more sensitive than standard tissue culture, which may take as long as 2 to 3 weeks to recover this virus.

3. VZV has a worldwide distribution. Disease is more common in temperate regions, with annual epidemics in the late winter and spring. The virus is spread by the respiratory route and is highly infectious, with approximately 90% of nonimmune household contacts and 10 to 35% of nonimmune classroom contacts becoming infected. Over 90% of children in temperate regions are infected by their 10th birthday.

4. In general, varicella causes much more severe illness in adults than in children. Immunocompromised children and nonimmune, pregnant women also are more prone to complications with this virus than is the general population. The severe illness seen with VZV in these patient populations is due in large part to the significant morbidity and mortality associated with varicella pneumonia. Other complications include hepatitis, arthritis, glomerulonephritis, encephalitis, and cerebellar ataxia. In addition, secondary bacterial infections of the skin lesions, as was seen in this case (cellulitis of the right hand), can also occur. These bacterial infections are most commonly caused by *Streptococcus pyogenes* and *Staphylococcus aureus*. VZV infections are associated with *S. pyogenes*-induced necrotizing fasciitis, as VZV skin lesions have been well recognized as an important portal of entry for *S. pyogenes*. Reye's syndrome, with encephalopathy, elevated transaminase levels, and elevated serum ammonia levels, can occur in children with varicella or influenza who take aspirin. It should be remembered that patients with VZV infection can have a prodrome characterized by fever, malaise, headache, and abdominal pain which is indistinguishable from many other viral illnesses. Therefore, infants and children with febrile illnesses should not be given aspirin.

5. Acyclovir is beneficial in treating varicella in both immunocompetent and immunocompromised children and adults. Because of its expense, the use of this agent has been controversial in immunocompetent children, but it is clearly indicated in immunocompromised children and in adults.

6. Herpes zoster (shingles) is a reactivation of a latent VZV infection. The dorsal root ganglia are latently infected following primary infections. Reactivation is most common in elderly patients but can be seen in all age groups. Typically, skin lesions appear in a single dermatomal distribution innervated by the specific dorsal root or extramedullary cranial ganglia where VZV had been latent. Pain often occurs with the rash and can persist even after the skin lesions heal. This persistent pain is called postherpetic neuralgia. It is the most important, most debilitating complication of herpes zoster. Rarely, skin lesions disseminate beyond the primary dermatome involved. In immunosuppressed patients, however, complicating viremia can occur, with dissemination to extradermatomal skin sites, lungs, liver, and the central nervous system. This condition, with extradermatomal sites of infection, is called disseminated zoster. Patients with zoster are also infectious although apparently not as infectious as patients with varicella.

7. Patients with chicken pox are very contagious. Secondary cases are frequently more severe. The increased severity is believed to be due to high viral inoculum. Hospitalized patients with chicken pox must be placed in respiratory isolation, and strict infection control measures regarding skin contact (hand washing, use of gloves and gowns, etc.) must be implemented.

Hospital workers without a history of chicken pox or those already known to be seronegative for VZV should not come into contact with VZV-infected patients. Seronegative health care workers who do come in contact with these infected patients should not have contact with other patients, especially immunocompromised ones, for a minimum of 2 weeks after exposure, the incubation period of this viral infection.

8. A live, attenuated varicella vaccine is available for use in the United States for all children over 12 months of age. Because it is a live virus vaccine, it should not be used in certain immunocompromised individuals, including children with leukemia, lymphoma, congenital immunodeficiency, or symptomatic HIV infection and children receiving high-dose immunosuppressive drugs for treatment of malignancy, nephrosis, or severe asthma. Current recommendations call for the vaccine to be given as a single dose in children 12 months to 13 years of age. Adolescents and adults with no previous evidence of disease should receive two doses of the vaccine 4 to 8 weeks apart since it is not as immunogenic in that population. The vaccine is very efficacious, vaccine failures are rare, and it has been shown to be particularly effective at preventing severe VZV disease. Post-licensure vaccine safety surveillance using the Vaccine Adverse Event Reporting System (VAERS) of the Centers for Disease Control and Prevention has shown the vaccine to be remarkably safe. Both vaccine-associated and natural infections have been noted postvaccination. Serious infections and deaths due to infection caused by the vaccine strain have been observed but are quite rare (1 death/1,000,000 doses of vaccine administered).

Two major questions remain unanswered concerning universal varicella vaccination. First, will the vaccine strain cause zoster in immunocompetent patients later in life? Limited data suggest that it may, but that the rates and severity of zoster are reduced compared with those in individuals who have natural disease. Second, will immunity wane in adults as natural disease declines, resulting in an at-risk population? Since adults are most vulnerable to severe varicella disease, this is a legitimate concern. Twenty-year follow-up data suggest that immunity persists, but these studies were done in settings where natural disease continues to be common, offering the opportunity for immunized individuals to receive a "booster" effect from exposure to infected individuals.

In immunodeficient patients or high-risk patients, such as VZV-seronegative pregnant women who cannot receive the vaccine, varicella-zoster immunoglobulin (VZIg) can be used prophylactically if these susceptible patients have been exposed to either varicella or zoster. A small number of these patients have been inadvertently given varicella vaccine instead of VZIg, including 19 pregnant women. There was no evidence of congenital varicella infection, although some infants were reported to have congenital malformations such as Down's syndrome.

REFERENCES

1. **Arvin, A. M.** 1996. Varicella-zoster virus. *Clin. Microbiol. Rev.* **9**:361–381.

2. **Wise, R. P., M. E. Salive, M. M. Braun, G. Terracciano Mootrey, J. F. Seward, L. G. Rider, and P. R. Krause.** 2000. Postlicensure safety surveillance for varicella vaccine. *JAMA* **284**:1271–1279.

CASE 35

The patient was a 9-year-old female who was brought to her pediatrician in February because of fever and rash for 2 days. She also had a headache, sore throat, and mild cough. There were no gastrointestinal symptoms. No one else in the household was ill, but she had a classmate with a similar illness.

On examination she was alert and in mild distress. Her temperature was 38.3°C, pulse rate was 110 beats/min, blood pressure was 90/60 mm Hg, and respiratory rate was 40/min. She had a mild conjunctivitis. Her posterior pharynx was injected, and petechiae were present on her soft palate. The buccal mucosa was injected with scattered raised papular lesions. She had a macular rash on her trunk, face, and arms (Fig. 1). Her chest radiograph was normal. A throat swab was sent for culture, and blood was drawn for viral serologic examination. Subsequently, the throat culture was read as negative for group A beta-hemolytic streptococci. Acute- and convalescent-phase (obtained 2 weeks later) serum specimens confirmed the clinical diagnosis, and the school nurse was notified.

1. What is the differential diagnosis in an individual who presents with the symptoms cited in this case, with specific emphasis on the skin rash? What is the agent of this patient's infection?

2. How is the diagnosis of this infection usually made?

3. Describe the typical clinical course of this infection, and name three complications which can occur.

4. How can infection with this virus be prevented? What is the current epidemiologic status of this disease in the United States? Worldwide?

5. From a public health perspective, supplementation with what vitamin has been shown to reduce mortality due to infection with this pathogen?

6. How would you manage her case? Are specific treatments available?

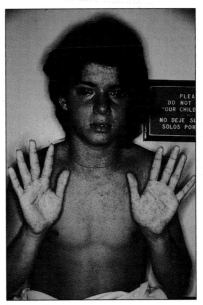

Figure 1

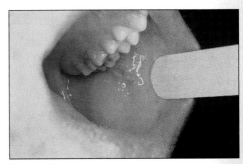

Figure 2

CASE DISCUSSION

1. In general, the differential diagnosis is quite large in patients with fever and rash, so it is important to focus on the specific type of rash. In this patient the rash was diffuse and macular (Fig. 1). Macular or maculopapular rashes are seen with viral infections due to measles virus, rubella virus, human herpesvirus type 6 (which causes roseola), Epstein-Barr virus, enteroviruses, cytomegalovirus, and parvovirus B19. Other types of infections associated with this type of rash include meningococcal infection, salmonellosis, mycoplasma infection, Rocky Mountain spotted fever, secondary syphilis, and infective endocarditis. Coexisting enanthemas (involvement of mucous membranes) can help to narrow down the differential diagnosis. This patient had a rash that was readily recognizable (i.e., measles) from its specific appearance and accompanying findings of coryza, conjunctivitis, pharyngitis, and palatal petechiae. The measles virus is a paramyxovirus (a single-stranded RNA virus). Humans are the only natural host, though nonhuman primates can catch the illness from humans.

2. The diagnosis of measles is usually made on clinical grounds, with laboratory diagnostic procedures playing a secondary role. A nasopharyngeal aspirate stained with fluorescein-labeled antibody detects the measles virus directly in clinical specimens, allowing same-day confirmatory laboratory diagnosis of infection.

Immunoglobulin M (IgM)-based tests are also available for the rapid diagnosis of measles. Serum samples for IgM antibody should be collected after day 3 of the rash since IgM antibody may not be detectable before this time. The virus can be isolated in tissue culture, but isolation is difficult and unreliable for the diagnosis of measles.

3. The measles virus, which is spread by the respiratory route, is the most easily transmitted human infectious agent. After exposure, the incubation period of measles is 10 to 14 days. Typically, patients initially develop fever, cough, coryza, conjunctivitis, sore throat, and headaches. Several days later a generalized morbilliform rash appears. Koplik's spots, which are pathognomonic for measles, may be seen. These are small bluish-gray lesions on a red base, which appear on the buccal mucosa (Fig. 2).

The virus multiplies in the upper respiratory tract and conjunctiva. Viremia then develops, and after this viremia the patient experiences fever, constitutional symptoms, and rash. Leukopenia usually accompanies acute infection.

In the upper respiratory tract, edema and loss of cilia as a result of the measles infection can predispose to secondary bacterial invasion, which can lead to bacterial pneumonia and otitis media. Pneumonia with respiratory failure is a potentially life-threatening complication of measles. In the developing world, secondary diarrheal disease and respiratory infections are often seen, especially in malnourished infants and children. The combination of measles and other infectious agents has a much higher mortality rate than does either disease alone. The mechanism by which measles predisposes to acquisition of other infectious agents is believed to be due to the immunosuppressive properties of this virus. This period of immunosuppression may last for weeks to months, during which time these children will continue to be at increased risk for other infectious agents.

The most severe complication is encephalitis, which develops in 1 of every 1,000 to 2,000 cases. This develops 1 to 14 days after the rash. A high proportion of patients with encephalitis are left with neurologic sequelae. Subacute sclerosing panencephalitis (SSPE) is an extremely rare complication of early infection and usually is seen in patients less than 2 years old. SSPE is a persistent encephalitic infection that is distinct from the encephalitis that may complicate acute measles infection. The virus that causes SSPE may be a defective measles virus or another measles virus variant. SSPE has an insidious onset, usually manifested by behavior problems. The disease progresses over a period of weeks to months, resulting in severe neurologic dysfunction, including seizure activity, loss of motor function, coma, and eventually death.

4. Measles can be prevented with live, attenuated vaccine. This vaccine is administered after maternally acquired immunity wanes, usually at 9 months (in the developing world) to 12 to 15 months (in developed countries). In addition to vaccination, infection may be prevented by the use of passive immunity (immunoglobulin). This is usually administered to persons at significant risk of severe measles following specific exposure. Immunoglobulin can be used in babies less than 1 year old and in children with cancer and/or specific defects in cell-mediated immunity in whom vaccination with a live, attenuated vaccine would be contraindicated.

In 1998 there were approximately 1 million deaths attributed to measles worldwide. Failure to deliver at least one dose of measles vaccine to all infants remains the primary reason for these deaths. Three regions of the World Health Organization (WHO) have targeted elimination of measles: the American Region, the Eastern Mediterranean Region, and the European Region. Global vaccination coverage was estimated at 72% in 1998. In 1998, 14 countries reported measles vaccination coverage of below 50%. By comparison, in the United States in 1999, only 100 confirmed cases of measles were reported. A total of 33 of these cases were imported, and 33 others were linked to imported cases. In the United States, there is 98% vaccine coverage among children entering school. In 48 states and the District of Columbia, a second dose of measles vaccine is required for school entry. Clearly, these are the types of public health measures that are needed globally to make worldwide eradication a reality.

5. Deficiency of vitamin A is known to increase the mortality due to both measles and diarrheal illnesses in children. Administering vitamin A to children with measles can reduce the mortality rate in individual cases by as much as 50% and decrease the severity of measles complications. From a public health perspective, institution of a program that ensures that children have periodic supplementation with vitamin A has been shown to reduce measles mortality. Fortification of bread and margarine with vitamin A is done in wealthier countries, and fortification of sugar has been successful in Latin America. Many developing countries either have or are planning to have programs to combat vitamin A deficiency.

6. This patient should be managed with supportive therapy. No specific antiviral treatment is available. The school nurse and local health department should be notified about this case (as was done) as soon as possible so that necessary control measures can be implemented.

REFERENCES

1. **Abramson, O., R. Dagan, A. Tal, and S. Sofer.** 1995. Severe complications of measles requiring intensive care in infants and young children. *Arch. Pediatr. Adolesc. Med.* **149:**1237–1240.

2. **Centers for Disease Control and Prevention.** 1999. Global measles control and regional elimination, 1998–1999. *Morb. Mortal. Wkly. Rep.* **48:**1124–1130.

3. **Centers for Disease Control and Prevention.** 2000. Measles—United States, 1999. *Morb. Mortal. Wkly. Rep.* **49:**557–560.

4. **Hersh, B. S., G. Tambini, A. C. Nogueira, P. Carrasco, and C. A. de Quadros.** 2000. Review of regional measles surveillance data in the Americas, 1996–99. *Lancet* **355:**1943–1948.

5. **Sommer, A.** 1997. Vitamin A prophylaxis. *Arch. Dis. Child.* **77:**191–194.

CASE 36 The patient was a 23-year-old male with a 3-year history of pain and itching of the toes of both feet and his left palm and fingers. Small raised, red lesions were visible on his left fingers. Peeling and scaling were observed on his feet and left palm. In the past 3 months, he had lost the nails from his great toes. Otherwise, he was in his usual state of good health, which included training for triathlons (swimming 1 hour a day, running, and biking). He had no systemic complaints. He had been treated for this condition twice in the past with an appropriate agent, the first time for 3 months during which time his condition improved. He stopped for 5 months, his condition worsened, and he again took the agent for 6 months, stopping after becoming ill following drinking "some" beer. His present illness was his third episode of the disease. Culture of the organism recovered from skin scrapings of his feet is shown in Fig. 1. Microscopic morphology is seen in Fig. 2.

1. What do you think is wrong with this patient? What organism was causing his infection?

2. Even though this patient had a chronic infection, he had no systemic complaints. Is this consistent with the pathogenesis of the organism causing his infection? Explain your answer.

3. How do you think this person got this infection? Why do you think it keeps recurring?

4. What are the limitations of therapy for this type of infection?

5. What is a Wood's lamp examination? Would it have been useful diagnostically in this patient? Explain.

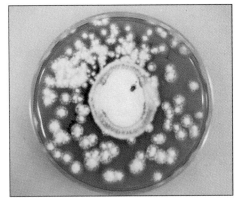

Figure 1

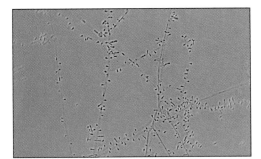

Figure 2

CASE DISCUSSION

CASE 36

1. This patient had tinea pedis or "athlete's foot." Because one of his hands was also infected, he had a variant of this condition known as "one-hand-two-feet" disease. Tinea pedis is an infection caused by a group of fungi called the dermatophytes. There are three genera of dermatophytes, *Trichophyton*, *Microsporum*, and *Epidermophyton*. The organism causing his infection is *Trichophyton rubrum* (Fig. 1). Microscopic examination of the organism shows microconidia being produced on the sides of the hyphae (Fig. 2). Macroconidia are rarely seen with this genus.

2. The finding in this patient of chronic infection without systemic complaints such as fever or malaise is consistent with infections with dermatophytes. Infections with these organisms are superficial. They are localized to keratinized, nonliving tissue such as keratinized skin, nails, and hair.

3. Spread of *T. rubrum* is human to human. Typically the organism is shed from infected individuals in hair and skin scales and is obtained by direct contact with this infected material. The organism survives best in moist environments. It is frequently found on pool decks and shower rooms, two areas that this individual frequented.

T. rubrum causes 80 to 90% of chronic dermatophyte infections. Patients who have chronic *T. rubrum* infection frequently have a negative delayed-type hypersensitivity reaction (negative skin test) to trichophyton-derived antigens, indicating no or poor cell-mediated immunity to this organism. Animal studies have shown the central role that cell-mediated immunity plays in resolution of dermatophyte infection. Therefore, this patient would probably be one of the 10 to 20% of individuals who are trichophyton skin test negative and at risk for developing chronic *T. rubrum* infection.

Dermatophytes are either anthropophilic (from humans), geophilic (from soil), or zoophilic (from animals). When humans are infected by geophilic and zoophilic species, there is usually a more intense inflammatory response than is the case for infection with anthropophilic species that have more successfully adapted to humans. Anthropophilic dermatophytes may have evolved from zoophilic dermatophytes. *T. rubrum*, an anthropophilic species, rarely infects animals other than humans. By contrast, *Microsporum canis* is commonly transmitted to humans from cats and dogs.

4. Therapy for dermatophytes can be either topical or systemic. Topical therapy works reasonably well for skin infections but is of little value for hair or nail infections, which require systemic therapy. Successful therapy of nail infections may require months. Discontinuation of therapy before then may result in relapses in patients, especially those with poor cell-mediated immunity to *T. rubrum*. Griseofulvin is safe with few side effects. However, it is known to interact with alcohol. This interaction can result, as occurred in this patient, in nausea and vomiting. Oral terbinafine, itraconazole, and fluconazole have also been used in the therapy of dermatophyte infections. These drugs are now more commonly prescribed than is griseofulvin. They are present in the distal nail plate within a few weeks of the initiation of therapy. It may be due to the persistence of these drugs in the nail for several months following the end of treatment that there is a higher mycological cure rate (on the order of 60 to 80%) with these drugs than there is with griseofulvin. The persistence of the drug in the nail bed is also the basis for the use of "pulse therapy" with itraconazole.

5. A Wood's lamp is an ultraviolet (UV) lamp used to examine hair infected with *Microsporum* species. *Microsporum*-infected hair will show green fluorescence under the Wood's lamp. Hair infected with *Trichophyton* sp. does not show green fluorescence. Since this patient did not have tinea capitis (infection of the hair and scalp), such an examination would not have been of value.

REFERENCES
1. **Aly, R.** 1999. Ecology, epidemiology and diagnosis of tinea capitis. *Ped. Infect. Dis. J.* **18**:180–185.

2. **Gupta, A. K., and R. K. Scher.** 1998. Oral antifungal agents for onychomycosis. *Lancet* **351**:541–542.

3. **Weitzman, I., and R. C. Summerbell.** 1995. The dermatophytes. *Clin. Microbiol. Rev.* **8**:240–259.

CASE 37 This 12-year-old girl was in her normal state of good health when she developed a fever of several days duration. She had no localizing symptoms, except for the development of a large rash on her back (Fig. 1). Her history was notable in that she lived in Connecticut near the New York State border and had recently been walking through tall grass where her sister was taking horseback riding lessons.

1. With what organism was she infected? What disease did she have?

2. What in her history is suggestive of this disease? How is this disease transmitted?

3. How, in the absence of a characteristic rash, is the diagnosis of this disease established?

4. She was appropriately treated with antibiotics and did well. What complications can occur in patients with this disease, particularly those in whom there is no treatment or inadequate therapy?

5. What efforts can be taken to prevent this illness?

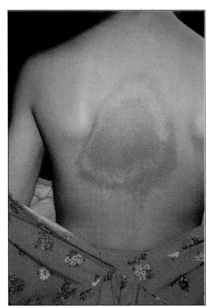

Figure 1

CASE DISCUSSION

CASE 37

1. This patient was infected with the spirochete *Borrelia burgdorferi*, which is the etiologic agent of Lyme disease. Her symptom of a nonspecific fever is consistent with Lyme disease, but it is the presence of the characteristic erythematous annular rash, referred to as erythema migrans, demonstrated in Fig. 1, that is diagnostic. The rash typically has a target-like appearance with expanding borders. Some patients will have these lesions at multiple sites. This patient had one rather prominent lesion.

2. The patient lives in Connecticut, a state with a very high incidence of Lyme disease. In fact, the disease was initially described in (and named for) a town in Connecticut. Other regions in which Lyme disease is endemic include other areas in the northeastern United States, Minnesota, Wisconsin, and northern California. *B. burgdorferi* is spread to humans by ticks of the genus *Ixodes*. In the northeastern United States, the white-footed mouse appears to be the primary reservoir of *B. burgdorferi*, which is present in the mouse's bloodstream. This mouse is also the preferred host for the *Ixodes scapularis* tick, formerly known as *Ixodes dammini*, the major vector of this spirochete. In other geographic locales, other *Ixodes* species act as major vectors.

Given that the patient was walking through tall grass, the sort of environment where ticks are likely to be found, she may well have had an ixodid tick attach to her during this time. The tick is frequently found in wooded areas, but also can be found in grassy areas. All three stages in the life cycle of the tick, i.e., larva, nymph, and adult, can feed on a human host, but only the nymph and adult stages can transmit the disease. Nymphs and adults infected after feeding on a *B. burgdorferi*-infected mouse pass the organism to humans during the blood meal, probably by regurgitating the spirochetes into the wound. Transfer of the spirochetes from the infected *I. scapularis* ticks to humans appears to require 36 to 48 hours of attachment. Ticks removed before that time probably do not transmit the spirochete. However, the nymph stage of the tick is extremely small (described as the size of a pencil point), so the tick bite may go unnoticed. Less than half of patients with documented Lyme disease are able to recall a tick bite.

3. The case definition of Lyme disease that is used for surveillance purposes is the presence of an erythema migrans rash greater than or equal to 5 cm in diameter or laboratory confirmation of infection with objective evidence of musculoskeletal, neurologic, or cardiovascular manifestations of Lyme disease. Unfortunately, the organism itself is difficult to grow from clinical specimens and requires complex media that are not available in most clinical laboratories, making culture a low-yield procedure. Because of this, isolation of this organism from clinical specimens is not routinely attempted. Although the polymerase chain reaction is used in some settings, such as in cases of arthritis and central nervous system disease, its low sensitivity limits its usefulness. The current laboratory recommendation is a two-test approach for the serologic diagnosis of Lyme disease. The serum specimen should first be tested using either an enzyme immunoassay (EIA) or an indirect immunofluorescent assay (IFA). Positive or equivocal specimens should then be tested with the more specific immunoglobulin G (IgG) and IgM Western blot (immunoblot). The sensitivity and specificity of the serologic tests vary in relation to the time in the course of the illness during which the specimen was obtained, with the tests being more accurate later in the disease course. The performance of different laboratories in Lyme serology testing varies greatly, with both false-positive and false-negative results occurring with increased frequency in certain laboratories.

4. Clinically, patients who are infected with *B. burgdorferi* may present with a clinical picture of headache and spinal fluid with a lymphocytic pleocytosis. Other neurologic symptoms and signs, such as cranial nerve VII palsy, peripheral neuropathy, meningoencephalitis, and subacute encephalopathy, may occur. Involvement of joints with clinical arthritis occurred in the original outbreak in Old Lyme, Connecticut, and continues to occur in cases of Lyme disease. Cardiac involvement with conduction defects and consequent arrhythmias also occurs in some patients. Finally, coinfection with another tick-borne pathogen, such as *Babesia microti*, the major etiologic agent of babesiosis, has been reported to occur in patients with Lyme disease.

5. Prevention of Lyme disease is the same as prevention of other tick-borne diseases. In endemic areas, the use of appropriate clothing, including long pants, long-sleeved shirts, and closed-toe shoes, is important when exposure to ticks may occur. The use of tick repellents, including the chemical *N,N*-diethyl-*m*-toluamide (DEET) on skin and clothing and permethrin on clothing is an additional precaution. Finally, examination of the skin after walking in an environment in which tick exposure is suspected allows for the removal of ticks before they are able to transmit *B. burgdorferi*.

In addition to the above methods of preventing tick bites, an available option in the prevention of Lyme disease included an FDA-approved Lyme disease vaccine based upon recombinant *B. burgdorferi* lipidated outer-surface protein A (rOspA) as an immunogen. This vaccine was removed from the American market by its manufacturer in February 2002 as a result of decreased sales. Results of a large-scale, randomized, controlled trial of safety and efficacy of this vaccine in persons aged 15 to 70 years residing in disease-endemic areas of the northeastern and north-central United States indicated that the vaccine was safe and efficacious when administered on a three-dose schedule of 0, 1, and 12 months.

The serodiagnosis of Lyme disease following vaccination may be problematic and remains of importance despite the discontinuation of the vaccine, given that significant numbers of people in areas where Lyme disease is endemic have received the vaccination. Vaccine-induced anti-rOspA antibodies routinely cause positive results on an enzyme-linked immunosorbent assay (ELISA), indicating exposure to *B. burgdorferi*. It is usually possible through careful interpretation of the results of immunoblots to discriminate between *B. burgdorferi* infection and previous rOspA immunization. Although vaccination is expected to elicit antibody to OspA only, natural infection results in the production of antibody to additional diagnostic antigen bands (to non-OspA borrelial antigens) seen in immunoblots. Another way to overcome this problem is by the use of an ELISA to a synthetic peptide, C(6), based on an immunodominant conserved region of the *B. burgdorferi* variable surface antigen. This ELISA has, in preliminary studies, been shown to be both sensitive and specific for Lyme disease infection, and to be negative in patients who have received the Lyme vaccine but in whom infection with *B. burgdorferi* has not occurred.

REFERENCES

1. **Centers for Disease Control and Prevention.** 1996. Recommendations for the use of Lyme disease vaccine. Recommendations of the Advisory Committee on Immunization Practices (ACIP). *Morb. Mortal. Wkly. Rep.* **48**(RR-7):1–17. (Erratum, **48**:833.)

2. **Centers for Disease Control and Prevention.** 1995. Recommendations for test performance and interpretation from the Second National Conference on Serologic Diagnosis of Lyme Disease. *Morb. Mortal. Wkly. Rep.* **44**:590–591.

3. **Liang, F. T., A. C. Steere, A. R. Marques, B. J. Johnson, J. N. Miller, and M. T. Philipp.** 1999. Sensitive and specific serodiagnosis of Lyme disease by enzyme-linked immunosorbent assay with a peptide based on an immunodominant conserved region of *Borrelia burgdorferi* vlsE. *J. Clin. Microbiol.* **37**:3990–3996.

4. **Nadelman, R. B., and G. P. Wormser.** 1998. Lyme borreliosis. *Lancet* **352**:557–565.

5. **Spach, D. H., W. C. Liles, G. L. Campbell, R. E. Quick, D. E. Anderson, Jr., and T. R. Fritsche.** 1993. Tick-borne diseases in the United States. *N. Engl. J. Med.* **329**:936–947.

6. **Thanassi, W. T., and R. T. Schoen.** 2000. The Lyme disease vaccine: conception, development, and implementation. *Ann. Intern. Med.* **132**:661–668.

CASE 38

The patient was a 6-year-old female from North Carolina. She was in her usual state of good health until 10 days prior to admission, when she had a tick removed from her scalp (Fig. 1). She developed a sore throat, malaise, and a low-grade fever 8 days after tick removal. She was seen by her pediatrician when she began developing a pink, macular rash, which started on her palms and lower extremities and spread to cover her entire body. The pediatrician's diagnosis was viral exanthem. One day prior to admission, she developed purpura, emesis, diarrhea, myalgias, and increased fever. On the day of admission, she was taken to her local hospital emergency room because of mental status changes and was admitted. Her physical examination was significant for diffuse purpura; periorbital, hand, and foot edema; cool extremities with weak pulses; and hepatosplenomegaly. Her laboratory studies were significant for a Na^+ level of 125 mmol/liter, platelet count of 26,000/µl, white blood cell (WBC) count of 14,900/µl, hemoglobin level of 8.8 g/dl, and greatly increased coagulation times. Ampicillin and chloramphenicol therapy was begun, and she was intubated and transferred to our institution; she died soon after arrival.

1. Which infectious agents are spread by ticks? Was the observation that a tick had been removed from her scalp important in this case?

2. What is the etiologic agent of this infection? What physical and laboratory findings are consistent with this infection?

3. Which condition(s) does her physical findings on admission suggest? List three organisms that can cause these types of physical findings.

4. Were her family members at increased risk for this infection? Explain.

5. Which specific test(s) is available for diagnosis of this infection?

Figure 1

CASE DISCUSSION

CASE 38

1. Ticks are vectors for *Borrelia burgdorferi* (the agent of Lyme disease), other *Borrelia* species (which cause relapsing fever), *Francisella tularensis* (the agent of tularemia), *Babesia microti* (the agent of babesiosis), *Rickettsia rickettsii* (the etiologic agent of Rocky Mountain spotted fever [RMSF]), human ehrlichiosis, two viral diseases, Colorado tick fever and Powassan encephalitis, and other rickettsial and viral diseases not found in the United States. *R. rickettsii* is endemic in the state of North Carolina, with Oklahoma or North Carolina reporting the largest number of RMSF cases on a yearly basis. Ninety percent of cases occur between the months of April and September, a period when ticks are most actively feeding. For *R. rickettsii* to be transmitted to humans, the infected tick must be attached for a minimum of 6 to 10 hours, and more than 24 hours may be required for transmission to occur. The incubation time after tick exposure ranges from 2 to 14 days, with a median of 7 days. The patient's development of symptoms 8 days after tick exposure is consistent with *R. rickettsii* infection. The tick removed from this child was *Dermacentor variabilis*, the common dog tick. This tick is known to transmit *R. rickettsii*, *F. tularensis*, and *Ehrlichia chaffeensis*.

2. The most likely etiology of her infection is *R. rickettsii*, which causes RMSF. The physical finding of a skin rash, in the presence of fever, myalgias, vomiting, and diarrhea, with a history of a tick bite, is highly suggestive of RMSF. Hyponatremia (low Na$^+$ level) is commonly seen in RMSF, as are low platelet counts and increased coagulation times, both of which are manifestations of disseminated intravascular coagulation (DIC), an often fatal complication of this infection.

This case took place in North Carolina. The failure of her pediatrician to recognize that she had RMSF most certainly contributed to her demise. Patients with a skin rash (seen in 90% of patients with RMSF), fever, and tick exposure in North Carolina have RMSF until proven otherwise. Mortality with RMSF is approximately 1 to 3% with timely, appropriate antimicrobial therapy with either doxycycline or chloramphenicol (used in children) but may reach as high as 25% in untreated individuals.

3. The finding of cool extremities with weak pulses is indicative of shock. Edema in this case is the result of increased vascular permeability, suggesting damage to endothelial cells, a well-known mechanism of *R. rickettsii*-induced pathologic changes. Purpura in the setting of septic shock indicates that the patient is suffering from DIC. Her platelet count and coagulation times are consistent with DIC. Almost any aerobic, gram-negative organism can cause septic shock and DIC; these include *Neisseria meningitidis*, *Pseudomonas aeruginosa*, members of the family Enterobacteriaceae, *Pasteurella multocida*, *F. tularensis*, *Vibrio vulnificus*, and *Haemophilus influenzae*.

Streptococcus pneumoniae, *Staphylococcus aureus*, and *R. rickettsii* can also cause septic shock. In the southwestern United States, *Yersinia pestis* should also be considered in the differential diagnosis of patients with septic shock and DIC.

4. RMSF is a tick-borne disease. There is no evidence of person-to-person spread. However, family clusters of RMSF have been described. In one such cluster, five family members developed disease with two dying of fulminant infection. A hyperendemic focus of ticks was found on the wooded property of the family. None of the ticks were found to be infected with *R. rickettsii*, but another rickettsial species (*R. amblyommii*) thought to be nonpathogenic for humans was found. Since only a small percentage of ticks (0.5%) are typically infected with *R. rickettsii*, these findings are not surprising. Therefore, it would not be surprising but it would be unusual if another family member developed RMSF near the time this child did.

Increasing human intrusion into wooded habitats infested with large numbers of ticks along with better diagnostic techniques for tick-borne diseases such as RMSF, *Ehrlichia* spp., and *B. burgdorferi* have resulted in the recognition of tick-borne diseases as important emerging infectious diseases. In a recent study of patients with a history of tick bite and unexplained fever in North Carolina, 16 of 35 patients had evidence of either RMSF or ehrlichiosis.

5. *R. rickettsii* can be detected directly in tissue biopsy specimens by using direct fluorescent-antibody (DFA) assay or other techniques. The DFA assay has the advantage of being very rapid and, in skilled hands, very specific. However, its sensitivity is very much dependent on the quality of the tissue biopsy. A negative test does not rule out this diagnosis. This test is available in only a limited number of laboratories. Immunohistochemical techniques can also be used to detect this organism in fixed tissues.

PCR (polymerase chain reaction) has also been applied to the detection of *R. rickettsii* but is not yet widely available. PCR is more sensitive in biopsies of the skin rash than it is from samples of blood. This is not surprising since the organism is an obligate, intracellular parasite that specifically infects endothelial cells, a cell type likely to be found in biopsy material but not in blood. The organism can also be isolated from blood by inoculation of guinea pigs, embryonated eggs, or tissue culture using the shell vial technique. However, cultivation of this organism is extremely dangerous and is attempted only in a few highly specialized laboratories.

Serologic tests are the most widely used diagnostic tests for detection of RMSF. At the University of North Carolina Hospitals, in an area in which RMSF is endemic, both a latex agglutination test and an indirect fluorescent-antibody assay (IFA) are used. The latex agglutination test is highly specific and has very good sensitivity.

False-negative results do occur, so alternative serologic tests should be available. The IFA can be used to confirm the latex agglutination test and also can be used as a primary diagnostic test. One of the major problems with the serologic tests is that early in the disease course, the individual may not have mounted a sufficiently strong immune response to cause a positive serologic test result. Follow-up serologic tests 1 to 4 weeks later may prove positive. In patients in whom RMSF is a distinct possibility, a negative serologic test result should not preclude the use of antimicrobial therapy since a fulminant, fatal disease course with this organism, as was seen in this case, is not unusual.

REFERENCES

1. **Carpenter, C. F., T. K. Gandhi, L. K. Kuo, G. R. Corey, S.-M. Chen, D. H. Walker, J. S. Dumler, E. Breitschwerdt, B. Hegarty, and D. J. Sexton.** 1999. The incidence of ehrlichial and rickettsial infections in patients with unexplained fever and recent history of tick bite in central North Carolina. *J. Infect. Dis.* **180:**900–903.

2. **Jones, T. F., A. S. Craig, C. D. Paddock, D. B. McKechnie, J. E. Childs, S. R. Zaki, and W. Schaffner.** 1999. Family cluster of Rocky Mountain spotted fever. *Clin. Infect Dis.* **28:**853–859.

3. **Paddock, C. D., P. W. Greer, T. L. Ferebee, J. Singleton, Jr., D. B. McKechnie, T. A. Treadwell, J. W. Krebs, M. J. Clarke, R. C. Holman, J. G. Olson, J. E. Childs, and S. R. Zaki.** 1999. Hidden mortality attributable to Rocky Mountain spotted fever: immunohistochemical detection of fatal, serologically unconfirmed disease. *J. Infect. Dis.* **179:**1469–1476.

4. **Thorner, A. R., D. H. Walker, and W. A. Petri, Jr.** 1998. Rocky Mountain spotted fever. *Clin. Infect. Dis.* **27:**1353–1360.

CASE 39

The patient was a 27-year-old male student who presented with ulcerated, raised lesions on his neck, back, and foot. The lesions began as raised red areas that spontaneously drained, ulcerated, healed, and then began to "grow" again. On presentation, the neck lesion had a raised, erythematous border with an ulcerative crater in the center (Fig. 1). His history was significant for travel in the Middle East the previous summer. He worked on an Israeli irrigation project in June and July and spent a week in the Sinai Peninsula in August, followed by several weeks along the Nile in Egypt. During his travel in Israel and Egypt, he experienced bites from multiple flying insects. While in Egypt, he suffered 8 days of watery diarrhea and a 20-lb (ca. 9-kg) weight loss and first noticed the neck lesion. This occurred 5 months prior to presentation. At presentation, he had a normal complete blood count and normal eosinophil count. The infecting organism was detected by direct examination and culture of the neck lesion (Fig. 2).

1. What is the organism causing his illness? Why doesn't he have systemic symptoms?

2. What is the natural history of his illness? Was the presence of multiple lesions unusual?

3. How is this disease spread? How can infection be prevented?

4. Why are infections with this organism typically seen on the head and neck?

5. How is this infection typically diagnosed?

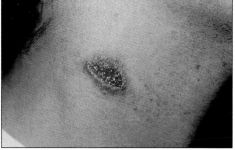

Figure 1

Figure 2

CASE 39

CASE DISCUSSION

1. The lesion and the organism detected by culture are both consistent with the protozoan *Leishmania* sp. There are three major disease syndromes caused by this parasite: cutaneous, mucocutaneous, and visceral leishmaniasis. Conventional wisdom concerning *Leishmania* contends that the different forms of the disease are caused by different species of the parasite. The relationship of specific species causing different disease manifestations is believed to be related to the temperature range of the individual parasites, with species unable to grow at the core human body temperature causing cutaneous disease, while those capable of growing at higher temperature cause visceral disease. Recent cases of atypical visceral disease caused by a variant of *L. tropica*, which typically causes cutaneous disease, have led to the reexamination of that idea.

This patient had Old World cutaneous leishmaniasis. This form of leishmaniasis is found in Mediterranean, Middle Eastern, and African countries and is caused by two species, *L. tropica* and *L. major*.

The question of why the patient did not have systemic symptoms may appear confusing since the patient had severe diarrhea with a 20-lb weight loss at the time his first skin lesion appeared. He in fact had systemic symptoms at that time. However, it is more likely that his symptoms were due to "traveler's diarrhea" contracted while visiting Egypt. The concurrent diarrhea and lesion appearance were a coincidence. Since his diarrhea had resolved 5 months previously, it clearly was not part of his current illness. This type of coincidence, which occurs frequently in medicine, can cloud the diagnostic picture.

Since this patient had cutaneous leishmaniasis, in which the disease process is limited to the skin, the absence of systemic symptoms is expected.

2. The natural history of this disease is consistent with the patient's clinical presentation. After infection, it takes approximately 1 week to 6 months for lesions to appear. In *L. tropica* infection the ulcers have raised borders and are often moist appearing. They usually spontaneously heal within several months, although they may result in disfiguring scars on the face. Once the lesions are healed, the individual will have lifelong immunity.

When multiple lesions are observed with this disease, they usually are the result of multiple bites by the disease vector. Metastatic lesions from a single lesion can also occur.

3. Old World leishmaniasis is spread by the bite of the sand fly belonging to the genus *Phlebotomus*. If the organism infecting the patient is *L. major*, it is believed that rodents act as a reservoir for this parasite and that the life cycle is completed in and transmitted by the insect vector. If the organism infecting the patient is *L. tropica*, transmission is from human to human via the sand fly. The infective phase, the flagellated promastigote form, multiplies in the gut of the insect and is injected into the skin during the insect's blood meal. In the skin of humans, the organism is taken up by macrophages where it survives and replicates within a cytoplasmic vacuole in the amastigote form. Cell-mediated immunity is key to the elimination of this parasite.

The sand fly's habitat is in rodent burrows and stone, concrete, or earthen cracks or crevices. The insect is typically found in shaded areas. It is a nocturnal feeder, so preventive measures are most important in the evening and at night. They include the use of insect repellents, wearing long-sleeved shirts and pants, and using insecticides in sleeping areas. The insect's small size (3 mm) makes screens and mosquito netting of limited protective value.

4. The sand fly cannot bite through clothes. Therefore, it can feed only on exposed areas of the skin. This typically is the head and neck in individuals who wear clothing that covers the trunk and extremities or sleep under bed clothing. The patient discussed here reported that he frequently wore only shorts and received multiple insect bites, explaining why he had lesions on his back and legs.

5. Cutaneous leishmaniasis is often diagnosed clinically by physicians who are experienced with this infection. However, this disease is not endemic in the United States and thus few physicians here will have seen an actual cause. Since cutaneous leishmaniasis can be confused with other conditions, including leprosy, tertiary syphilis, dimorphic fungal infections, and noninfectious causes, laboratory support of this diagnosis is helpful. In this case, the amastigotes were observed in a Giemsa-stained biopsy of the lesion. In addition, the organism was grown on culture using triple N medium, an enriched medium designed for the recovery of *Leishmania*. Recovery of the organism on culture is necessary if speciation of the organism, which is done by isoenzyme analysis, is desired. Speciation is not important for therapeutic reasons since this disease is self-limiting. PCR (polymerase chain reaction) techniques are being developed for speciation but are not yet as discriminatory as isoenzyme analysis.

REFERENCES
1. **Alvar, J., C. Cañavate, B. Gutiérrez-Solar, M. Jiménez, F. Laguna, R. López-Vélez, R. Molina, and J. Moreno**. 1997. *Leishmania* and human immunodeficiency virus: the first ten years. *Clin. Microbiol. Rev.* **10:**298–319.

2. **Klaus, S., and S. Frankenburg.** 1999. Cutaneous leishmaniasis in the Middle East. *Clin. Dermatol.* **17:**137–141.

3. **Magill, A. J., M. Grogl, R. A. Gasser, Jr., W. Sun, and C. N. Oster.** 1993. Visceral infection caused by *Leishmania tropica* in veterans of Operation Desert Storm. *N. Engl. J. Med.* **328:**1383–1387.

4. **Salman, S. M., N. G. Rubeiz, and A.-G. Kibbi.** 1999. Cutaneous leishmaniasis: clinical features and diagnosis. *Clin. Dermatol.* **17:**291–296.

FIVE

Central Nervous System Infections

INTRODUCTION TO SECTION V

Infections of the central nervous system (CNS) are infrequent compared with the other infections we have discussed thus far, but they are very important because of the high mortality rates and the serious sequelae associated with them, including learning, speech, and motor skills disorders; seizures; and hearing and sight loss. The most frequent CNS infections are meningitis, encephalitis, and abscess. Intoxication caused by tetanus and botulinum toxins can affect the CNS, causing spastic or flaccid paralysis, but these diseases are quite rare in the developed world.

There are two major forms of meningitis, **septic** and **aseptic.** Septic meningitis is typically caused by bacteria. The cerebrospinal fluid (CSF) is usually cloudy, with over 1,000 white blood cells per μl with neutrophils predominating; increased protein levels due to inflammation; and decreased glucose due in part to metabolism by white blood cells. Aseptic meningitis can be caused by viruses, fungi, or *Mycobacterium tuberculosis*. In aseptic meningitis, the CSF is more likely to be grossly "clear" due to a lower white blood cell count, typically in the range of 100 to 500 per μl. Except for very early in the disease course, the predominant cell type is mononuclear, primarily lymphocytes. CSF glucose levels are frequently normal but may be decreased in over half of the patients with fungal or mycobacterial infections. CSF protein levels are frequently normal except with *M. tuberculosis*, where they are typically elevated.

Bacterial meningitis is most common in the very young, the very old, and the immunocompromised. Group B streptococci are the most common cause of neonatal meningitis (newborns to 2 months). *Listeria monocytogenes* is another organism that causes neonatal disease. It also is an important agent of meningitis in the immunosuppressed. Gram-negative enteric bacilli, including *Escherichia coli*, *Klebsiella pneumoniae*, and *Citrobacter* spp., may also cause neonatal meningitis. Congenital syphilis, which may manifest itself during the neonatal period, frequently will have a CNS component, neurosyphilis. Bacterial meningitis is most commonly seen in children 2 months to 5 years of age. Until recently, *Haemophilus influenzae* type b was the most common cause of bacterial meningitis in this age group. The widespread use of conjugated *H. influenzae* type b vaccine has resulted in a dramatic decline in the incidence of this disease. *Streptococcus pneumoniae* and *Neisseria meningitidis* are now the leading causes of meningitis in this age group and the elderly.

Recent trials of a conjugated 7-valent *S. pneumoniae* vaccine in children less than 2 years of age have shown that the vaccine has an efficacy of greater than 95% in preventing invasive pneumococcal disease, suggesting that pneumococcal meningitis will be much less common in this patient population once vaccination is widespread. Individuals with head trauma are also at risk for developing bacterial meningitis. The organisms most frequently associated with this type of bacterial meningitis are coagulase-negative staphylococci (especially in patients with CNS shunts or who have undergone neurosurgical procedures), *Staphylococcus aureus*, and *Pseudomonas aeruginosa*. In patients who have had meningitis more than once with *S. pneumoniae*, it is worth seeking a previously unrecognized anatomic defect in the skull due to what may have seemed to be minimal trauma, in some cases in the distant past. *M. tuberculosis* meningitis is seen primarily in children and the immunosuppressed.

Viral meningitis is typically caused by the enteroviruses other than the polioviruses. It is a disease seen primarily during the summer months in infants and young children. Herpes simplex virus can cause a usually benign meningitis associated with primary genital tract infections. This is not to be confused with herpes simplex encephalitis, which can occur in neonates and during reactivation of latent infection in adults. This form of herpes infection can produce necrotic lesions in the brain resulting in long-term sequelae or death.

Fungal meningitis is seen primarily, but not exclusively, in the immunocompromised. It is of particular importance in AIDS patients, in whom *Cryptococcus neoformans* is far and away the most important cause.

Encephalitis due to infectious agents is due primarily to viruses. Herpes simplex virus causes probably the most common form of viral encephalitis encountered in the developed world. Arthropod-borne viruses such as West Nile, Eastern equine, Western equine, St. Louis, and La Crosse encephalitis viruses are encountered in the United States. In many eastern states, an epidemic of rabies in animals is continuing. In 2000, four cases of human rabies were reported in the United States. As in essentially all cases of human rabies, they were fatal.

Parasites may also cause CNS infection. The most frequently encountered parasite causing CNS infections in the developed world is *Toxoplasma gondii*. Encephalitis due to this organism occurs primarily in AIDS patients and represents reactivation of latent infection. One of the most common causes of a clinical presentation of CNS infection in the developing world is cerebral malaria. A major cause of adult onset of seizures in certain areas of the developing world where pork is a source of protein in the diet is cysticercosis. This disease occurs when eggs of the pork tapeworm *Taenia solium* are ingested. The parasite is unable to complete its life cycle in humans, and cyst-like lesions occur throughout the body, including the brain. An ameba, *Naegleria fowleri*, causes a rare and frequently fatal form of meningoencephalitis. It is found in individuals who swim in warm fresh water during the summer months.

Brain abscesses occur through direct extension from a contiguous site such as an infected paranasal sinus, following trauma, or by hematogenous spread from another infected site. Typically, patients with abscesses due to hematogenous spread have either endocarditis or a lung abscess. **Septic emboli,** which are small blood clots containing infectious agents, are released from the primary infection site and enter the blood-stream. The emboli lodge in capillaries in the brain causing localized hemorrhage and producing sites for the initiation of infection, which evolve into brain abscesses. The organisms most frequently causing brain abscess in immunocompetent individuals are *S. aureus* and organisms usually found in the oropharynx, including the viridans group streptococci, *Actinomyces* spp., and anaerobic bacteria. In immunocompromised individuals, *Aspergillus* spp., *Mucor*, *Rhizopus*, and *Nocardia* spp. can cause brain abscesses. In trauma patients, *S. aureus* and gram-negative rods are frequently seen. In diabetic patients (especially those with ketoacidosis), rhinocerebral zygomycosis due to *Mucor*, *Rhizopus*, and other fungi within the *Zygomycetes*, can extend from the sinuses into the brain, causing extensive necrosis.

TABLE 5 CENTRAL NERVOUS SYSTEM INFECTIONS

ORGANISM	GENERAL CHARACTERISTICS	PATIENT POPULATION	DISEASE MANIFESTATION
Bacteria			
Actinomyces spp.	Branching, gram-positive bacilli, usually anaerobic	Individuals with aspiration pneumonia	Brain abscess
Bacillus anthracis	Spore-forming gram-positive bacillus	Individuals with severe anthrax infection; victims of bioterrorism	Meningitis
Citrobacter spp.	Enteric gram-negative bacillus	Neonates	Meningoencephalitis with abscess
Clostridium botulinum	Toxin-producing, anaerobic, gram-positive bacillus	Infants; adults who ingest botulinum toxin	Botulism, flaccid paralysis
Coagulase-negative staphylococci	Catalase-positive, gram-positive cocci	Individuals with foreign bodies, e.g., shunts or bolts	Meningitis
Escherichia coli	Lactose-fermenting, gram-negative bacillus	Neonates	Meningitis
Group B streptococci (*Streptococcus agalactiae*)	Catalase-negative, gram-positive cocci	Neonates, immunocompromised adults	Meningitis
Haemophilus influenzae type b	Gram-negative, pleomorphic bacillus	Unvaccinated children	Meningitis
Listeria monocytogenes	Catalase-positive, gram-positive coccobacillus	Neonates, adults with cell-mediated immunity defect	Meningitis, rhomboencephalitis
Mycobacterium tuberculosis	Acid-fast bacillus	Children; patients with AIDS	Tuberculous meningitis, CNS[a] tuberculomas
Neisseria meningitidis	Oxidase-positive, gram-negative diplococcus	All ages; outbreaks in college students and military	Meningitis
Nocardia spp.	Aerobic, partially acid-fast branching bacilli	Individuals with pulmonary or cutaneous nocardiosis	Brain abscess
Oral streptococci (*S. sanguis, S. mutans*, etc.)	Alpha-hemolytic, gram-positive cocci	Individuals with aspiration pneumonia, endocarditis	Brain abscess

(continued next page)

TABLE 5 CENTRAL NERVOUS SYSTEM INFECTIONS *(continued)*

ORGANISM	GENERAL CHARACTERISTICS	PATIENT POPULATION	DISEASE MANIFESTATION
Prevotella sp., *Porphyromonas* sp.	Anaerobic, gram-negative bacilli	Individuals with aspiration pneumonia	Brain abscess
Pseudomonas aeruginosa	Oxidase-positive, gram-negative bacillus	Individuals with head trauma or foreign bodies	Meningitis
Staphylococcus aureus	Catalase-positive, gram-positive coccus	Individuals with head trauma or foreign bodies	Meningitis, brain abscess
Streptococcus pneumoniae	Catalase-negative, gram-positive coccus	Primarily young children and elderly	Meningitis
Fungi			
Aspergillus spp.	Acute-angle, septate hyphae in tissue	Neutropenia with invasive aspergillosis	Brain abscess
Coccidioides immitis	Dimorphic mold	Geographically limited to Lower Sonoran life zone; increased rate of disseminated disease in non-Caucasians, immunocompromised, especially AIDS	Meningitis
Cryptococcus neoformans	Encapsulated, round yeast	Cell-mediated immunity defect, especially AIDS	Meningitis, cryptococcoma
Mucor sp., *Rhizopus* sp.	Ribbon-like, aseptate hyphae in tissue	Diabetics, neutropenic individuals	Necrotizing encephalitis, rhinocerebral zygomycosis
Parasites			
Acanthamoeba sp.	Ameba	Immunocompromised or immunocompetent	Granulomatous amebic encephalitis or keratitis
Naegleria fowleri	Ameba	Individuals who dive into warm, fresh water	Fatal amebic meningoencephalitis
Plasmodium falciparum	Delicate, ring forms in red blood cells	Individuals who visit or reside in areas where malaria is endemic	Cerebral malaria

Taenia solium	Larval cyst	Individuals who ingest *T. solium* eggs	Seizures, calcified or noncalcified lesions in brain; calcified lesions in muscle
Toxoplasma gondii	Large cysts in tissue	Cell-mediated immunity defect, especially AIDS patients	Encephalitis, abscess
Viruses			
Arthropod-borne viruses (arboviruses), including Eastern equine encephalitis virus, St. Louis equine encephalitis virus, West Nile virus	Both enveloped and nonenveloped ssRNA[b]	Children and adults bitten by viral arthropod vector	Encephalitis; fatality rate depends upon specific virus and age of individual
Echovirus/coxsackievirus	Nonenveloped ssRNA	Children and adults during summer months	Aseptic meningitis
Herpes simplex virus	Enveloped dsDNA[c]	Neonates, individuals with primary genital herpes, individuals with primary or recurrent herpes infection	Necrotizing encephalitis; benign, aseptic meningitis; necrotizing hemorrhagic encephalitis
Human immunodeficiency virus (HIV)	Enveloped RNA retrovirus	AIDS patients	AIDS-associated dementia; predisposes to other CNS infections
Polioviruses	Nonenveloped ssRNA	Nonvaccinated individuals; live, attenuated vaccine, especially in the immunocompromised	Polio paralysis
Rabies virus	Enveloped ssRNA	Individuals bitten or scratched by nonvaccinated, rabid dog, cat, or other mammal; bat contact	Rabies

[a]CNS, central nervous system.

[b]ssRNA, single-stranded RNA.

[c]dsDNA, double-stranded DNA.

CASE 40 The patient was a 4-year-old female with a fraternal twin. Her medical history was unremarkable and her vaccination status was appropriate for her age. During the previous week, she and her twin and two younger siblings had had upper respiratory tract infections with nasal congestion. On the night prior to admission, she developed fever and left ear pain. She was examined and found to have a bulging left eardrum consistent with otitis media. She was begun on oral amoxicillin/clavulanic acid. Early the next afternoon, she continued to have fever and was irritable. Later that afternoon, she also complained of a headache and neck pain and was brought to the emergency department. On physical examination, she was lethargic, listless, and had a stiff neck. There were no focal neurologic findings, but it was noted that she had a rash on her right cheek which the physician thought might be an early stage of a petechial rash. He performed a lumbar puncture. The fluid was cloudy with a white blood cell (WBC) count of 6,750/µl (normal, 0 to 3 cells/µl), 95% being neutrophils, and 189 red blood cells/µl. Her cerebrospinal (CSF) protein level was 170 mg/dl (normal, 15 to 45 mg/dl) and CSF glucose level was 25 mg/dl (normal, 50 to 75 mg/dl). A Gram stain of the CSF is shown in Fig. 1. The culture of the organism recovered from CSF is shown in Fig. 2. Of note, her serum immunoglobulin (Ig) levels (IgG, IgA, IgM) were normal.

1. What pathologic process is going on in this child? What organism do you think is infecting this patient?

2. What is the significance of her having otitis media in the development of her infection? What is the major virulence factor(s) produced by this organism? When this organism does not produce one of these virulence factors, what types of infection does it typically cause?

3. What is the significance of her developing a petechial rash? What organism that causes bacterial meningitis frequently has this type of

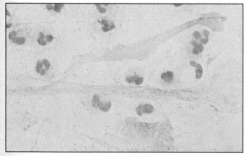

Figure 1

Figure 2

rash associated with it? What are the pathophysiologic events responsible for the formation of this rash?

4. Since 1991, how has the epidemiology of invasive infection caused by this organism changed? Why did this change occur?

5. How do her vaccination status and immunoglobulin levels help you understand this case?

CASE DISCUSSION

CASE 40

1. The lumbar puncture and other clinical findings indicate that this child had bacterial meningitis. She had a very high CSF WBC count of 6,750/µl with 95% neutrophils. Both findings are consistent with bacterial meningitis. In cases of fungal, viral, and mycobacterial meningitis, cell counts usually are in the range of 50 to 500 WBC/µl, with a predominance of mononuclear rather than polymorphonuclear cells. In bacterial meningitis, CSF glucose levels are generally decreased while protein levels are markedly increased. Decreased CSF glucose is due in part to its utilization by CSF polymorphonuclear cells. Increased CSF protein levels are due to inflammation and alterations in the blood-brain barrier. The occurrence of irritability, evolving to include lethargy and a stiff neck, is consistent with the clinical presentation of bacterial meningitis.

The patient is infected with *Haemophilus influenzae*. Serotyping of the organism revealed it to be serotype f. Prior to 1990, *H. influenzae* type b was the most common cause of bacterial meningitis in children from 6 months to 5 years of age. With the development and widespread use of a conjugated vaccine against serotype b *H. influenzae*, the number of cases of meningitis and other invasive infections (bacteremia, septic arthritis, buccal cellulitis, epiglottitis) due to this organism has declined dramatically (see answer to question 4 for further details).

2. The initial stage in the development of bacterial meningitis in infants and young children is frequently an upper respiratory tract illness, characterized by a "runny nose" or otitis media with a low-grade fever, due to an encapsulated organism, in this particular case, an unusual *H. influenzae* serotype, type f. From the upper respiratory tract, *H. influenzae* type f invades the bloodstream. In the absence of opsonic antibodies, the capsule surrounding this bacterium allows it to evade phagocytosis. By a process that is not clearly understood, the organism is able to cross the blood-brain barrier and infect the meninges. Many of the pathologic events that follow, including inflammation and edema, are due to the host's release of cytokines (tumor necrosis factor, interleukins) in response to the presence of the lipo-oligosaccharide found in the outer membrane of these gram-negative bacteria.

Nonencapsulated *H. influenzae* organisms typically cause infections of the mucous membranes in the respiratory tract. These infections include otitis media (as was seen in this patient), sinusitis, conjunctivitis, bronchitis, and occasionally pneumonia. Capsule seems to be important for the development of invasive disease such as bacteremia since immunocompetent individuals rarely develop invasive disease with nonencapsulated organisms. Bacteremic pneumonia due to nonencapsulated *H. influenzae* has been reported in immunosuppressed individuals.

3. Petechiae are small (<3 mm in diameter) skin lesions which usually indicate hemorrhage in capillaries and venules near the skin surface. Petechial lesions are frequently associated with endotoxin-induced disseminated intravascular coagulation (DIC) and septic shock. Although a discussion of the entire cascade of events that lead to DIC is beyond the scope of this text, one of the key features of DIC is clotting in small vessels resulting in hemorrhage which is manifested by petechial lesions, as was seen in this child. Petechiae are most commonly seen in cases of bacteremia and meningitis caused by *Neisseria meningitidis*. They are also frequently seen in patients with *Rickettsia rickettsii* infections. The finding of petechial lesions can be an ominous sign prognostically and requires prompt, aggressive medical intervention. Petechial lesions in patients with disseminated *H. influenzae* infections are unusual.

4. Beginning in 1991, a sharp decline in invasive disease due to *H. influenzae* type b (Hib) was seen in the United States. In 1987, invasive Hib disease incidence was estimated at 88 per 100,000 per year. By 1995, the incidence of disease had dropped to 1.6 cases per 100,000 per year. Why the dramatic decline? Much of the decline can be attributed to the widespread use of Hib vaccine. First used in this country in 1986, Hib vaccine has resulted in a decline in the incidence of Hib invasive disease. The first widely used vaccine was a polysaccharide vaccine made from the polyribosylphosphate (PRP) capsule of the type b organism. Although this vaccine was safe and effective in children greater than 2 years of age, it was not effective in children less than 2 years of age, the age group in whom invasive Hib disease was most prevalent. The reason for its ineffectiveness in this population was that, like all polysaccharide antigens, this vaccine induced a T-cell-independent response. Children less than 2 years old respond poorly, if at all, to T-cell-independent antigens. Additional doses of the PRP vaccine did not induce a booster effect, and this vaccine did not prevent respiratory carriage, facilitating spread of the organism to nonimmune populations. As a result, there was only a modest decline in the number of cases of invasive Hib disease after widespread use of the polysaccharide vaccine.

Clearly, an improved vaccine strategy was necessary. In 1991, the first of four conjugated Hib vaccines was licensed in the United States. In the conjugated vaccine, the Hib capsular polysaccharide was coupled to a protein to form a complex of the two molecules. The four different proteins used as conjugates were diphtheria toxoid, a genetically modified diphtheria toxoid, tetanus toxoid, and meningococcal outer membrane protein. Because of its protein component, this "new" antigen is processed differently by the immune system than the polysaccharide antigen was alone. This process requires the interaction of two types of lymphocytes, the T-helper cell and the antibody-producing B cell. This interaction is mediated by the T-helper cell recognizing a specific

B cell. The T-helper cell binds to this B cell and releases cytokines which, in turn, stimulate this B cell to proliferate and produce antibodies. An antigen that stimulates this elegant series of events is known as a T-cell-dependent antigen. Protein or protein conjugates are T-cell-dependent antigens.

The conjugated vaccine does not have the drawbacks of the polysaccharide vaccine. Because it is composed of a T-cell-dependent antigen, infants as young as 2 months generate an immune response to it. It shows a booster effect so immunity can be enhanced by using multiple doses, and most but not all studies suggest that it reduces nasopharyngeal carriage of Hib.

Although the vaccine has been highly successful in industrialized countries where it has been widely used, Hib invasive disease remains an important health problem globally. The vaccine is currently too expensive to be used in the developing world. It is estimated that this vaccine prevents >95% of cases of invasive Hib infection in the industrialized world. However, from a global perspective, Hib vaccination is estimated to prevent only 5 to 8.5% of cases of invasive Hib disease. In the developing world, Hib meningitis is estimated to have a mortality of 30% and long-term neurologic sequelae, including partial or complete hearing loss, and learning disabilities are estimated to occur in approximately 30% of survivors. Clearly, worldwide Hib vaccine programs would be a relatively inexpensive way to have a significant impact on the health of the world's children.

One of the theoretical concerns about this vaccine is that other serotypes of *H. influenzae* would take over the ecological niche of type b in the body with a concomitant rise in cases of meningitis due to other *H. influenzae* serotypes such as type f, as seen in this case. Although a slight rise has been seen, it has been insignificant compared to the decline in cases due to *H. influenzae* type b.

5. By the age of 4 years, this child would have received four doses of the conjugated Hib vaccine, at 2, 4, 6, and 12 to 15 months. However, because she developed invasive disease, it was feared that she had not made a protective immune response against this organism. One of the most common reasons for failure to mount an effective immune response is immunodeficiency. Since this child had normal immunoglobulin levels, it was felt that she did not have an obvious deficiency in the humoral arm of her immune system. Her immune system work-up was not pursued further when it was determined she was infected with the type f strain of *H. influenzae*. Antibodies to type b antigen have not been shown to protect against type f infections.

REFERENCES

1. **Peltola, H.** 2000. Worldwide *Haemophilus influenzae* type b disease at the beginning of the 21st century: global analysis of the disease burden 25 years after the use of polysaccharide vaccine and a decade after the advent of conjugates. *Clin. Microbiol. Rev.* **13:**302–317.

2. **Perdue, D. G., L. S. Bulkow, B. G. Gellin, M. Davidson, K. M. Peterson, R. J. Singleton, and A. J. Parkinson.** 2000. Invasive *Haemophilus influenzae* disease in Alaskan residents aged 10 years and older before and after infant vaccination programs. *JAMA* **283:**3089–3094.

3. **Tunkel, A. R., and W. M. Scheld.** 1993. Pathogenesis and pathophysiology of bacterial meningitis. *Clin. Microbiol. Rev.* **6:**118–136.

CASE 41

This 19-year-old student was in his usual state of health until the evening prior to admission, when he went to bed with a headache. He told his mother that he felt feverish, and on the following morning his mother found him in bed, moaning and lethargic. He was brought to the emergency room, where he appeared toxic and drowsy but oriented. His temperature was 40°C, his heart rate was 126/min, and his blood pressure was 100/60 mm Hg. His neck was supple. He had an impressive purpuric rash (Fig. 1), not blanching, most prominent on the trunk, legs, and wrists. A Gram stain of material taken from one of the patient's skin lesions is shown in Fig. 2. His white blood cell count was 26,000/µl with 25% band forms. The platelet count was 80,000/µl.

Blood cultures were obtained, a lumbar puncture was performed, and the patient was begun on intravenous ceftriaxone. Cerebrospinal fluid (CSF) glucose, protein, and white blood cell count were normal, and CSF bacterial culture was negative. Blood cultures grew the organism seen in Fig. 2.

1. What organism was causing this patient's illness? Is the finding of a normal CSF profile, without evidence of meningitis, commonly observed in infection with this organism? Explain your answer.

2. Is this organism ever part of the normal oropharyngeal flora? Explain your answer.

3. Which immunologic abnormalities predispose individuals to infection with this organism?

4. Which serogroup(s) causes illness? The serogroup is based on antigen from which part of the bacterium?

5. Which prophylactic strategies are useful for large populations?

6. Which prophylactic strategies can be used for exposed individuals?

7. What types of antibiotic resistance have been found in this organism?

8. What is a purpuric rash, and which virulence factor plays a central role responsible for its appearance?

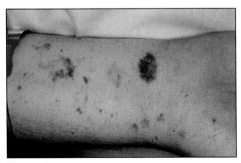

Figure 1

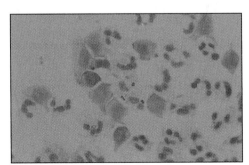

Figure 2

CASE DISCUSSION

1. The clinical presentation and the presence of gram-negative diplococci growing in the blood and seen on the Gram stain of a skin lesion strongly indicated that the etiologic agent of this infection was *Neisseria meningitidis*. *Neisseria gonorrhoeae* is also a gram-negative diplococcus, but the patient's clinical picture is more suggestive of infection with *N. meningitidis* than with *N. gonorrhoeae*. The two organisms can be differentiated in the clinical laboratory by biochemical means; *N. gonorrhoeae* oxidizes glucose, while *N. meningitidis* oxidizes both glucose and maltose.

Bacteremic disease with *N. meningitidis* does not always result in the development of meningitis. Meningococcemia without meningitis is a well-recognized syndrome caused by this organism. Cases of meningococcemia can vary clinically from mild disease to overwhelming sepsis. In patients with fulminant meningococcemia, the disease course from initial symptoms to death can be measured in hours. These patients may die before developing signs and symptoms of meningitis. Similarly, the absence of a cerebrospinal pleocytosis in the setting of central nervous system infection appears to be a risk factor for a poor outcome in cases of invasive meningococcal disease.

2. *N. meningitidis* is usually considered to be part of the normal oropharyngeal flora and can be found in a significant minority (2 to 10%) of healthy people. During epidemics of meningococcal disease in institutionalized populations, such as new recruits to the military, colonization rates may increase dramatically.

3. Most people who are colonized with this organism mount a humoral immune response to it. These individuals produce bactericidal antibodies, which appear to be protective for invasive disease. The very small percentage of patients who do not make bactericidal antibodies in response to colonization by this organism are at high risk for the development of invasive disease. Some patients who make antibodies have deficiencies in the terminal components of the complement pathway, and therefore these antibodies may not be bactericidal, nor can the alternative complement pathway be triggered. This complement deficiency places them at risk for disseminated disease as well, although these patients often present with less severe disease. Finally, asplenic individuals are thought to be at increased risk for infection with this organism, but the data supporting this conclusion are not as convincing as are the data for overwhelming pneumococcal infection following splenectomy.

4. The *N. meningitidis* serogroups most commonly associated with meningitis in the United States are types A, B, C, Y, and W135. The two most frequently isolated serogroups are B (50%) and C (20%). Typically, groups A and C are thought of as epidemic strains because of their association with epidemics, whereas group B isolates are

most likely to cause sporadic cases. Cases due to group B are most frequent because of the rarity of epidemics of *N. meningitidis* groups A and C in the United States. The serogroups are based on the biochemical structure of the capsular polysaccharide that surrounds the organism. Nonencapsulated organisms rarely cause invasive disease, indicating that encapsulation is critical to the pathogenicity of the organism. This patient's blood isolate was identified as belonging to group C.

5. Vaccination is the mainstay of prophylactic strategies for large, at-risk populations. A quadrivalent vaccine for groups A, C, Y, and W135 is the formulation currently available in the United States. Each dose consists of 50 μg of the four purified bacterial capsular polysaccharides. For both adults and children, vaccine is administered subcutaneously as a single 0.5-ml dose. The antibody responses to each of the four polysaccharides in the quadrivalent vaccine are serogroup-specific and independent. Currently no vaccine is available for group B meningococci because of the poor immunogenicity of this capsular polysaccharide. It is thought that the poor immunogenicity is due to the alpha 2-8-linked *N*-acetylneuraminic acid (NANA; sialic acid) being structurally and antigenically homologous with polysialosyl glycopeptides of neural cell adhesion molecules found in fetal and newborn human brain.

The specific indications for the use of the vaccine include certain high-risk groups, such as persons who have terminal complement component deficiencies and those who have anatomic or functional asplenia. Research, industrial, and clinical laboratory personnel who are exposed routinely to *N. meningitidis* in solutions that may be aerosolized also should be considered for vaccination. Vaccination with the quadrivalent vaccine may benefit travelers to countries in which *N. meningitidis* is hyperendemic or epidemic. Epidemics of meningococcal disease regularly occur in that part of sub-Saharan Africa known as the "meningitis belt," which extends from Senegal in the west to Ethiopia in the east. Epidemics in the meningitis belt usually occur during the dry season (i.e., from December to June); thus, vaccination is recommended for travelers visiting this region during that time. Though it is outside the "meningitis belt," epidemics of meningococcal disease have also been associated with pilgrimage to the Hajj in Saudi Arabia. In 2000, a total of more than 300 cases of invasive meningococcal disease due to serogroup W135 occurred in people from at least a dozen countries who attended the Hajj, and in their close contacts.

The issue of routine vaccination of college students is less clear. Based upon available data, it appears that those college students who live in dormitories are at an increased risk of invasive meningococcal disease compared with other persons their age. Nevertheless, because the risk for meningococcal disease among college students is low, vaccination of all college students, all freshmen, or only freshmen who live in dormitories or residence halls is not likely to be cost-effective for society as a whole.

Despite these data, an increasing number of colleges and universities are requiring meningococcal vaccination as a condition of enrollment.

Routine vaccination of children aged less than 2 years with the quadrivalent meningococcal polysaccharide vaccine is not recommended because of its relative ineffectiveness and its relatively short duration of protection in this age group.

6. Both vaccination and chemoprophylaxis may be in order for exposed individuals, especially health care workers who come into close contact with respiratory secretions of infected individuals. Rifampin has historically been the drug of choice for antimicrobial prophylaxis. It penetrates well into respiratory secretions and is well tolerated. The purpose of chemoprophylaxis is twofold, first to protect the individual receiving the drug, and then to eliminate nasopharyngeal carriage of the organism to limit its spread in the general population. However, cases of meningococcal meningitis in patients given rifampin have been reported. These isolates were found to be rifampin resistant. In addition, rifampin will not eliminate carriage in 10 to 20% of colonized individuals. It is worth noting that treatment with penicillin does not eradicate nasopharyngeal carriage of the organism.

There are three practical points concerning rifampin prophylaxis, First, patients should be informed that it causes secretions to turn orange. Urine, breast milk, and tears will be affected, and contact lenses can be permanently stained. Second, pharmacies in rural areas may not stock this drug, making it inconvenient to obtain and thus adversely affecting compliance. Third, in the setting of a case or an outbreak of meningococcal disease, antibiotic prophylaxis with rifampin is often given unnecessarily to many individuals who did not have intimate contact with a case but, because of the panic surrounding the event, contact a health care provider who is not able to properly evaluate the situation. Arrangements should be made to ensure that all close contacts can obtain this drug. Other agents that have been used for antibiotic prophylaxis include ciprofloxacin and ceftriaxone. These antibiotics are 90 to 95% effective in reducing nasopharyngeal carriage of *N. meningitidis* and are suitable alternatives to rifampin for chemoprophylaxis. Problems with these alternative therapies include the facts that ciprofloxacin is not normally given to children owing to concerns about growing cartilage, and ceftriaxone can only be given parenterally. An advantage of ciprofloxacin as compared with rifampin is the likelihood that a pharmacy will stock ciprofloxacin (only a single dose is used in adult patients) as compared with rifampin.

7. Antibiotic resistance in *N. meningitidis* has been reported for sulfonamides, which in the past were commonly used to eradicate nasopharyngeal carriage. High-level resistance to rifampin is due to mutations in the gene for the beta subunit of RNA polymerase. Resistance to penicillin G is usually on the basis of the production of altered penicillin-binding proteins and has been found in a number of countries. Of greater concern is the report of rare strains able to produce β-lactamase. Other antibiotics to which *N. meningitidis* has been reported to be resistant include tetracycline and chloramphenicol. The report of chloramphenicol resistance is worrisome, as this drug is commonly used to treat meningococcal infections in developing countries.

8. Purpuric skin lesions can be manifestations of disseminated intravascular coagulation (DIC). Petechial lesions are pinpoint, purplish-red lesions that are caused by hemorrhage in the intradermal vascular bed. Purpuric lesions are similar to petechial lesions but are larger, probably representing coalescence of a number of petechial lesions. Although many different events can initiate DIC, the endotoxin found in the outer membrane of *N. meningitidis* is a well-recognized mediator of DIC.

REFERENCES

1. **Centers for Disease Control and Prevention.** 2000. Prevention and control of meningococcal disease. Recommendations of the Advisory Committee on Immunization Practices (ACIP). *Morb. Mortal. Wkly. Rep.* **49**(RR-7):1–10.

2. **Centers for Disease Control and Prevention.** 2000. Meningococcal disease and college students. Recommendations of the Advisory Committee on Immunization Practices (ACIP). *Morb. Mortal. Wkly. Rep.* **49**(RR-7):11–20.

3. **Centers for Disease Control and Prevention.** 2000. Serogroup W-135 meningococcal disease among travelers returning from Saudi Arabia—United States, 2000. *Morb. Mortal. Wkly. Rep.* **49**:345–346.

4. **Galimand, M., G. Gerbaud, M. Guibourdenche, J. Y. Riou, and P. Courvalin.** 1998. High-level chloramphenicol resistance in *Neisseria meningitidis*. *N. Engl. J. Med.* **339**:868–874.

5. **Malley, R., S. H. Inkelis, P. Coelho, W. C. Huskins, and N. Kuppermann.** 1998. Cerebrospinal fluid pleocytosis and prognosis in invasive meningococcal disease in children. *Pediatr. Infect. Dis. J.* **17**:855–859.

6. **Pollard, A. J., and M. Levin.** 2000. Vaccines for prevention of meningococcal disease. *Pediatr. Infect. Dis. J.* **19**:333–344.

CASE 42

The patient was a 3½-week-old male who was born at term by cesarean section. At birth he had a left diaphragmatic hernia that was repaired soon thereafter. He required intubation at that time and continued to require respiratory support. Over a 24-hour period, the infant developed bulging anterior fontanelles, increased respiratory and heart rates, wide fluctuations in blood pressure, and difficulties maintaining adequate tissue perfusion, and his peripheral white blood cell (WBC) count increased from 6,300 to 13,700/μl. The child began to have focal seizures as well. A cerebrospinal fluid (CSF) examination showed 3,900 WBC/μl with 92% neutrophils, glucose level of 2 mg/dl, and protein level of 350 mg/dl. Gram stain of the child's CSF is shown in Fig. 1. The organism from the CSF is shown in Fig. 2.

1. What is your diagnosis for this patient? Is it consistent with his physical and laboratory findings? Explain.

2. What is the most likely organism causing his infection? What other organism has similar Gram stain and colonial morphology? What simple, rapid test would you use to distinguish these two organisms?

3. There are two forms of this infection in neonates. Compare and contrast these two forms. Which form does this patient have?

4. Beside infections in neonates, what other populations are at risk for invasive infection with this organism?

5. Describe the key virulence factor produced by the infecting organism and discuss its role in pathogenesis of infection.

6. Vaccines are currently under development for this organism. Describe the components that you would include in this vaccine. Who should receive this vaccine? Why would they receive it?

7. Since vaccines against the organism are not currently available, discuss strategies for prevention of neonatal infections with this organism. How effective have they been in preventing early-onset disease? How effective have they been in preventing late-onset disease?

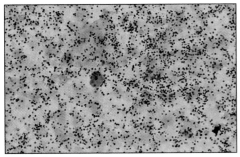

Figure 1

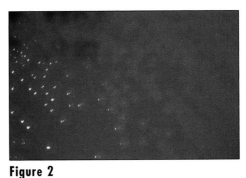

Figure 2

CASE DISCUSSION

1. This patient has bacterial meningitis. The physical finding of bulging anterior fontanelles is due to inflammation and increased intracranial pressure caused by infection. His increased respiratory and heart rates, fluctuations in blood pressure, and difficulties in maintaining adequate tissue perfusion are all signs of sepsis and are often seen in individuals with bacterial meningitis. Seizures are common in patients with meningitis. His CSF findings of 3,900 WBC/μl with a neutrophil predominance, low glucose, and high protein are all consistent with bacterial meningitis. The finding of gram-positive cocci (Fig. 1) in the microscopic examination of CSF is conclusive evidence of bacterial meningitis.

2. The bacteria which most frequently cause neonatal meningitis are the group B streptococci (GBS). The Gram stain and colonial morphology seen in Fig. 1 and 2 are consistent with GBS. However, *Listeria monocytogenes*, a much less frequent cause of neonatal infection, can be confused with GBS on Gram stain, even though *L. monocytogenes* is classified as a gram-positive coccobacillus. On sheep blood agar both organisms have very similar colonial morphology and both are weakly beta-hemolytic. The clinical disease these organisms cause is indistinguishable. Accurate identification is important when therapeutic choices are being made because cephalosporin therapy is not effective against *L. monocytogenes*. Accurate identification is also important in understanding the epidemiology of disease caused by these two organisms. Accurately distinguishing these organisms can be easily accomplished by the catalase test. Catalase is an enzyme which catalyzes the following reaction: $H_2O_2 \rightarrow 2H_2O + O_2$ (gas). Catalase activity can be detected by smearing the test organism on a glass slide and placing a drop of hydrogen peroxide (H_2O_2) on the smear. A bubbling reaction due to the release of O_2 occurs immediately if the organism produces catalase. *L. monocytogenes* is catalase positive; GBS is catalase negative. The patient's isolate was catalase negative; further testing including serogrouping confirmed it to be GBS.

3. Both GBS and *L. monocytogenes* can cause both early- and late-onset infections in neonates. In early-onset disease, the organism is spread vertically from mother to infant, as evidenced by the mother and child having both the same serotype and the same genotype. In the mother, the GBS or *L. monocytogenes* infection may manifest itself as a mild febrile illness, with the mother having "flu-like" symptoms, or she may be an asymptomatic carrier. The neonate is generally infected in utero or, in some instances, during passage through a colonized birth canal. Neonates with early-onset disease often are ill at birth or become infected in the first 3 days of life. They typically are low-birth-weight babies, often born to mothers who have not received prenatal care and have prolonged rupture of membranes. The major focus of infection is the lungs, which are poorly developed and poorly functioning in low-birth-weight babies. The children are often septic. Mortality in early-onset disease is 4 to 6%. Approximately 75% of GBS neonatal infections are early onset.

Late-onset disease usually occurs between 10 and 14 days of age but can be seen up to 3 months after birth. Late-onset disease is typically seen in full-term infants. The epidemiologic link between mother and child is more tenuous in late-onset disease, with only 50% of the children having the same serotype as that colonizing their mothers. A major focus of infection in late-onset disease is the central nervous system, with meningitis being more common than it is in early-onset disease. Mortality is not as high in late-onset disease, but it is still significant. Surviving infants may have neurologic sequelae, including blindness, hearing loss, and developmental and educational delay.

4. Given the great emphasis on preventing GBS in neonates, it is not surprising that 70% of invasive GBS infections occur outside of the neonatal period. Of those, about 90% occur in nonpregnant adults, with 6% in pregnant women and 3% in children from 3 months to 14 years of age. Diabetes mellitus is the most common risk factor for invasive GBS. Invasive GBS is 14 times more likely to occur in diabetics than in persons of the same age without diabetes. GBS bacteremia in diabetics is frequently secondary to cellulitis or foot ulcers. These ulcers are frequently seen as a result of peripheral vascular disease, which is common in diabetics. Other underlying conditions associated with invasive GBS are cardiovascular disease, solid cancers, alcoholism, and cirrhosis. The disease is much more common in those 65 years of age or older, and the mortality in that population is 15%, three times as high as the 5% seen in neonates. Along with early-onset disease, GBS infections during pregnancy can be responsible for septic abortion and stillbirth, as well as bacteremia, chorioamnionitis, and endometritis.

5. The key virulence factor of GBS is the capsule. There are multiple serotypes, with type III, the most commonly encountered serotype in neonates, being responsible for 36% of early-onset and 71% of late-onset disease. Other serotypes commonly seen in human disease include Ia, Ib, II, and V. Type V is an important cause of invasive disease in adults and is responsible for approximately one-third of those cases. Like *Haemophilus influenzae* type b capsule, GBS capsule inhibits complement-mediated phagocytosis, allowing the organism to evade the immune system. Capsule-specific antibodies reverse this inhibition.

6. Currently, the GBS vaccine is experimental. The most important component of any vaccine against GBS would be capsular polysaccharide. Because cross-immunity is not conferred among GBS serotypes, a vaccine should contain capsular polysaccharide from the common GBS serotypes (see answer to question 5). Polysaccharides are

T-cell-independent antigens. Primary and even secondary immune response may be poor in some individuals. Coupling the polysaccharides to a protein carrier molecule produces an antigen that elicits a T-cell-dependent immune response, resulting in a more predictable and protective immune response to the polysaccharide antigen. Tetanus toxoid coupled with different capsular polysaccharide serotypes has been shown to elicit a protective response in animals and in humans.

The target population in which disease must be prevented is the fetus/neonate. This population cannot be effectively vaccinated. However, the mother, who can pass protective immunoglobulin G (IgG) transplacentally, can be vaccinated. Animal studies have shown protection in neonates challenged with GBS whose mothers have been previously vaccinated.

The conjugate vaccine might also be useful in diabetics, who are at a much greater risk for infections than the general population.

7. Until GBS vaccines are proven safe and efficacious, prevention of GBS infection in the neonatal period is dependent upon the use of prophylactic antibiotics. Currently there is agreement that intrapartum intravenous administration of penicillins, or clindamycin or erythromycin in penicillin-allergic mothers, is a successful prophylactic strategy for GBS neonatal infection.

In 1996, the American Academy of Pediatrics, the American College of Obstetricians, and the Centers for Disease Control and Prevention jointly issued guidelines as to which pregnant women should receive intrapartum antibiotics. Between 15 and 35% of pregnant women are colonized with GBS in the genitourinary tract. Therefore it is not necessary to give intrapartum antibiotics to all women during labor. It is recommended that women be screened by culture at 35 to 37 weeks of gestation for the presence of GBS in their vagina and rectum. If positive, then they should be offered the choice of intrapartum antibiotics. Women who have had a previous child with GBS neonatal infection or have had GBS bacteriuria during pregnancy should be given intrapartum antimicrobial agents. In addition, women without prenatal care who deliver at less than 37 weeks of gestation, have intrapartum fever of ≥38°C, or have rupture of membranes for longer than 18 hours should also receive intrapartum antibiotics.

The result of this strategy for preventing early-onset GBS disease has been impressive. Since these recommendations, there has been a 65% decline in the number of cases of early-onset GBS disease in the United States. This result has been seen despite the fact that there is not yet universal compliance with these guidelines as well as 20% resistance to erythromycin and 15% resistance to clindamycin among GBS isolates. GBS remains universally susceptible to penicillin. There is concern, however, as the rate of early-onset sepsis due to *Escherichia coli*, especially isolates resistant to ampicillin, appears to have increased among very-low-birthweight infants. This may be due to an increase in the use of antibiotics as a result of these guidelines. Intrapartum antibiotic therapy has not had any impact on the incidence of late-onset GBS disease.

REFERENCES

1. **Baker, C. J., L. C. Paoletti, M. A. Wessels, H.-K. Guttormsen, M. A. Rench, M. E. Hickman, and D. L. Kasper.** 1999. Safety and immunogenicity of capsular polysaccharide-tetanus toxoid conjugate vaccines for group B streptococcal types 1a and 1b. *J. Infect. Dis.* **179:**142–150.

2. **Centers for Disease Control and Prevention.** 2000. Early onset group B streptococcal disease—United States, 1998–1999. *Morb. Mortal. Wkly. Rep.* **49:**793–796.

3. **Schrag, S. J., S. Zywicki, M. M. Farley, A. L. Reingold, L. H. Harrison, L. B. Lefkowitz, J. L. Hadler, R. Danila, P. R. Cieslak, and A. Schuhat.** 2000. Group B streptococcal disease in the era of intrapartum antibiotic prophylaxis. *N. Engl. J. Med.* **342:**15–20.

4. **Schuhat, A.** 1998. Epidemiology of group B streptococcal disease in the United States: shifting paradigms. *Clin. Microbiol. Rev.* **11:**497–513.

5. **Stoll, B. J., N. Hansen, A. A. Fanaroff, L. L. Wright, W. A. Carlo, R. A. Ehrenkranz, J. A. Lemons, E. F. Donovan, A. R. Stark, J. E. Tyson, W. Oh, C. R. Bauer, S. B. Korones, S. Shankaran, A. R. Laptook, D. K. Stevenson, L.-A. Papile, and W. K. Poole.** 2002. Changes in pathogens causing early-onset sepsis in very-low-birth-weight infants. *N. Engl. J. Med.* **347:**240–247.

CASE 43

The patient was a 3½-month-old male who presented in August with a 2-week history of diarrhea which abated with oral rehydration. One week later, he developed a fever with a temperature of 39.2°C and respiratory symptoms. He was found to have some wheezing and right otitis media. He was treated with Pediazole for presumed *Chlamydia* infection. He continued to have fever and developed irritability and vomiting. He returned to the clinic and was admitted.

On physical examination he was irritable and had a temperature of 36.6°C. He had tachycardia with a pulse of 180 beats/min. His blood pressure was normal. His fontanelles were normal. His neck was supple. His tympanic membranes were dull and distorted bilaterally. The rest of his examination was unremarkable. Laboratory tests showed anemia with a hemoglobin level of 10.4 g/dl; the white blood cell (WBC) count was 9,300/µl with 60% lymphocytes. Electrolyte levels were normal. A lumbar puncture was done, and the cerebrospinal fluid (CSF) revealed a WBC count of 75/µl with 72% neutrophils, 8% lymphocytes, and 20% monocytes; the glucose level was 60 mg/dl, and the protein level was 22 mg/dl (both normal). A Gram stain was negative for bacteria, and few polymorphonuclear leukocytes (PMNs) were present.

CSF samples were sent for viral cultures. Intravenous ceftriaxone and ampicillin were begun empirically for presumed bacterial meningitis. One day later his anterior fontanelle was full. A head computed tomogram (CT) scan was normal. On the second hospital day his condition had improved and his anterior fontanelle was less full. Blood, urine, and CSF bacterial cultures were negative. He was discharged on the fourth hospital day to complete a 10-day course of intramuscular ceftriaxone on an outpatient basis.

After discharge, his CSF viral culture became positive.

1. Does this patient have meningitis? Explain your answer.

2. Which type of virus is most likely to be causing this infection?

3. Describe the transmission and pathogenesis of infection with viruses of this group.

4. Give some examples of specific clinical syndromes associated with particular virus serotypes in this group.

5. Describe the treatment and prevention of these viral infections.

6. What issues will be of importance as the campaign to eradicate a viral infection found in this genus nears success?

7. This child received a 10-day course of antimicrobial agents even though he had a viral infection. How might this inappropriate use of antimicrobial agents be either shortened or eliminated?

CASE DISCUSSION

CASE 43

1. The finding of >3 white blood cells per µl in CSF is abnormal and may be indicative of meningitis. The patient was treated for bacterial meningitis because of the predominance of neutrophils in the CSF. Early in the course of viral meningitis (the first 24 to 48 hours), neutrophils can be the predominant cell type. However, mononuclear cells are predominant later in the disease. The normal CSF glucose level also argues against bacterial meningitis. This CSF picture may also be seen in mycobacterial and fungal meningitis, both of which would be very unusual in an immunocompetent infant. Finally, a positive CSF viral culture confirms the diagnosis of viral meningitis.

2. Viruses of the enterovirus genus (coxsackieviruses, echoviruses, enteroviruses, and polioviruses) characteristically cause aseptic meningitis during late summer and early fall. Because of poliovirus vaccine usage, it is most likely that this patient had infection due to either a coxsackievirus or an echovirus. These single-stranded RNA viruses are all members of the *Picornaviridae.* This patient's viral culture was positive for a coxsackievirus.

3. Enteroviruses are typically transmitted by the fecal-oral route, but some serotypes (notably coxsackievirus A24 and enterovirus 70) may be spread by respiratory secretions or fomites. Most enteroviral infections are asymptomatic.

Enteroviruses are acid stable; after replication in the oropharynx they can survive transit through the stomach. Replication then occurs in the lymphoid tissue of the gastrointestinal tract, after which the viruses can pass to the bloodstream. Spread to the liver, spleen, lymph nodes, and specific target organs (meninges, heart, and skin) can result. The virus can then replicate in those organs, resulting in clinical disease.

4. Many enteroviruses can cause aseptic meningitis. The incidence is highest in children less than 1 year old, such as this patient. Myocarditis, pericarditis, and pleurodynia are associated with group B coxsackievirus. Polioviruses most commonly cause an inapparent infection but may cause nonparalytic poliomyelitis, which is clinically similar to aseptic meningitis, or paralytic poliomyelitis. In paralytic poliomyelitis, flaccid paralysis occurs following a prodromal illness. Herpangina (vesicular oral ulcers) and hand-foot-and-mouth disease are classically caused by group A coxsackieviruses, although hand-foot-and-mouth disease may also be caused by enterovirus 71. A variety of coxsackieviruses and echoviruses have been associated with benign viral exanthems, mimicking measles or rubella.

5. The management of patients with enteroviral infections includes supportive care. Infection control measures are recommended for patients and their families to interrupt transmission of virus to others who may be susceptible. These measures include hand washing. This is of particular importance in day care centers, where agents that are spread by the fecal-oral route (e.g., hepatitis A virus, *Giardia lamblia*, *Cryptosporidium* spp., *Shigella* spp.) are of particular concern.

   Vaccines, both inactivated and live, attenuated preparations, are available for the prevention of poliovirus infections. These are made from poliovirus-1, poliovirus-2, and poliovirus-3. In immunocompetent individuals, the live, attenuated vaccine developed by Sabin was the vaccine of choice in the United States until relatively recently *OPV* and remains in use in many other countries. The attenuated virus in this vaccine replicates in the gastrointestinal tract and can be shed in feces. Its efficacy is greater than 95%. However, this vaccine has caused clinical disease in both immunocompetent and immunocompromised individuals. Vaccine-associated paralytic poliomyelitis is rare, occurring once for every 2.5 million doses of oral (live attenuated) vaccine administered. The greatest risk of vaccine-associated disease is in immunocompromised children or in contacts of recently vaccinated children who have waning immunity and are exposed to the virus via these children. Since 1979, the only indigenous cases of polio reported in the United States have been associated with the use of the live oral polio vaccine. Because of this, the Advisory Committee on Immunization Practices now recommends exclusive use of inactivated poliovirus vaccine for routine childhood polio vaccination in the United States.

6. As the worldwide eradication of natural poliovirus infection nears, it will be important to define what strains, if any, should be stored in reference laboratories. This has been a contentious issue for smallpox virus, which is a potential agent of bioterrorism and biowarfare. Other issues relating to poliovirus as global eradication approaches include the change from the live attenuated oral polio vaccine to the inactivated polio vaccine, as has been discussed above for the United States. One of the concerns with the use of the oral live vaccine is the possibility of the continued circulation of neurovirulent strains derived from the vaccine strains. Another issue is the need for continued vigilance as cases are sought. Cases of acute flaccid paralysis are investigated as potential markers of undiagnosed poliomyelitis. It will be essential to continue this practice until the global eradication of poliovirus infection has been confirmed.

7. Because of the high mortality and morbidity associated with bacterial meningitis, the physician caring for this child decided to treat him with a full course of an antibacterial agent. This decision was based on evidence of bacterial meningitis (abnormal cell count and some clinical evidence of meningitis) and the lack of evidence for the diagnosis of the more benign and self-limited viral meningitis due to the enteroviruses. The reason this diagnosis could not be established more quickly is that enteroviruses grow slowly in tissue culture and certain types may not grow at all.

Molecular diagnostic techniques such as polymerase chain reaction (PCR) show promise of providing the diagnosis of enteroviral meningitis via detection of the virus in CSF within 12 to 24 hours. Currently CSF culture for enteroviruses may take 7 to 10 days to become positive. PCR at present is not widely available for the detection of enteroviruses, and it is expensive. Nevertheless it holds the promise of being a valuable tool in the diagnosis of enteroviral meningitis, and a number of published reports suggest that the use of PCR will help to reduce or eliminate the inappropriate use of antibacterial agents in this clinical setting and to shorten the length of hospitalization.

REFERENCES

1. **Centers for Disease Control and Prevention.** 2000. Poliomyelitis prevention in the United States. Updated recommendations of the Advisory Committee on Immunization Practices (ACIP). *Morb. Mortal. Wkly. Rep.* **49**(RR-5):1–22.

2. **Centers for Disease Control and Prevention.** 1998. Deaths among children during an outbreak of hand, foot, and mouth disease—Taiwan, Republic of China, April–July 1998. *Morb. Mortal. Wkly. Rep.* **47**:629–632.

3. **Centers for Disease Control and Prevention.** 2000. Enterovirus surveillance— United States, 1997–1999. *Morb. Mortal. Wkly. Rep.* **49**:913–916.

4. **Sawyer, M. H.** 1999. Enterovirus infections: diagnosis and treatment. *Pediatr. Infect. Dis. J.* **18**:1033–1039.

CASE

44

The patient was a 38-year-old HIV-positive male with a CD4 count of 80/μl. The patient had a 1-week history of progressively worsening headache, photophobia, lethargy, and fevers to 38.5°C. On the morning of his admission, he became confused, disoriented, and ataxic, having fallen three times. On physical examination, he was lethargic and could only answer a few questions before falling asleep. His vital signs were all within normal limits. Chest examination and radiograph were normal. He had a head computed tomogram (CT) scan which was also normal. Because of his declining mental status and history of headache and photophobia, a lumbar puncture was done. The cerebrospinal fluid (CSF) revealed 32 white blood cells per μl with 89% lymphocytes and 6% monocytes, a glucose level of 22 mg/dl, and a protein level of 89 mg/dl. Gram stain of his CSF is shown in Fig. 1. Serum and CSF tests for the presence of a specific antigen were positive. The organism that was recovered from his CSF and blood is shown growing on a sheep blood agar plate in Fig. 2.

↓ glu
↑ pro
Bacterial

1. What is the organism most likely to be causing his illness? Are his CSF parameters (cell count and chemistries) consistent with infection with this organism? What other organisms are frequently seen causing central nervous system infections in this patient population?

2. What virulence factor does this organism produce, and what is its role in the pathogenesis of this disease?

3. What is the specific antigen that was found in his serum and CSF? Explain two different ways this antigen test is used in managing HIV patients. What other organism will give a positive reaction in this test?

4. How did this patient become infected? Beginning in 1996, what changes have occurred in the epidemiology of infection with this organism? Why have these changes occurred?

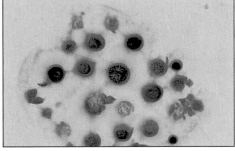

Figure 1

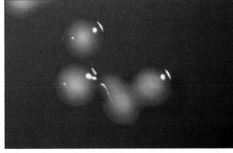

Figure 2

5. Three months later, the patient again presented with symptoms consistent with his initial illness. A CSF Gram stain obtained at this time is seen in Fig. 3. His CSF antigen titer on his first admission had been 1:100,000. It had dropped to 1:200 after therapy and on his latest admission was 1:1,600. How do you interpret his CSF antigen titers?

6. How should this patient be managed to prevent future infections with this organism?

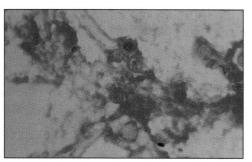

Figure 3

CASE DISCUSSION

CASE 44

1. The Gram stain reveals round yeast cells which are gram variable. The most common such agent causing central nervous system infection in HIV-infected individuals with CD4 counts of <200/μl is *Cryptococcus neoformans*. Both the Gram stain and the colonial morphology seen in Fig. 1 and 2 are consistent with this organism. Further biochemical characterization, including a rapid urease test, confirmed this organism as *C. neoformans*.

CSF cell counts of <100/μl are commonly seen in HIV-infected patients with cryptococcal meningitis. CSF glucose levels are frequently normal but may be low as was seen here. CSF protein levels are also frequently normal but may be elevated as was seen in this patient.

Toxoplasma gondii and HIV both are common causes of central nervous system infection in HIV-infected patients. Both cause an encephalopathic picture. A variety of other agents must be considered when *C. neoformans* and *T. gondii* have been ruled out in these patients. Some of the more important of these include *Mycobacterium tuberculosis*, *Nocardia* spp., *Treponema pallidum* (neurosyphilis), the herpes viruses (cytomegalovirus and herpes simplex virus), dimorphic fungi including *Histoplasma capsulatum* and *Coccidioides immitis*, and encapsulated bacteria.

2. On Gram stain, it is often possible for the skilled observer to detect a capsule surrounding the yeast. The capsular material often stains gram negative while the yeast cell stains gram positive. This explains the gram-variable appearance of this organism. Alternatively, the capsule can be nicely demonstrated using India ink, a negative staining technique. With this method, the yeast cell is seen in a dark background surrounded by a clear "halo." This halo is due to the inability of the ink to penetrate the capsule surrounding the cell.

The cryptococcal capsule is antiphagocytic. In animal studies, the capsular polysaccharide has been shown to inhibit cell-mediated immunity as well. Finally, soluble cryptococcal capsular polysaccharide has been shown to activate the alternative complement pathway. In patients with a high concentration of circulating capsular polysaccharide, this may result in depletion of complement, causing inefficient opsonization and reduced phagocytosis of this organism.

3. The specific antigen that was found in this patient's CSF and serum was the capsular polysaccharide of *C. neoformans*. Capsular polysaccharide is detected by reacting these body fluids with latex particles coated ("sensitized") with antibodies specific for this antigen. Alternatively, enzyme-linked immunosorbent assays have been developed for the quantitative detection of this antigen. Most laboratories use latex agglutination as their method to detect cryptococcal antigen.

These cryptococcal antigen detection methods can be used in two ways. One is diagnostically. An HIV-infected patient who presents with central nervous system symptoms such as headache and lethargy and has cryptococcal antigen detectable in his or her serum is at high risk for having cryptococcal meningitis. In HIV-infected patients with culture-proven cryptococcal meningitis, >99% have a positive cryptococcal antigen test. The second way in which the antigen test can be used is to follow response to therapy. Cryptococcal antigen can be quantitated in both serum and CSF by serially diluting these body fluids and determining the highest dilution (most dilute) which produces a positive agglutination reaction. This test can be used to follow the patient, with titers falling with successful therapy. Increases in titer may herald relapse (see answer to question 5 for further details). Patients with sepsis due to *Trichosporon beigelii* and *Capnocytophaga canimorsus* can be falsely positive for cryptococcal antigen.

4. Cryptococcal meningitis usually begins as an asymptomatic pulmonary infection. The patient is infected with *Cryptococcus* by inhaling it from the environment. The organism's natural habitat is soil; it grows particularly well in pigeon droppings and other bird guano. Areas with large pigeon populations, such as urban parks, are places where exposure to this organism may be increased. Asymptomatic pulmonary infection may progress to fungemia and meningitis in the individual with defects in cell-mediated immunity, especially HIV-positive patients, the population in whom this infection is most frequently seen.

The widespread use of highly active antiretroviral therapy (HAART) began in the United States in 1995. HAART is a combination of antiretroviral drugs that usually includes a protease inhibitor. The effect of this therapy is to preserve immune function in some HIV-infected patients and to reconstitute it in others. The end result has been a reduction of opportunistic infections and a significant decrease in morbidity and mortality in HIV-infected patients. We saw a marked decline in the number of cases of cryptococcal meningitis in our patient population at the University of North Carolina Hospitals beginning in the last quarter of 1994, a time period when many of our patients had begun to receive HAART in clinical trials. Most HIV patients who develop cryptococcal meningitis today are patients who have undiagnosed HIV infections. They typically have CD4 counts <200/µl and often have CD4 counts of <100/µl.

5. In an HIV-infected patient who has been treated for cryptococcosis with a corresponding drop in CSF or serum antigen titers, a rising cryptococcal antigen level coupled with clinical symptoms may herald a relapse of his cryptococcal infection. In addition to being used to determine cryptococcal antigen levels, the CSF specimen should be cultured. Microscopic examination of CSF may be misleading in diagnosing active infection since nonviable yeast cells can remain visible in the CSF for weeks to months. Recent studies of cryptococcal meningitis relapse have shown that in most patients, the same genotype causes the initial infection and the relapse. In some patients, the genotype causing relapse is different from the genotype causing the initial infection. This second observation can be explained in two ways. One, the patient did not relapse but has developed a "new" infection. Since most patients do not make antibodies to cryptococcal antigens, reinfection with a different genotype of *C. neoformans* would not be surprising. Alternatively, it has been shown that patients may be infected with multiple genotypes. In the initial infection one genotype could be predominant while a second genotype could be predominant in the relapse.

6. Guidelines have been developed for the prevention of relapse of cryptococcal meningitis. The strategy is twofold. First, HAART should be used in HIV-infected patients. HIV-infected patients who develop cryptococcal meningitis have, as a rule, CD4 counts of <200/µl, and the risk of developing initial infection or relapse increases as the CD4 count declines below 200. HAART has been shown to successfully reconstitute the immune response in many individuals, as evidenced by CD4 counts rising above 200/µl. In addition to attempts to reconstitute the immune system or prevent its decline, HIV-infected patients who develop cryptococcal meningitis receive antifungal prophylaxis for life. Oral fluconazole is the drug of choice for this purpose. There are three major problems with lifelong prophylactic antimicrobial therapy. First, fluconazole is very expensive, although by preventing relapses it could be argued that this therapy is cost-effective. Second, compliance is an issue, especially when the patients must take very complex drug regimens including multiple antiretroviral agents. Most relapses of cryptococcal meningitis are due to failure to comply with either HAART, fluconazole prophylaxis, or both. Finally, development of drug resistance must also be considered as a potential problem. Fluconazole-resistant strains have been recovered from HIV patients who received fluconazole prophylactically. Although fluconazole resistance in this patient population is currently rare, it is likely to increase.

REFERENCES
1. **Buchanan, K. L., and J. W. Murphy.** 1998. What makes *Cryptococcus neoformans* a pathogen? *Emerg. Infect. Dis.* **4:**71–83.

2. **Mondon, P., R. Petter, G. Amalfitano, R. Luzzati, E. Concia, I. Polacheck, and K. J. Kwon-Chung.** 1999. Heteroresistance to fluconazole and voriconazole in *Cryptococcus neoformans. Antimicrob. Agents Chemother.* **43:**1856–1861.

3. **Palella, F. J., Jr., K. M. Delaney, A. C. Moorman, M. O. Loveless, J. Fuhrer, G. A. Satten, D. J. Aschman, S. D. Holmberg, and the HIV Outpatient Study Investigators.** 1998. Declining morbidity and mortality among patients with advanced human immunodeficiency virus infection. *N. Engl. J. Med.* **338:**853–860.

4. **Powderly, W. G.** 2000. Prophylaxis for opportunistic infections in an era of effective antiretroviral therapy. *Clin. Infect. Dis.* **31:**597–601.

5. **Saag, M. S., R. J. Graybill, R. A. Larsen, P. G. Pappas, J. R. Perfect, W. G. Powderly, J. D. Sobel, and W. E. Dismukes, for the Mycoses Study Group Cryptococcal Subgroup.** 2000. Practice guidelines for the management of cryptococcal disease. *Clin. Infect. Dis.* **30:**710–718.

6. **Sullivan, D., K. Haynes, G. Moran, D. Shanley, and D. Coleman.** 1996. Persistence, replacement, and microevolution of *Cryptococcus neoformans* strains in recurrent meningitis in AIDS patients. *J. Clin. Microbiol.* **34:**1739–1744.

CASE 45

The patient was a 21-year-old migrant farmworker who was 27 weeks pregnant. She presented with complaints of fever, headache, chills, frequency and urgency of urination, decreased appetite, and a 1-day history of diarrhea and decreased fetal movement. On physical examination, she had a temperature of 38.3°C, abdominal tenderness, and tachycardia. Her chest was clear on auscultation, and no cervical discharge or tenderness was seen on pelvic examination. Her laboratory studies were significant for a white blood cell count of 21,300/μl. A cervical specimen assayed for *Chlamydia trachomatis* using the ligase chain reaction was negative. Cervical culture for *Neisseria gonorrhoeae* was not done. Two blood cultures were drawn, and the patient was begun on ampicillin-sulbactam and gentamicin.

The next morning the patient complained of right costovertebral tenderness and abdominal pain. On ultrasound, there was no fetal movement, and intrauterine fetal demise was suspected. Labor was induced, and a stillborn infant was delivered vaginally. Cultures of blood, placenta, and umbilical cord all grew the organism seen in Fig. 1 and 2.

1. What organisms do you think are likely infecting this patient? If you learned that the organism was catalase positive, how would that help you decide what organism was infecting this patient?

2. What is the significance of headache in this patient's history? What is the natural history of this disease in pregnancy?

3. What other patient populations are at risk for infection with this organism?

4. How is this organism spread? What special characteristic of this organism may be important in its spread?

5. This organism is classified as a facultative intracellular organism. Briefly describe how this organism is able to evade the host's immune system and survive intracellularly?

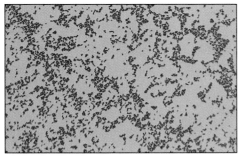

Figure 1

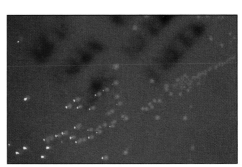

Figure 2

CASE

45 CASE DISCUSSION

1. The Gram stain and colonial morphology of the organism infecting the mother and her child are consistent with two organisms that are frequently associated with fetal demise, group B streptococci (GBS) and *Listeria monocytogenes*. On Gram stain, the organisms, for all practical purposes, are indistinguishable even though textbooks indicate that they are morphologically distinct, the GBS appearing as cocci and *L. monocytogenes* as bacilli. In clinical specimens, GBS frequently appear as gram-positive diplococci while *L. monocytogenes* appear as gram-positive, diphtheroid-like bacilli. These organisms can appear so much alike on Gram stain that even a skilled microscopist may not be able to distinguish them. The colonial morphology of these organisms on a 5% sheep blood agar plate is identical, with both organisms producing grayish white colonies surrounded by a narrow zone of beta-hemolysis (see Fig. 2).

Although the organisms have many similar characteristics, they can be easily distinguished in the laboratory based on the catalase test: *L. monocytogenes* is catalase positive while GBS are negative. It is important to identify these organisms accurately for both therapeutic and epidemiological reasons. (See case 42 for further discussion.) The disease course is similar for these two organisms, with the highest morbidity and mortality seen in low-birth-weight children who have poorly developed lungs. The incidence of neonatal infection with GBS is much higher than that for *L. monocytogenes*. Because early-onset disease is quite rare with *L. monocytogenes*, there is no organized attempt to screen women for this organism prenatally at 35 to 37 weeks of gestation as there is for GBS. There also are no available clinical trial data that show that intrapartum antibiotic therapy prevents early-onset disease with *L. monocytogenes*.

2. A prominent feature of the clinical spectrum of *L. monocytogenes* disease is meningitis. Headache is an important clinical symptom of meningitis, as it is of many clinical syndromes. Meningitis due to *Listeria* is almost always seen in neonates or severely immunocompromised patients. The headache seen in the mother is part of the flu-like illness which is commonly seen in pregnant women who are bacteremic with *L. monocytogenes*. The disease is usually self-limiting in pregnant women but can have devastating consequences for their unborn children.

The organism, which can be vertically transmitted, causes chorioamnionitis, resulting in septic abortion, stillbirth (as was seen in this case), premature birth, and serious infections in the early neonatal period (0 to 3 days), including pneumonia, bacteremia, and meningitis.

Like GBS, *L. monocytogenes* can also cause late-onset disease in infants. Late-onset neonatal disease occurs after the first week of life up to the third month. These children tend to have been born at term, and meningitis is more common in late-onset disease.

3. Besides pregnant women and their unborn children, serious *Listeria* infections are usually limited to immunocompromised individuals. In particular, organ transplant recipients, patients with AIDS or malignancy (especially lymphoma and chronic lymphocytic leukemia), and those receiving corticosteroids are at high risk for developing serious infections with this organism. Meningitis is a prominent feature of the disease spectrum in these patients. Mortality rates with systemic listeriosis are estimated to be 20%, with most fatalities in newborns and the immunocompromised.

4. Listeriosis is clearly a food-borne infection. Dairy products (particularly soft cheeses), undercooked chicken, and prepared meats such as hot dogs and cold cuts have all been implicated as vehicles for the transmission of *Listeria*. One striking example of the ability of *Listeria* to cause food-borne outbreaks occurred in the Latino community in Los Angeles in 1985. In this outbreak, it was noted that there was a high rate of fetal and neonatal infections and death due to *Listeria* infection in Latinos that was not occurring in other populations in the city. Through careful epidemiologic study, it was learned that a soft cheese referred to as "Mexican-style cheese," which was sold and consumed primarily in the Latino community, was contaminated with *L. monocytogenes*. Further studies showed that this cheese was frequently contaminated with unpasteurized milk, and this unpasteurized milk was the ultimate source of the *Listeria*.

Outbreaks of *Listeria* infection due to contaminated foods may not be readily recognized because small numbers of cases (less than 50) may be spread throughout the United States. In addition, the median incubation period of invasive disease (bacteremia, meningitis) with this organism is 3 weeks. By the time cases are recognized, the potential food source may no longer be available for culture. In addition, implicated foods such as hot dogs or cold cuts may be distributed from a central location under many different brand names to locations throughout the country, making the connection between a particular food and a specific supplier more difficult. Surveillance programs that monitor food-borne diseases, such as the CDC-sponsored FoodNet, play an important role in recognizing these outbreaks and in limiting their spread by effecting producer recalls of tainted products.

The reason why *L. monocytogenes* can be transmitted by foods that are almost always refrigerated before consumption is due to the fact that this organism can grow at 4°C. This is an unusual characteristic for an organism that causes human disease. Most human pathogens grow in a temperature range between 20 and 37°C. Another human pathogen, *Yersinia enterocolitica*, can also grow at 4°C. Not surprisingly, the vehicle of transmission for outbreaks of disease due to this organism has also been dairy products and processed meats.

L. monocytogenes is one of the leading causes of serious food-borne illnesses. It is estimated that there are as many as 500 deaths annually in the United States that are due to this organism. Recently, *L. monocytogenes* food-borne outbreaks of febrile gastroenteritis in immunocompetent individuals have been described. This disease entity appears to be different from invasive disease seen in immunosuppressed and pregnant women. The incubation period is short (12 to 26 hours) and gastroenteritis symptoms are prominent.

5. *L. monocytogenes* produces a cell surface virulence factor called internalins which allows it to invade epithelial cells. After invasion, the organism is found in a vacuole within the host cell. The organism produces a second virulence factor, listeriolysin O. Listeriolysin O lyses the vacuole, allowing the organism to enter the cytoplasm of the host cell where it can multiply. There, a third virulence factor, the cell surface protein ActA, mediates the binding of host cell actin to one end of the bacteria. This actin "tail" propels the bacterium to the periphery of the cell where the bacterium induces the formation of protrusions called filopodia. These filopodia are taken up by adjacent cells where this infective process can begin anew. Because the bacterial cells are never exposed to the extracellular environment, humoral immunity plays little if any role in the immune response to this organism. Rather, cell-mediated immunity is central to control of infection with this organism, explaining why individuals with defects in cell-mediated immunity, such as transplant recipients and AIDS patients, have increased risk of invasive disease with this organism.

REFERENCES

1. **Aureli, P., G. C. Fiorucci, D. Caroli, G. Marchiaro, O. Movara, L. Leone, and S. Salmaso.** 2000. An outbreak of febrile gastroenteritis associated with corn contaminated with *Listeria monocytogenes*. *N. Engl. J. Med.* **342:**1236–1241.

2. **Finlay, B. B., and S. Falkow.** 1997. Common themes and molecular pathogenicity revisited. *Microbiol. Mol. Biol. Rev.* **61:**136–169.

3. **Linnan, M. J., L. Mascola, X. D. Lou, V. Goulet, S. May, C. Salminen, D. W. Hird, L. Yonekura, P. Hayes, R. Weaver, A. Audurier, B. D. Plikaytis, S. L. Fannin, A. Kleks, and C. V. Broome.** 1988. Epidemic listeriosis associated with Mexican-style cheese. *N. Engl. J. Med.* **319:**823–828.

4. **Mead, P. S., L. Slutsker, V. Dietz, L. F. McCraig, J. S. Breese, C. Shapiro, P. M. Griffin, and R. V. Tauxe.** 1999. Food-related illness and death in the United States. *Emerg. Infect. Dis.* **5:**607–625.

5. **Southwick, F. S., and D. L. Purich.** 1996. Intracellular pathogenesis of listeriosis. *N. Engl. J. Med.* **334:**770–776.

C A S E

46

This 29-year-old man, who was originally from the Cape Verde Islands, was in his usual state of good health until the day of admission, when his family noted that he could not speak and could not move the right side of his body. He was taken to the hospital emergency department. Neurologic examination was notable for aphasia. On ophthalmologic exam, the optic nerves appeared normal. The patient also complained of hearing voices. His recent medical history was notable for negative HIV antibody test results. A magnetic resonance imaging (MRI) scan was performed; it was remarkable for the presence of a right frontal ring-enhancing lesion (Fig. 1), a 1-cm ring-enhancing lesion with focal edema in the left temporal lobe, and multiple small, thin-walled lesions less than 5 mm in diameter (not shown here).

A lumbar puncture was performed. The opening pressure was within normal limits and the cerebrospinal fluid (CSF) demonstrated no cells, a normal glucose level, and a normal protein level.

1. What is the differential diagnosis of this patient's intracranial process?

2. Serum and CSF were sent for serologic studies to the Centers for Disease Control and Prevention (CDC), where the clinical diagnosis was confirmed serologically. What is the likely diagnosis?

3. Are there any additional tests that might help to establish this diagnosis while the serologic studies are pending?

4. Which parasite causes this infection?

5. How do people become infected with this parasite? What in this patient's history indicates that he is at increased risk for this infection?

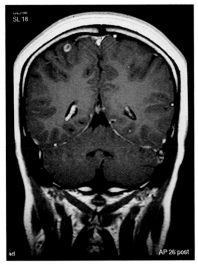

Figure 1

CASE DISCUSSION

CASE 46

1 and 2. This patient's signs and symptoms are consistent with a seizure and postictal neurologic abnormalities. Hearing voices can be explained by the temporal lobe location of the lesion. The presence of multiple intracranial lesions is consistent with a noninfectious process, such as cancer with metastases to the brain, as well as with several infectious processes. Multiple brain lesions would be consistent with central nervous system toxoplasmosis in an immunocompromised host. This patient had a negative serology for HIV and had no other clinical history suggesting immunosuppression, so toxoplasmosis was less likely. Central nervous system tuberculosis occasionally presents with tuberculomas in the brain, and cannot be ruled out on the basis of the information provided in the history. A negative purified protein derivative (PPD) skin test would be helpful in decreasing the probability of the diagnosis of tuberculosis. Multiple brain abscesses are another possibility, but none of these possibilities would adequately explain the calcifications seen on radiologic examination (see the answer to question 3). As part of this patient's evaluation, serologic studies were performed by the CDC, which confirmed the diagnosis as cysticercosis, a parasitic infection due to the larval form of the tapeworm *Taenia solium*. Both serum and CSF should be sent for serologic studies in the setting of suspected cerebral cysticercosis.

3. An X-ray study of the soft tissues will frequently demonstrate "rice grain" calcifications (Fig. 2) in skeletal muscle that are essentially diagnostic of cysticercosis. An additional test that is of low yield but suggestive of the diagnosis of cysticercosis is the presence of eosinophils in the CSF, a finding that can also occur in a number of other conditions. This patient, however, did not have any white blood cells in his CSF, so there would not be any CSF eosinophils.

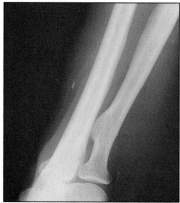

Figure 2

4. Infection with the larval forms of the parasite *T. solium*, the pork tapeworm, causes cysticercosis. Infection with the beef tapeworm, *Taenia saginata*, does not cause cysticercosis in humans.

5. Humans can be infected in three ways with the pork tapeworm. They can be infected through ingestion of undercooked pork containing cysticercus larvae or by ingestion of food or water contaminated with human feces containing *T. solium* eggs, or they can be autoinfected with eggs after defecation and failure to properly clean their hands. Following the ingestion of contaminated undercooked pork, cysticercus larvae are liberated in the stomach. The larval form migrates to the small intestine and over a period of months develops into the adult tapeworm. The worms reproduce there, and eggs are excreted in the feces. When *T. solium* eggs are ingested, either by autoinfection from the hands of infected individuals or in fecally contaminated food or water, gastric acid and pancreatic enzymes cause the release of oncospheres (motile larvae) that penetrate the intestinal wall and are disseminated in the blood, from which they can encyst in a variety of tissues including the brain, eyes, and skeletal muscle. In this form of the infection, the parasite's life cycle cannot be completed and the larval form never develops into the adult worm. These larval forms eventually die, and calcification occurs around the degenerating larval form. The formation of tissue cysts only occurs following ingestion of the eggs and does not follow ingestion of cysts. Cysticercosis is common in areas of Mexico and other Latin American countries with poor sanitation and in many other parts of the world, such as the Cape Verde islands. It is a major cause of adult-onset seizure disorders in these areas. This patient's geographic location and his adult-onset seizure disorder are significant factors to consider in the differential diagnosis.

Although this patient had edema, which is indicative of inflammation, the survival of cysts within the central nervous system may occur without clinical manifestations if there is no inflammation around the parasites. In such settings, the cysts are not calcified but can be demonstrated with magnetic resonance imaging (MRI) scans. Ideally, a determination should be made whether a patient has inactive infection, which does not merit antiparasitic therapy, or active disease, in which case antiparasitic therapy may be indicated.

REFERENCES

1. **García, H. H., O. H. Del Brutto, and the Cysticercosis Working Group in Peru.** 1999. Heavy nonencephalitic cerebral cysticercosis in tapeworm carriers. *Neurology* **53:**1582–1584.

2. **García, H. H., and O. H. Del Brutto.** 2000. *Taenia solium* cysticercosis. *Infect. Dis. Clin. North Am.* **14:**97–119, ix.

3. **White, A. C., Jr.** 1997. Neurocysticercosis: a major cause of neurological disease worldwide. *Clin. Infect. Dis.* **24:**101–113.

CASE 47

The patient was a 36-year-old HIV-positive male with a CD4 count of 60/µl. He previously had *Pneumocystis carinii* pneumonia for which he continued to receive prophylaxis with aerosolized pentamidine. At presentation he also had intermittent diarrhea due to *Cryptosporidium*. He was on a study protocol evaluating the efficacy of the combination of AZT (zidovudine) and 3TC (lamivudine) as antiretroviral therapy.

He presented with complaints of headaches, weakness, and difficulty maintaining his balance while walking. A friend who brought the patient to the emergency room related that the patient had seemed quite agitated over the previous few days, his speech was slurred, and he made several inappropriate comments. The friend brought the patient to the emergency room when the patient failed to recognize him.

On physical examination, the patient was afebrile with normal vital signs. He was not oriented to time or place, his speech was slurred and inappropriate, and he was unable to count backward from 100 by 7. A lumbar puncture was performed which revealed 68 white blood cells per µl with 78% lymphocytes and 20% monocytes, a protein level of 67 mg/dl, and a glucose level of 55 mg/dl. A cryptococcal antigen test of the cerebrospinal fluid was negative. Imaging of the brain revealed multiple ring-enhancing lesions (Fig. 1).

1. What organism do you think is causing this patient's symptoms? How would you confirm this diagnosis?

2. Do you believe this represents an acute infection or a reactivation of a latent one? Explain your answer.

3. Name two ways individuals can become infected with this organism. What stage of the parasite is found in each transmission mode? How could each mode of transmission be prevented?

4. What other populations are at increased risk for debilitating or life-threatening infections with this organism?

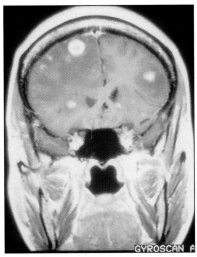

Figure 1

CASE DISCUSSION

1. The differential diagnosis of ring-enhancing lesions seen on head computed tomogram (CT) scan includes both infectious and noninfectious diseases. The presence of this type of lesion indicates that there has been some breakdown in the normally intact blood-brain barrier.

Of noninfectious causes, malignancy is the most common process. The presence of multiple ring-enhancing lesions would suggest the possibility of metastatic cancer (for example, squamous cell carcinoma of the lung) with spread to multiple sites within the brain. However, in a patient with AIDS, a far more likely malignancy would be central nervous system (CNS) lymphoma. This must be considered in this patient's differential diagnosis.

Of infectious causes, a process that can spread hematogenously to multiple sites within the brain is a possibility. Such processes would include septic emboli from a vegetation on a heart valve, as would be seen in infective endocarditis, from other endovascular infections, or from lung abscesses. These processes can result in multiple brain abscesses with resulting loss of the blood-brain barrier, which would be seen as ring-enhancing lesions on head CT. It would be important to know if this patient was actively injecting illicit drugs or if he had other infections that could result in infective endocarditis.

CNS infection with *Mycobacterium tuberculosis* may result in CNS tuberculomas, and, similarly, infection with *Cryptococcus neoformans* may result in the presence of cryptococcomas. Although both infections occur with an increased frequency in patients with AIDS, these are not extremely common causes of ring-enhancing CNS lesions.

The most common cause of ring-enhancing CNS lesions in patients with AIDS is toxoplasmosis, caused by *Toxoplasma gondii*. The diagnosis of toxoplasmosis can be established by performing a brain biopsy. (Figure 2 shows a positive biopsy from another patient with toxoplasmosis.)

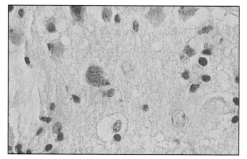

Figure 2

2. *Toxoplasma* infection in AIDS patients is typically due to reactivation of a latent parasitic infection rather than an acute infection. The initial infection, which is often not noticed, may have occurred years prior to the patient's infection with HIV and subsequent suppression of cell-mediated immunity.

3. Infection with *T. gondii* most commonly occurs by ingestion of infective oocysts. An infected cat, the definitive host for *T. gondii*, sheds unsporulated oocysts that require several days before they convert to infective oocysts. Ingestion of infective oocysts in the environment (from cat feces) or of trophozoites present in cysts in undercooked or raw meat is the means by which primary infections typically occur. Alternatively, transplacental infection, in which a fetus acquires *T. gondii* during the parasitemia that occurs during the mother's primary toxoplasma infection, can occur as well. These infections are potentially fatal, especially in the first trimester of pregnancy. Transmission of *T. gondii* has also been documented to occur as a result of organ transplantation and blood transfusions.

Prevention of toxoplasmosis includes eating meat that has been properly cooked (to kill any viable cysts) and, particularly for pregnant women, only changing the litter box within 24 hours after the cat has defecated. This does not give the oocysts time to become infectious. The use of gloves and hand washing when handling cat litter is important to prevent fecal-oral spread of the oocysts. If the pregnant woman is a gardener, wearing gloves while working with soil is another important preventive measure, because cats may defecate there as well. In some countries, such as Austria and France, prenatal screening and treatment for toxoplasmosis are performed, though the risk-benefit ratio of such a program is unclear.

4. In addition to AIDS patients, infants who acquire congenital toxoplasmosis may have a poor outcome. Clinically, the infected infant may have cerebral calcifications, seizure disorder, hydrocephalus, or ocular involvement. The disease may be fatal, or it may result in learning disabilities.

Another group of patients at high risk for toxoplasmosis is organ transplant recipients, particularly heart and heart-lung transplant recipients. In addition to reactivation, which may occur due to immunosuppression that is used to prevent organ rejection, an explanation for the increased risk in this patient population is that a solid organ, such as the transplanted donor heart, may contain viable trophozoites within cysts. Needless to say, in transplanting the heart, a large amount of muscle mass that may be latently infected with viable trophozoites will be transferred to the organ recipient. This risk in cardiac transplant patients is high enough that many centers routinely give such patients prophylactic anti-*Toxoplasma* therapy beginning at the time of transplant.

Other patients at increased risk for reactivation resulting in toxoplasmosis are those with lymphoma and leukemia who are immunosuppressed as a result of their underlying disease as well as cytotoxic drug therapy.

REFERENCES

1. **Centers for Disease Control and Prevention.** 2000. Preventing congenital toxoplasmosis. *Morb. Mortal. Wkly. Rep.* **49**(RR-2):57–75.

2. **Lebech, M., O. Andersen, N. C. Christensen, J. Hertel, H. E. Nielsen, B. Peitersen, C. Rechnitzer, S. O. Larsen, B. Nørgaard-Pedersen, E. Petersen, and the Danish Congenital Toxoplasmosis Study Group.** 1999. Feasibility of neonatal screening for toxoplasma infection in the absence of prenatal treatment. *Lancet* **353**:1834–1837.

3. **Quality Standards Subcommittee, American Academy of Neurology.** 1998. Evaluation and management of intracranial mass lesions in AIDS. *Neurology* **50**:21–26.

CASE 48

The patient was a 34-year-old male with a history of tobacco and alcohol abuse (12 cans of beer per day). Two months prior to admission he was seen at an outside hospital, where he was found to have a necrotic lesion in his right upper lobe on chest X ray. He was PPD (purified protein derivative) negative, and three sputum cultures done for *Mycobacterium* spp. were negative. He had no risk factors for HIV infection. Four weeks prior to admission, he presented with fever, productive cough, night sweats, chills, and a 10-lb (4.5-kg) weight loss. He was treated with amoxicillin-clavulanic acid for 14 days. Fever, chills, and night sweats decreased. On admission, he presented with a palpable, firm right chest wall mass (4 by 4 cm), which was aspirated for diagnostic purposes. The aspirated material was dark green and extremely viscous. Two days later, the nurses found him urinating on the wall of his room. Because this was unusual behavior for this patient, it was decided to perform a computed tomogram (CT) scan of the head; the scan revealed multiple ring-enhancing lesions. The patient was taken to surgery and the brain lesions were aspirated. A Gram stain of the organisms recovered from the brain lesion aspirate showed a branching, beaded gram-positive rod (Fig. 1). Figure 2 shows the colonial morphology of the organism recovered from the brain lesion aspirate after 4 days of incubation.

1. What do the lesions in the patient's head probably indicate? What therapy does this patient need?

2. Two genera of organisms are likely on the basis of the Gram stain. What test could be performed on the aspirate from the brain lesions to determine the genus to which this organism belonged? Why is it important to distinguish between these two genera?

3. Explain the pathogenesis of his brain lesions.

4. Which lifestyle factor(s) predisposed this patient to develop this infection?

5. What is the usual outcome of patients with brain lesions caused by this organism?

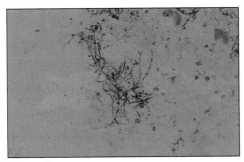

Figure 1

Figure 2

CASE DISCUSSION

1. Ring-enhancing lesions in the brain usually are associated with infection but can also indicate the presence of a tumor or an infarction. This patient had surgical aspiration of what proved to be multiple brain abscesses. Surgical drainage has two specific purposes. One is therapeutic. The surgeon's mantra "to heal, use steel," where the abscess is drained by aspiration, is key to a successful clinical outcome. Abscesses, especially large ones as were seen in this patient, may not respond to antimicrobial therapy alone. The reason for this is that antimicrobial agents probably will not penetrate to the center of these purulent lesions at high enough levels to kill or inhibit the growth of the infecting organism(s).

The other purpose of the surgical procedure is diagnostic. By determining the exact etiology of the infection, the most appropriate antimicrobial therapy can be used.

2. Organisms which on Gram stain appear as branching, beaded gram-positive rods and frequently cause infection are limited to two genera, *Actinomyces* and *Nocardia*. They can be differentiated by a modified acid-fast stain. *Nocardia* spp. are partially acid fast, whereas *Actinomyces* spp. are not. In addition, *Nocardia* spp. are strict aerobes, whereas *Actinomyces* spp. prefer anaerobic growth conditions. This patient's specimen grew *Actinomyces* sp. from both chest wall and brain aspirates. A modified acid-fast stain done on the material aspirated from the brain was negative.

The "molar tooth" appearance of the colonies (Fig. 2) is consistent with *Actinomyces* sp. However, the early colonial morphology of *Nocardia* sp. can mimic that of *Actinomyces*. After more extended incubation, *Nocardia* sp. colonies will begin to have a "velvety" appearance and a strong "dirt-like" odor.

It is important to differentiate these organisms for therapeutic reasons. Both require therapy lasting for months. The sulfa drugs are used to treat *Nocardia* while penicillin G is the drug of choice for *Actinomyces*.

3. Brain abscesses are unusual. They occur most commonly secondarily to lung abscesses, dental infections, endocarditis, sinusitis, head trauma, and meningitis. Central nervous system infection with *Actinomyces* presents as abscesses in two-thirds of patients. These abscesses are most frequently secondary to lung abscesses (as was seen in this case) or dental infection. There are three major routes by which bacteria infect the brain and cause abscess. One is by direct extension from a contiguous infected site such as sinusitis or dental infections. The second is by direct introduction of bacteria following trauma. The third is via septic emboli (small blood clots contaminated with microorganisms) that enter the circulation after breaking away from the primary lesion, typically endocarditis or lung abscesses. These emboli can enter the circulation of the brain and actually block the capillary bed in the brain tissue. This blockage results in hemorrhage and local infection called cerebritis, which becomes walled off forming an abscess. Most brain abscesses are caused by organisms that are part of the oral microflora. *Actinomyces* is a member of the oral microflora. The most frequently recovered organisms from brain abscesses are the viridans group streptococci and anaerobic gram-negative rods such as *Fusobacterium* spp., *Prevotella* spp., *Porphymonas* spp., and *Bacteroides* spp. They either infect directly or, because they are the most frequent cause of both endocarditis (viridans group streptococci) and lung abscess (viridans group streptococci and anaerobic gram-negative rods), indirectly. Multiple species of both facultative and strictly anaerobic oral bacteria are frequently seen in brain abscesses. In two-thirds of *Actinomyces* brain abscesses, it is the only organism recovered.

4. This patient has a history of excessive alcohol consumption. Excessive alcohol consumption can result in unconscious states. Clearance of oral secretions is believed to be depressed as a result of a decreased gag reflex and/or the alcohol-induced stupor. As a result, oropharyngeal flora, which are usually easily cleared, can infect the lower airways. In addition, vomiting can lead to aspiration of gastric acid and enzymes, which can cause localized tissue damage and create an ideal focus for pulmonary infection. The end result can be a lung abscess followed by a secondary brain abscess. In this case, the patient had an aspiration pneumonia and lung abscess due to *Actinomyces*. His clinical course in many ways was consistent with tuberculosis, with findings of night sweats, fevers, significant weight loss, and productive cough. However, his initial work-up ruled out that infection. Unfortunately, his oral antibacterial regimen, assuming he was compliant, was not sufficient to prevent development of a lung abscess and the subsequent brain abscess.

5. Central nervous system infection with *Actinomyces* has both high morbidity and high mortality. Mortality is close to 30% with this infection and is frequently only recognized at autopsy. In those who survive, 55% will have some type of neurological sequelae.

REFERENCES
1. **Mathisen, G. E., and J. P. Johnson.** 1997. Brain abscess. *Clin. Infect. Dis.* **25:**763–769.
2. **Smego, R. A., Jr.** 1987. Actinomycosis of the central nervous system. *Rev. Infect. Dis.* **9:**855–865.

CASE 49 The patient was a 6-month-old male who presented with a 3-day history of increasing hypotonia and a 1-day history of dehydration. Three days prior to admission, the mother had noticed decreased suck while breast-feeding. The next day, it took him 1 hour to take his normal amount of breast milk. He normally took only 15 minutes. During the evening 2 days prior to admission, the parents noticed the infant had generalized weakness with decreased movement and difficulty sitting up. That evening the child would not breast-feed. The morning prior to admission, the parents noted the infant was increasingly floppy and took only 1 oz of breast milk (versus his usual 4½ oz) and 1 oz of juice. He had trouble with gurgling in the back of his throat, very poor head control, and increased floppiness. They took him to a local physician, who noted severe hypotonia and dehydration. The parents were told to take their son to the local emergency room. On review of systems, it was noted that the child began to be constipated 4 weeks previously and had had only two small stools over the last 6 days. His diet consisted of breast milk, occasional sweet potatoes, and rice cereal. Abnormal findings on physical examination consisted of generalized hypotonia with head lag. Cerebrospinal fluid (CSF) findings were within normal limits. The patient was admitted to the pediatric intensive care unit (PICU). Because of increasing respiratory difficulty, he was intubated. He remained on the ventilator for 6 days, was extubated, and was discharged to home 11 days after admission.

1. What is the condition this child has? What clinical clues are present which help you make this diagnosis?

2. What is the organism causing this condition? Briefly describe the epidemiology and pathogenesis of this disease.

3. How is the laboratory diagnosis of this disease made?

4. How is this form of the disease typically treated?

5. What other forms of disease can be seen with this organism? Describe the similarities and differences in these different forms of the disease.

6. Why is there increased concern about this organism among governmental agencies such as the Department of Defense, the Centers for Disease Control and Prevention, and the Federal Bureau of Investigation?

CASE DISCUSSION

1. This patient has infant botulism. This diagnosis must be considered in any infant who has the constellation of symptoms seen in this child. Because infant botulism is rare, the clinical manifestations seen are frequently attributed to sepsis, central nervous system infections, or other, more esoteric diagnoses such as Guillain-Barré syndrome or inborn errors of amino acid metabolism. The misdiagnosis of sepsis may further complicate the disease course of these children. (See answer to question 4 for further details.) The disease is characterized by descending paralysis. The initial signs of the disease include constipation; poor suck (which is typically noticed by breast-feeding mothers); and increasing hypotonia, frequently first characterized by the child's inability to hold up its head and by ptosis (drooping of the upper eyelids).

2. The patient is infected with *Clostridium botulinum. C. botulinum* produces a neurotoxin called botulinum toxin, which, on a weight basis, is the most potent biologic toxin known. There are seven different types of botulinum toxin designated A to G, with types A, B, and E being responsible for most human disease. Two other *Clostridium* species, *C. baratii* and *C. butyricum*, have been shown to cause human disease via the production of botulinum toxin.

Infant botulism is the most commonly seen form of botulism in the United States. The highest incidences of disease are in California, due primarily to type A botulinum toxin, and in the Delaware Valley area of Pennsylvania and New Jersey, involving primarily type B botulinum toxin. Toxin A-producing strains are the predominant type in soil in California while toxin B-producing strains predominate east of the Mississippi River. This disease occurs sporadically; no outbreaks of infant botulism have been reported. In this disease, spores of the organisms are ingested either in foodstuffs (e.g., honey) or from dust. The peak incidence of infant botulism is at 2 to 4 months of age, with formula-fed babies developing disease earlier than breast-fed babies, perhaps reflecting differences in bowel flora seen in those two populations. It is important to note that most cases of infant botulism occur around the time of weaning. At this time, the infant's bowel flora undergoes significant change and it is this change that is believed to provide an ecological niche for the growth of and toxin production by this organism. An adult form of "infant botulism" has also been postulated since the source of some cases of adult botulism cannot be identified. These individuals typically have had gastrointestinal surgery, have chronic gastrointestinal disease such as inflammatory bowel disease, or have had recent antibiotic treatment. A toxin-producing organism can be recovered from the gastrointestinal tract of these patients.

C. botulinum produces toxin in the gastrointestinal tract which is absorbed into the bloodstream and binds irreversibly to the presynaptic nerve endings. The toxin, like many exotoxins, consists of two chains or subunits designated A (enzymatically active) and B (binding). For the toxin to be biologically active, it must be cleaved by a protease. Once cleaved, the B subunit binds to neurons and forms a pore in the neuronal membrane through which the A subunit can enter the cell. There it blocks the release of acetylcholine by preventing its exocytosis. Acetylcholine release is necessary for the excitation of the muscle fibers. This blockage of neurotransmitter release results in the flaccid paralysis seen in this disease.

3. The mouse lethality assay is the standard test used to detect botulinum toxin in serum, foodstuffs, and stool specimens. In this test, serum, food, or stool filtrates are injected intraperitoneally into mice and the animals are observed over a 96-hour period for the development of paralysis. Typically, the paralysis begins in the hind legs and eventually results in the death of the animal. To ensure that this paralysis is specifically due to botulinum toxin, control mice are injected with an aliquot of the serum or stool specimen which has been preincubated with polyvalent antiserum raised against the different botulinum toxin types. Animals receiving this mixture will not develop paralysis if the paralysis is due to botulinum toxin.

An alternative to this assay is an enzyme immunoassay (EIA). Unfortunately, the EIA is not nearly as sensitive as the mouse bioassay and may not be sensitive enough to detect all cases of botulism. Polymerase chain reaction (PCR) for detection of botulinum toxin genes is currently under development, but its accuracy in the diagnosis of botulism from clinical specimens has not been adequately studied.

In infant botulism and adult forms of infant botulism, the organism may be isolated from stool, but culture lacks the sensitivity of the mouse assay and is only an adjunct procedure useful for studying the molecular epidemiology of these organisms.

4. Death from botulism is typically due to respiratory arrest from paralysis of the diaphragmatic muscles involved in breathing. As a result, mechanical ventilatory support is central to therapy for all forms of botulism. Typically, ventilatory support is required for 2 to 8 weeks while the affected nerve endings regenerate.

Botulinum antitoxin has been shown to reduce the mortality of botulism in patients with food-borne and wound forms of the disease. Because the botulinum antitoxin is of equine origin, there has been reluctance to use this product in infants because of concerns about anaphylactic reactions. Currently, a human-derived antitoxin is under investigation for use in infants. Its efficacy has not been reported to date.

Many patients with infant botulism are initially believed to have sepsis. As a result, they may receive empiric antimicrobial therapy, which includes an aminoglycoside. Aminoglycosides are contraindicated in patients with botulism because they have been shown to potentiate the activity of the toxin. Therefore, in patients with ptosis or hypotonia, the use of aminoglycosides should be avoided until the diagnosis of botulism can be ruled out.

5. There are two other forms of naturally occurring botulism, food-borne and wound botulism. In food-borne botulism, the toxin is pre-formed in food, typically canned goods or smoked fish (especially type E toxin) or meat, and ingested with the food. This disease is an intoxication rather than an infection, as is seen in both infant and wound botulism where toxin is produced in the host. Food-borne botulism tends to cause outbreaks of disease. These outbreaks are most commonly associated with improperly home-canned vegetables and usually affect only family members. In these outbreaks, spores of *C. botulinum* are not killed by the canning or other food processing techniques. An anaerobic environment is produced, the spores vegetate, and the organism grows, producing toxin. When the food is then consumed, it is either not heated at all (smoked meats or fish) or not heated to temperatures sufficient to inactivate the toxin (canned vegetables and soups). Outbreaks in which large numbers of individuals become ill are almost always associated with commercially prepared foods. Diagnosis of this disease can be made by detecting toxin in stool, feces, or the implicated food. Alaska has the highest rate of food-borne botulism in the United States, due to the consumption of native Inuit foods such as fermented fish and fermented marine mammals, in which anaerobic conditions permit the growth of *C. botulinum* and the elaboration of its toxin.

In wound botulism, the organism is introduced into a wound with devitalized tissue. This dead tissue provides an anaerobic environment where the organism can grow and produce toxin that can enter the bloodstream. Wound botulism has been seen with increasing frequency in users of black tar heroin who inject the drug by "skin popping," i.e., intradermal injection of the drug. In wound botulism, the organism can be detected in the wound by culture in addition to detection of the toxin in serum. These patients should be treated with both antitoxin and antibiotics. Penicillin G is the antibiotic of choice. These patients may also require ventilatory assistance. Interestingly, two patients have been described in the literature who have had two episodes of wound botulism associated with intradermal drug use, and another individual has been reported to have had two episodes of food-borne botulism. This suggests that the initial toxin dose was not sufficient for them to mount a protective immune response that could prevent subsequent episodes.

6. Botulinum toxin has been "weaponized" by several countries. During the Gulf War, missiles with warheads containing botulinum toxin were reported to have been produced by Iraq, although there was never any evidence that these bioweapons were used. Botulinum toxin is recognized as a potential weapon of bioterrorists. The toxin in crude form is easily produced. Animal studies have shown that the toxin can enter the bloodstream following inhalation, making it theoretically possible to deliver this agent by aerosol. Contamination of various foods is another possible scenario by which this toxin could be used to attack a population. Because the medical management of botulism often requires ventilatory support, and the number of ventilators and the skilled individuals to support their use is limited, a successful bioterrorism attack on a large population with this toxin is of great concern to governmental agencies.

REFERENCES

1. **Fenicia, L., G. Franciosa, M. Pourshaban, and P. Aureli.** 1999. Intestinal toxemia botulism in two young people caused by *Clostridium butyricum* type E. *Clin. Infect. Dis.* **29:**1381–1387.

2. **Midura, T. F.** 1996. Update: infant botulism. *Clin. Microbiol. Rev.* **9:**119–125.

3. **Shapiro, R. L., C. Hatheway, and D. L. Swerdlow.** 1998. Botulism in the United States: a clinical and epidemiologic review. *Ann. Intern. Med.* **129:**221–228.

4. **Werner, S. B., D. Passaro, J. McGee, R. Schecter, and D. J. Vugia.** 2000. Wound botulism in California, 1951–1998: recent epidemic in heroin injectors. *Clin. Infect. Dis.* **31:**1018–1024.

SIX

Systemic Infections

INTRODUCTION TO SECTION VI

Systemic infections can be caused by many different infectious agents: bacterial, fungal, viral, and parasitic. One common finding for all systemic infections is the need for a **portal of entry**. The portal of entry can be via the skin (as in mosquito-borne diseases such as malaria), via the oral route (as in typhoid fever), via sexual contact (as in HIV infection), as a bloodborne pathogen (as in hepatitis B virus infection), via the respiratory tract (as in measles), and by vertical transmission via transplacental infection (as in congenital cytomegalovirus infection).

In many cases of systemic infection, **colonization** occurs prior to the dissemination of the infectious agent throughout the body. In some diseases (e.g., tetanus and diphtheria) the infection itself is caused by a noninvasive organism and the systemic symptoms are caused by the dissemination of a **toxin** that is responsible for the disease. In most cases, however, the etiologic agent is disseminated via the hematogenous route.

Patients may have certain risk factors or defects in host defenses that predispose them to specific types of infections. Examples of defects in host defenses that predispose to certain specific types of infections include breaches in the integrity of the skin (patients with burns, patients with invasive medical devices), defects in cell-mediated immunity (AIDS, corticosteroid use), defects in humoral immunity (hypogammaglobulinemia), decreased splenic function (splenectomy, sickle-cell disease), quantitative defects in neutrophils (neutropenia following chemotherapy), qualitative defects in neutrophils (chronic granulomatous disease, Chediak-Higashi syndrome), and deficiencies in the complement system. It is important to be able to recognize these risk factors when they are present and to understand the defect that predisposes the patient. Conversely, it is important to be able to suspect a specific defect in host defenses when a patient presents with a systemic infection.

Protection of the host from a systemic infection can occur as a result of acquired immunity due to a prior infection or due to a vaccination to that agent. Unfortunately, efficacious vaccines are not available for the majority of infectious agents, and in many diseases, infection does not lead to protective immunity.

Important agents of systemic infection are listed in Table 6. Please note that virtually all bacteria can potentially be isolated from the blood under circumstances of specific host defects, such as the presence of an intravenous catheter. Many of the etiologic agents listed have a particular organ tropism (such as the liver for hepatitis viruses) but may also cause systemic illness.

TABLE 6 SELECTED SYSTEMIC PATHOGENS

ORGANISM	GENERAL CHARACTERISTICS	SOURCE OF INFECTION	DISEASE MANIFESTATION
Bacteria			
Acinetobacter spp.	Glucose-nonfermenting, gram-negative bacilli	Exogenous	Nosocomial UTI,[a] nosocomial pneumonia, nosocomial and line-related bacteremia
Bacillus anthracis	Aerobic gram-positive bacillus	Exogenous; zoonosis; possible agent of bioterrorism	Cutaneous, pulmonary (with hemorrhagic mediastinitis), gastrointestinal, bacteremia, meningitis
Bartonella henselae	Fastidious, gram-negative bacillus	Exogenous; cats appear to be primary host	Cat scratch disease; bacillary angiomatosis (in immuno-compromised hosts)
Borrelia burgdorferi	Spirochete	Exogenous, tick to human	Lyme disease; rash, arthritis, nervous system and cardiac manifestations
Brucella spp.	Oxidase-positive, fastidious, gram-negative bacilli	Exogenous; zoonosis; possible agent of bioterrorism	Lymphadenopathy, hepatosplenomegaly, genitourinary, bone, and CNS[b] infection
Clostridium botulinum	Anaerobic gram-positive bacillus	Exogenous; improperly canned food; possible agent of bioterrorism	Botulism; flaccid paralysis with prominent cranial nerve symptoms
Clostridium perfringens	Anaerobic gram-positive bacillus	Exogenous	Wound infection, gas gangrene, bacteremia, gangrenous cholecystitis, food poisoning
Clostridium tetani	Anaerobic gram-positive bacillus	Exogenous	Tetanus
Coagulase-negative staphylococci	Catalase-positive, coagulase-negative, gram-positive cocci	Endogenous	Nosocomial and line-related bacteremia, prosthetic valve endocarditis
Corynebacterium diphtheriae	Aerobic gram-positive bacillus	Exogenous	Diphtheria
Enterobacter spp.	Lactose-fermenting, gram-negative bacilli	Endogenous	Community and nosocomial UTI, bacteremia, endocarditis
Enterococcus spp.	Catalase-negative, gram-positive cocci	Endogenous	Wound infections, nosocomial UTI, bacteremia, intra-abdominal infections

(continued next page)

TABLE 6 SELECTED SYSTEMIC PATHOGENS *(continued)*

ORGANISM	GENERAL CHARACTERISTICS	SOURCE OF INFECTION	DISEASE MANIFESTATION
Escherichia coli	Lactose-fermenting, gram-negative bacillus	Endogenous	Community and nosocomial UTI, bacteremia, intra-abdominal infections
Francisella tularensis	Fastidious, gram-negative bacillus	Exogenous; zoonosis; tick to human; direct contact with animal; inhalation; ingestion of contaminated food or water; possible agent of bioterrorism	Skin ulcer, lymphadenopathy, ocular involvement, bacteremia, pneumonia
Group A streptococci (*Streptococcus pyogenes*)	Catalase-negative, gram-positive cocci	Exogenous	Pharyngitis, cellulitis, bacteremia, scarlet fever, necrotizing fasciitis, pneumonia, post-streptococcal glomerulo-nephritis and rheumatic fever
Group B streptococci (*Streptococcus agalactiae*)	Catalase-negative, gram-positive cocci	Endogenous	Sepsis, meningitis, cellulitis, UTI (diabetics)
Klebsiella pneumoniae	Lactose-fermenting, gram-negative bacillus	Endogenous	Community and nosocomial UTI, bacteremia, intra-abdominal infections
Mycobacterium avium complex	Acid-fast bacilli	Exogenous	Disseminated disease
Mycobacterium tuberculosis	Acid-fast bacillus	Respiratory, may be exogenous (primary) or endogenous (reactivation)	Chronic pneumonia with or without cavitation, adenopathy, gastrointestinal involvement, pleural involvement, peritonitis, meningitis, bone infection, genitourinary infection, miliary tuberculosis
Neisseria gonorrhoeae	Oxidase-positive, gram-negative diplococcus	Exogenous, direct sexual contact; vertical (mother to child)	Urethritis, cervicitis, pharyngitis, pelvic inflammatory disease, proctitis, bacteremia, septic arthritis, conjunctivitis
Neisseria meningitidis	Oxidase-positive, gram-negative diplococcus	Exogenous	Meningitis, bacteremia, pneumonia
Pasteurella multocida	Oxidase-positive, gram-negative bacillus	Exogenous; zoonosis (often animal bite or scratch)	Cellulitis, bacteremia, osteo-myelitis, meningitis, pneumonia
Proteus mirabilis	Lactose-nonfermenting, gram-negative bacillus	Endogenous	Community and nosocomial UTI, bacteremia

Organism	Characteristics	Transmission	Clinical manifestations
Pseudomonas aeruginosa	Glucose-nonfermenting, oxidase-positive, gram-negative bacillus	Exogenous	Community and nosocomial UTI, nosocomial pneumonia, nosocomial bacteremia, chronic pulmonary infections in patients with cystic fibrosis
Rickettsia rickettsii	Rickettsial organism	Exogenous, tick to human	Rocky Mountain spotted fever
Salmonella typhi	Lactose-nonfermenting, gram-negative bacillus	Exogenous, human to human via contaminated food or water	Typhoid fever, bacteremia, intestinal disease, perforation of colon
Staphylococcus aureus	Catalase-positive, coagulase-positive, gram-positive coccus	Endogenous and exogenous	Skin infections, bacteremia, endocarditis, osteomyelitis, septic arthritis, pneumonia, food poisoning
Streptococcus pneumoniae	Catalase-negative, gram-positive coccus	Endogenous	Community-acquired pneumonia, sinusitis, otitis media, bacteremia, meningitis, endocarditis, septic arthritis, peritonitis
Treponema pallidum	Spirochete (does not Gram stain)	Exogenous, direct sexual contact; vertical (mother to child)	Primary, secondary, latent, and late syphilis; can affect any organ
Viridans group streptococci	Catalase-negative, gram-positive cocci	Endogenous	Dental caries, endocarditis, bacteremia, abscesses
Yersinia pestis	Gram-negative bacillus	Zoonosis; flea to human; person to person in pneumonic form; possible agent of bioterrorism	Localized lymphadenopathy (bubonic), high-grade bacteremia, pneumonia, meningitis
Fungi			
Aspergillus spp.	Molds with septate hyphae	Exogenous	Pneumonia, sinusitis, external otitis, allergic processes, disseminated infection
Blastomyces dermatitidis	Dimorphic mold	Exogenous	Pneumonia, meningitis, bone infection
Candida albicans	Yeast, often germ tube positive	Endogenous	Thrush, vaginal yeast infection, diaper rash, esophagitis, nosocomial bloodstream infection

(continued next page)

TABLE 6 **SELECTED SYSTEMIC PATHOGENS** *(continued)*

ORGANISM	GENERAL CHARACTERISTICS	SOURCE OF INFECTION	DISEASE MANIFESTATION
Candida spp., non-*albicans*	Yeasts, germ tube negative	Endogenous	Thrush, vaginal yeast infection, nosocomial bloodstream infection
Coccidioides immitis	Dimorphic mold	Exogenous	Pneumonia, meningitis, bone infection
Cryptococcus neoformans	Encapsulated yeast	Exogenous	Meningitis, pneumonia, bloodstream infection
Histoplasma capsulatum	Dimorphic mold	Exogenous	Pneumonia, disseminated infection
Paracoccidioides brasiliensis	Dimorphic mold	Exogenous	Ulcerative mucosal lesions in the mouth, nose, larynx and oropharynx, lung, skin, other organs
Penicillium marneffei	Dimorphic mold	Exogenous	Disseminated disease in immuno-compromised; skin lesions, lung involvement
Zygomycetes infection	Molds with aseptate hyphae	Exogenous	Pneumonia, sinusitis, invasive
Parasites			
Babesia microti	Can be seen on peripheral blood smear	Exogenous, tick to human	Babesiosis
Leishmania donovani	Amastigotes in tissue touch preparation	Exogenous (*Phlebotomus* sand flies)	Kala-azar
Plasmodium spp.	Can be seen on peripheral blood smear	Exogenous (*Anopheles* mosquitoes)	Malaria
Strongyloides stercoralis	Nematode	Exogenous; endogenous (autoinfection and hyperinfection)	Gastrointestinal, pulmonary (pneumonia, wheezing), disseminated in hyperinfection
Taenia solium	Tapeworm	Exogenous (consumption of pig meat; fecal-oral from humans passing eggs)	Gastrointestinal infection, cysticercosis (brain, muscles, other organs)
Toxoplasma gondii	Protozoan	Exogenous; endogenous (reactivation)	CNS, ocular, hepatic, pulmonary

Viruses

Coxsackieviruses	Nonenveloped, ssRNA[c]	Children and adults during summer months	Aseptic meningitis, myocarditis, hand-foot-and-mouth, pleuritis
Cytomegalovirus	Enveloped, dsDNA[d]	Immunocompromised and newborn	Pneumonia, hepatitis, gastro-intestinal ulcers, congenital infection
Dengue viruses	Enveloped, ssRNA	Exogenous (*Aedes* mosquitoes)	"Breakbone fever," headache, fever, rash, myalgia, sometimes hemorrhagic fever/shock
Epstein-Barr virus	Enveloped, dsDNA	Often present in saliva, exogenous	Mononucleosis, lympho-proliferative disorders
Filoviruses (Ebola virus, Marburg virus)	RNA virus	Reservoir in nature is unknown; nonhuman primates and humans have been sources; possible agent of bioterrorism	Hemorrhagic fever with high mortality
GB virus C (hepatitis G virus)	RNA virus	Blood-borne, sexual	Of unclear clinical significance
Hantaviruses	Enveloped, ssRNA	Rodent excreta	Pneumonia, hemorrhagic fever with renal dysfunction
Hepatitis A virus	Nonenveloped RNA virus	Fecal-oral transmission	Hepatitis
Hepatitis B virus	Enveloped DNA virus	Blood-borne, sexual, and vertical transmission	Hepatitis, chronic carriers, cirrhosis, hepatocellular carcinoma
Hepatitis C virus	RNA virus	Blood-borne pathogen	Hepatitis, chronic carriers, cirrhosis, hepatocellular carcinoma
Hepatitis D virus	RNA virus	Blood-borne pathogen, requires hepatitis B coinfection	Fulminant hepatitis, requires coinfection with hepatitis B virus
Hepatitis E virus	RNA virus	Fecal-oral transmission	Hepatitis, increased severity in pregnant women
Herpes simplex viruses	Enveloped, dsDNA viruses	Person to person, including via sexual contact; reactivation of latent infection; during passage through the birth canal	Genital, oral, ocular, encephalitis, neonatal infection, esophagitis (immunocompromised)
Human T-cell lymphotropic virus type 1	Enveloped RNA retrovirus	Blood-borne pathogen	Tropical spastic paraparesis, T-cell leukemia

(continued next page)

TABLE 6 **SELECTED SYSTEMIC PATHOGENS** (*continued*)

ORGANISM	GENERAL CHARACTERISTICS	SOURCE OF INFECTION	DISEASE MANIFESTATION
Human herpesvirus type 6	Enveloped, dsDNA	Person to person	Exanthem subitum (roseola)
Human herpesvirus type 8 (Kaposi's sarcoma–associated herpesvirus)	Enveloped, dsDNA	Person to person, including sexual transmission	Kaposi's sarcoma in HIV-infected individuals
Human immunodeficiency viruses (HIV-1 and HIV-2)	Enveloped RNA retroviruses	Blood-borne, sexual, and vertical transmission	AIDS
Mumps virus	Enveloped, ssRNA	Respiratory spread	Mumps, parotitis, pancreatitis, orchitis, meningitis
Parvovirus B19	Small DNA virus	Person to person, including vertical transmission	Erythema infectiosum, arthritis, transient aplastic crisis, hydrops fetalis
Rubella virus (German measles)	Enveloped, ssRNA	Person to person, including vertical transmission	Inapparent or subclinical infection in adults, birth defects in infants
Rubeola virus (measles)	Enveloped, ssRNA	Respiratory spread	Measles, pneumonia, encephalomyelitis, subacute sclerosing panencephalitis
Smallpox	Large DNA virus	Must be assumed to be due to bioterrorism or biological warfare; respiratory spread; direct contact	Prominent vesicular rash, up to 30% mortality
Varicella-zoster virus	Enveloped, dsDNA	Respiratory spread; direct contact; reactivation of latent infection	Chicken pox; zoster (may disseminate)
Yellow fever virus	Enveloped, ssRNA	Exogenous (*Aedes* mosquitoes)	Severe hepatitis, significant mortality rate

[a]UTI, urinary tract infection.

[b]CNS, central nervous system.

[c]ssRNA, single-stranded RNA.

[d]dsDNA, double-stranded DNA.

CASE 50 This 53-year-old man with a past medical history of non-insulin-dependent diabetes mellitus and hypertension was in his usual state of health until 4 days prior to admission when he developed fatigue, fever, chills, and a cough occasionally productive of green sputum. Over the 2 days prior to admission, he had drenching sweats, increasing dyspnea, and left-sided pleuritic chest pain. The patient had smoked 2 packs of cigarettes a day for 40 years.

His physical examination was notable for an increased respiratory rate of 22/min, and crackles were heard over the right middle, left middle, and left lower lung fields. A chest radiograph demonstrated right lower lobe, left lingular, and left lower lobe infiltrates.

A Gram stain of the patient's sputum contained >25 polymorphonuclear leukocytes per low-power field and 4+ (many) gram-positive diplococci. Culture of the sputum grew 4+ (many) *Streptococcus pneumoniae* as well as normal respiratory flora. One set (both bottles) of two sets of blood cultures drawn prior to the administration of antibiotics grew the organism shown on Gram stain in Fig. 1. The organism growing from a subculture of the blood is shown in Fig. 2. Further biochemical testing revealed the organism to be catalase positive and coagulase negative.

1. What is the blood culture isolate?

2. What is the significance of this patient's blood culture isolate?

3. How can one determine the clinical significance of this organism in a given patient?

4. What is the clinical impact of this type of blood culture isolate in terms of length of stay, antibiotic administration, and additional testing?

5. What can be done to prevent this type of blood culture isolate in a health care facility?

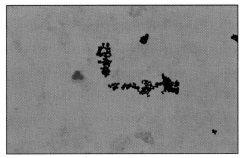

Figure 1

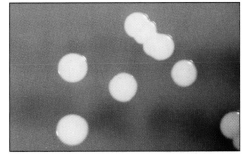

Figure 2

CASE

50

CASE DISCUSSION

1. Gram-positive cocci in clusters that are catalase positive are most likely staphylococci. The coagulase test helps to differentiate *Staphylococcus aureus*, which is coagulase positive, from the other staphylococci, which are often grouped together as "coagulase-negative staphylococci." Coagulase-negative staphylococci are a heterogeneous group of several different species including the urinary tract pathogen *Staphylococcus saprophyticus* as well as *Staphylococcus epidermidis* and many others.

This patient's blood isolate was not identified to the species level by the clinical microbiology laboratory but was identified as a coagulase-negative staphylococcus. In many cases, this information is sufficient for the clinician.

2. This patient's blood culture isolate is judged to be a contaminant. His clinical picture is most consistent with community-acquired pneumonia due to *S. pneumoniae*. Approximately two-thirds of patients with pneumococcal pneumonia do not have positive blood cultures for the pneumococcus. The recovery of coagulase-negative staphylococci from the patient's blood is not at all consistent with his clinical picture and should not lead his doctor to alter antibiotic therapy. The most common reason for a blood culture contaminant to be isolated is inadequate disinfection of the patient's skin prior to venipuncture.

In addition to coagulase-negative staphylococci, other organisms that are often contaminants include skin flora such as *Propionibacterium acnes*, *Micrococcus* spp., diphtheroids, *Bacillus* spp., and viridans group streptococci.

3. The clinical significance of a blood culture positive for a possible contaminant can only be determined by weighing all the clinical information available. For example, consider a patient who has a central venous line who develops fevers, erythema, and tenderness at the catheter site. Multiple sets of blood cultures are positive for coagulase-negative staphylococci, and the patient's central venous catheter line tip, when removed, grows >15 colonies of the same organism on culture. This patient is far more likely to have a real infection with this organism than is a patient similar to the one described in this case. The laboratory staff are often asked by clinicians, "Is this a contaminant or is it a real pathogen?" This question is not one that the laboratorian can answer. It is up to the patient's physician to weigh his or her clinical observations with the laboratory data to judge the significance of the patient's blood culture isolate.

When more than one set of blood cultures are positive for coagulase-negative staphylococci, it is more likely that the positive blood culture represents a true pathogen. The laboratory should, in this circumstance, help the clinician by providing susceptibility test results. If the results of these tests are the same for the different isolates, this is evidence that the different isolates are, in fact, the same organism, and is supportive evidence that it represents a true pathogen. A more definitive test that is available in some laboratories involves the use of molecular fingerprinting techniques such as pulsed-field gel electrophoresis to determine if the isolates are genetically distinguishable.

4. Patients with contaminated blood cultures often have their hospitalization stay increased by several days at an average cost of several thousand dollars per patient. Since anywhere from 1 to 6% of all blood cultures drawn may become positive with a contaminant, the excess cost to a health care institution in the course of a year due to contaminated blood cultures is staggering. While patients are being evaluated in the hospital for blood culture contaminants, they may receive unnecessary antibiotics and further diagnostic studies which entail some risk. Thus it is prudent for hospitals to make efforts to decrease blood culture contamination rates.

5. Strict attention to the method by which cultures are obtained is important. In particular, the use of tincture of iodine as the skin antiseptic has been shown to be superior to the use of an iodophor. In addition, some hospitals have phlebotomists who are specifically trained in proper techniques for obtaining blood for culture. Studies have shown that blood culture contamination rates are lower in these hospitals. Having specially trained staff to draw blood cultures should be cost-effective given the high costs associated with the management of patients who have contaminated blood cultures.

REFERENCES
1. **Bates, D. W., L. Goldman, and T. H. Lee.** 1991. Contaminant blood cultures and resource utilization. The true consequences of false-positive results. *JAMA* **265:**365–369.

2. **Strand, C. L., R. R. Wajsbort, and K. Sturmann.** 1993. Effect of iodophore vs. iodine tincture skin preparation on blood culture contamination rate. *JAMA* **269:**1004–1006.

3. **Weinbaum, F. I., S. Lavie, M. Danek, D. Sixsmith, G. F. Heinrich, and S. S. Mills.** 1997. Doing it right the first time: quality improvement and the contaminant blood culture. *J. Clin. Microbiol.* **35:**563–565.

CASE 51

This 39-year-old intravenous drug user (actively using cocaine on the date of admission) was admitted with cellulitis of the right arm after experiencing fevers for several weeks. He had been treated with outpatient antibiotics without relief of either associated chills or dizziness. Two sets of blood cultures were obtained on admission. A trans-thoracic echocardiogram demonstrated a 1-cm vegetation on the ventral surface of the aortic valve. The patient left the hospital against medical advice but was readmitted 2 days later for antimicrobial therapy.

Past medical history was notable for multiple hospital admissions for both cellulitis and abscesses primarily involving the patient's arm. He had had multiple drug rehabilitation treatment attempts without success.

Physical examination demonstrated a thin, unkempt man in no acute distress with multiple "needle track" marks on both his upper and lower extremities. No splinter hemorrhages or signs of embolic phenomena were noted on the extremities. Cardiac exam was notable for a grade II/VI systolic murmur best heard at the left sternal border. The spleen tip was palpable. The right arm had a 10-by-6-cm excoriated area with surrounding induration.

Gram stain of an organism detected in both sets of the blood cultures obtained at admission is shown in Fig. 1. Growth of the organism on a blood agar plate is shown in Fig. 2. The organism grew in broth containing 6.5% NaCl, hydrolyzed esculin in the presence of bile (i.e., was bile esculin positive), and was catalase negative.

1. What type of infection does this patient have?

2. What organisms frequently cause this type of infection in intravenous drug users (IVDUs)? What organism is causing his infection?

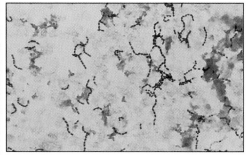

Figure 1

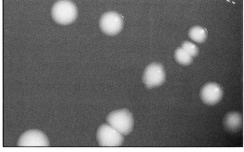

Figure 2

3. How does intravenous drug use predispose the patient to this type of infection? Briefly describe the pathogenesis of this infection. Describe what other organs may be secondarily infected and the mechanism by which secondary infections occur.

4. For what other infectious agents is this individual at increased risk?

5. When considering antimicrobial therapy for this infection, what general strategy should be employed?

6. What antimicrobial resistance problems have recently emerged involving this organism? What strategies have been employed to reduce the spread of these organisms?

CASE DISCUSSION

1. This patient has bacterial endocarditis. The keys to making this diagnosis are the detection by echocardiogram of a vegetation on his aortic heart valve and the presence of continuous bacteremia in his bloodstream as detected by his positive blood cultures. The use of criteria for the diagnosis of infective endocarditis has been advocated, and this patient would meet the clinical criteria of Durack et al. on the basis of the community-acquired enterococcal bacteremia and the presence of the vegetation. Physical findings consistent with endocarditis include enlarged spleen and the presence of a heart murmur. Twenty to 40% of patients with endocarditis have cutaneous findings due to embolic events secondary to endocarditis. This patient did not have these lesions.

2. The most common agents of bacterial endocarditis in IVDUs are *Staphylococcus aureus*, viridans group streptococci, *Candida albicans*, *Enterococcus* spp., and gram-negative bacilli including *Pseudomonas aeruginosa*. The organism description indicates that this patient was infected with an organism belonging to the genus *Enterococcus*. Further phenotypic characteristics would be required to determine to which species this organism belonged.

3. The pathogenesis of bacterial endocarditis is dependent upon damage to heart valves, which typically produces turbulent blood flow. Turbulence in blood flow may result in the deposition of platelets and fibrin, the initial stage in formation of vegetations. IVDUs do not use "sterile technique" when they inject drugs. Therefore, transient bacteremia with either skin flora (*S. aureus*, viridans group streptococci, enterococci, or *Candida* spp.) or environmental organisms (*P. aeruginosa*) may occur following drug injection. These organisms have been shown to adhere readily to thrombotic lesions on the heart valve. The adherent bacteria begin to grow, and platelet and fibrin deposition continues, resulting in an enlarging vegetation. As this vegetation continues to grow, small pieces containing fibrin, platelets, and bacteria may "break off," causing septic emboli. Septic emboli enter the bloodstream and can become lodged in the vascular bed, resulting in localized hemorrhage and infection. Common secondary infections due to septic emboli include brain, kidney, spleen, lung, and liver abscesses. Anatomically, IVDUs are more likely to have "right-sided" endocarditis affecting the tricuspid valve than are other people with infective endocarditis.

4. IVDUs often share needles or reuse needles used by others, exposing themselves to blood from other individuals. As a result, they are at increased risk for many blood-borne infectious agents. The most common and important agents acquired by this behavior are HIV, hepatitis B, C, and D viruses, and, in geographically appropriate areas, human T-cell lymphotropic virus type 1. Uncommon infections can also be acquired from the injectable drugs. As a result, tetanus due to infection with *Clostridium tetani* and other life-threatening soft-tissue clostridial infections are well documented in IVDUs.

5.   Enterococci are susceptible to very few antimicrobial agents. Most strains are susceptible only to vancomycin and ampicillin/penicillin G. Enterococci are often tolerant to these agents, meaning that the organisms are inhibited but not killed by the specific antimicrobial agent. This is problematic when treating patients with endocarditis because phagocytic cells provide little help in clearing the infection on the heart valve so that killing of organisms within vegetations is very much dependent upon antimicrobial activity. Studies in vitro and in animals have shown that aminoglycosides, although inactive alone at concentrations achievable in the bloodstream, greatly enhance the killing power of ampicillin/penicillin G or vancomycin when one of these cell wall-active agents is given in combination with the aminoglycoside. This enhancement of antimicrobial killing power when drugs are given in combination is known as synergy. Typically, either ampicillin or vancomycin is given in combination with gentamicin to treat enterococcal endocarditis. Major resistance problems have developed, however (see answer to question 6).

The other important problem when treating a patient with endocarditis is the poor penetration of antimicrobial agents into the infected vegetation. As a result, therapy for bacterial endocarditis is typically long-term, lasting 4 to 6 weeks.

6.   Drug resistance has become a major problem in *Enterococcus* spp. There are three major problems with acquired drug resistance in organisms belonging to this genus, in addition to the organism's well-known intrinsic resistance to cephalosporins, monobactams (aztreonam), clindamycin, and trimethoprim-sulfamethoxazole. Isolates have obtained genes which encode for resistance to ampicillin/penicillin G, aminoglycosides, and, most recently and perhaps most disturbingly, vancomycin. Resistance to ampicillin/penicillin G in enterococcal isolates is due to either production of β-lactamase or modification of penicillin-binding proteins (PBPs). Of these two mechanisms, modification of PBPs resulting in decreased binding of penicillins is more important clinically and occurs more frequently.

All strains of enterococci are resistant to aminoglycosides at concentrations achievable in serum, e.g., gentamicin MIC of 16 to 64 μg/ml. This is known as low-level resistance. Fortunately, as discussed in the answer to question 5, aminoglycosides can be used synergistically with cell wall-active agents. Unfortunately, strains of enterococci have been recognized that have high-level resistance to gentamicin (MIC ≥500 μg/ml). This high-level resistance is due either to modification of the aminoglycoside binding site on the ribosome or, more commonly, to the production of enzymes which modify and thus inactivate the aminoglycosides. Isolates that have high-level resistance to gentamicin generally possess high-level resistance to tobramycin and amikacin but not to streptomycin; however, high-level resistance to streptomycin has also been reported. When enterococci demonstrate high-level resistance to aminoglycosides, the synergy between cell wall-active agents and the aminoglycoside is lost.

Perhaps the most disturbing trend in enterococcal drug resistance is the development of resistance to vancomycin. Vancomycin has generally been thought of as the "drug of last resort" for multidrug-resistant gram-positive organisms. The emergence of vancomycin resistance has challenged that dogma. There have been five types of vancomycin resistance described in enterococci: VanA, VanB, VanC, VanD, and VanE. Vancomycin resistance is due to the production of enzymes that modify the vancomycin target, significantly reducing the ability of vancomycin to block cell wall synthesis. Vancomycin-resistant enterococci (VRE) are frequently resistant to ampicillin and high levels of aminoglycosides as well. When these isolates are detected in serious infections such as endocarditis, there are few options. New antibiotics that have recently been approved by the U.S. Food and Drug Administration which have good activity against enterococci include the streptogramin combination of quinupristin and dalfopristin, which is only active against *Enterococcus faecium,* and the oxazolidinone linezolid, which is active against both *E. faecium* and *Enterococcus faecalis.* Resistance in enterococci has already been reported for both of these agents.

It is well recognized that multidrug-resistant enterococci and in particular VRE are more common in patients who have received vancomycin previously, although it has been reported that the association with prior vancomycin use is confounded by the greater duration of hospital stay in patients with VRE. Therefore, judicious use of vancomycin is important. For example, vancomycin once was the drug of choice for treatment of *Clostridium difficile* colitis despite the fact that metronidazole has been shown to be a reasonable and much less expensive therapeutic choice. Because of concerns about VRE, most institutions tightly control the use of vancomycin to treat *C. difficile* disease, encouraging the use of metronidazole instead.

Because enterococci are part of the normal gut flora, the gastrointestinal tract is a frequent reservoir for VRE. Studies have shown that patients who carry drug-resistant organisms in their gastrointestinal tract frequently contaminate their environment. Therefore, strict infection control measures, including patient isolation and barrier nursing precautions (the wearing of masks, gloves, and gowns and the strict enforcement of hand washing), are essential in preventing nosocomial spread of this organism.

REFERENCES

1. **Bayer, A. S., A. F. Bolger, K. A. Taubert, W. Wilson, J. Steckelberg, A. W. Karchmer, M. Levison, H. F. Chambers, A. S. Dajani, M. H. Gewitz, J. W. Newburger, M. A. Gerber, S. T. Shulman, T. J. Pallasch, T. W. Gage, and P. Ferrieri.** 1998. Diagnosis and management of infective endocarditis and its complications. *Circulation* **98:**2936–2948.

2. **Carmeli, Y., M. H. Samore, and C. Huskins.** 1999. The association between antecedent vancomycin treatment and hospital-acquired vancomycin-resistant enterococci: a meta-analysis. *Arch. Intern. Med.* **159:**2461–2468.

3. **Durack, D. T., A. S. Lukes, and D. K. Bright, and the Duke Endocarditis Service.** 1994. New criteria for diagnosis of infective endocarditis: utilization of specific echocardiographic findings. *Am. J. Med.* **96:**200–209.

4. **Montecalvo, M. A., W. R. Jarvis, J. Uman, D. K. Shay, C. Petrullo, K. Rodney, C. Gedris, H. W. Horowitz, and G. P. Wormser.** 1999. Infection-control measures reduce transmission of vancomycin-resistant enterococci in an endemic setting. *Ann. Intern. Med.* **131:**269–272.

5. **Murray, B. E.** 2000. Vancomycin-resistant enterococcal infections. *N. Engl. J. Med.* **342:**710–721.

CASE 52

The patient was a 32-year-old Haitian male referred to the hospital with a 3-week history of fever, nausea, vomiting, and diarrhea. Four days after returning from Haiti, where he had seen unembalmed bodies at a funeral, he developed a temperature to 39.5°C, myalgias, constipation, and rectal pain. He was admitted to an outside hospital overnight and given intravenous (i.v.) cefotaxime. He was discharged on oral cephalexin. His symptoms recurred 2 weeks later, 1 week prior to admission, and his therapy was changed to metronidazole. Five days prior to admission, the patient developed fever, diarrhea with six watery stools per day, nausea, vomiting, and dark urine. On the day of admission, the patient passed out while walking to the bathroom.

On admission, he had a temperature of 37.7°C, supine pulse rate of 104 beats/min, and blood pressure of 115/75 mm Hg. The rest of his physical and laboratory findings were unremarkable. On the second hospital day, the patient became acutely agitated, pulled out his i.v. lines, and tried to leave the hospital. He claimed that a voodoo curse had been placed on him and that he was "already dead." Hepatitis and HIV serologic tests were negative. Blood and stool cultures were diagnostic. A Gram stain and growth on MacConkey and blood agars and a triple sugar iron slant of the organism causing the patient's illness are shown in Fig. 1, 2, and 3.

1. What was the etiologic agent of this infection?

2. Explain the significance of the patient's travel history in his developing this infection.

3. Why did the patient have a syncopal episode on his way to the bathroom? Would his vital signs on admission explain why he passed out? Is his bizarre behavior consistent with his infection?

4. Organisms were found in both blood and stool. Briefly explain the pathogenesis of the bacteremia.

Figure 1

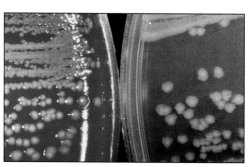

Figure 2

5. What other types of clinical specimens might be useful for culture in this disease?

6. If this person worked in the food industry, what action should be taken before he was allowed to return to work?

7. What antimicrobial resistance problems have recently emerged with this organism?

8. How can this infection be prevented in travelers?

Figure 3

CASE DISCUSSION

1. The infectious agent was *Salmonella typhi.* The organism is an enteric, lactose-nonfermenting, gram-negative rod (Fig. 1 and 2). A key biochemical characteristic of *S. typhi* is its ability to produce a small ring of H_2S on triple sugar iron agar slants at the top of the butt (Fig. 3). The vast majority of other *Salmonella* isolates will produce H_2S throughout the butt. *S. typhi* is the etiologic agent of typhoid fever.

2. In patients with febrile illnesses, a travel history can often be helpful, especially if patients have been to areas that have infectious agents not found in the United States (malaria, yellow fever, etc.). Individuals who travel to developing countries such as Haiti, with poor sanitation which can result in fecal contamination of water and food, are at increased risk for infections acquired by the fecal-oral route, such as typhoid fever. Typhoid fever is an unusual infection in the United States, with 80% of cases occurring in individuals who are infected while traveling in the developing world, where it is very common. At least some domestically acquired cases originate from food handlers who have recently traveled from the developing world as well. It should be noted that cerebral malaria also should be considered in patients who present with a history similar to this patient, i.e., travel to an area where malaria is endemic, no history of malaria prophylaxis, fever, diarrhea, and altered mental status.

3. Because of a 5-day history of diarrhea and vomiting, it is likely that the patient was dehydrated. Dehydration leads to depletion in intravascular volume. This decrease in volume results in hemodynamic changes such as postural (supine versus standing) changes in heart rate and blood pressure. When this patient stood up to go to the bathroom, his blood pressure dropped, causing him to pass out. His bizarre behavior can also be attributed to his typhoid fever. Approximately 20% of patients with typhoid fever will have some type of neuropsychiatric disturbance.

4. *S. typhi* is spread by ingestion of fecally contaminated food or water. A fairly high inoculum (10^6 organisms) is needed because these organisms are rapidly killed at pH 2, the pH of the normal stomach. The organisms which survive their transit through the stomach then multiply in the small intestine. They invade the intestinal mucosa of the ileum via M cells. M cells are specialized epithelial cells that play a role in gut mucosal immunity. Normally, antigens from the lumen of the gut are taken up by these cells and are then processed by antigen-presenting cells in the Peyer's patches. The organism has a series of genes encoding proteins involved in invasion of epithelial and epithelial-like cells. This series of genes is located in a region of the bacterial chromosome called a "pathogenicity island." The typhoid bacilli subvert the function of the M cells to invade the Peyer's patches. Within the Peyer's patches, they are phagocytized by macrophages where they survive and multiply. From the Peyer's patches, they can be carried to the bloodstream via the lymphatics.

5. Bone marrow culture is the specimen of choice in making the diagnosis of typhoid fever when routine blood and stool cultures are negative. Bone marrow cultures are positive in approximately 90% of patients, including patients in the early stages of appropriate antimicrobial therapy. Conventional blood cultures are positive in only 50% of patients, and stool cultures are positive in only 33%.

6. Humans are the only known host of *S. typhi*. Spread of this disease is typically from food and water that has been fecally contaminated by an infected individual, including asymptomatic carriers. Approximately 1 to 3% of patients who have typhoid fever will become chronic carriers of *S. typhi*. They excrete large numbers (10^6 CFU/ml) of *S. typhi* in their feces and can continue to do so for many years. Food workers who are carriers and do not practice good hygiene could contaminate the food they handle, spreading the organisms to large numbers of individuals. As a result, workers in the food industry who have had an *S. typhi* infection should have three negative stool cultures before being allowed to return to work. These cultures should be done over a period of at least 5 to 7 days to prevent sampling error.

In the carrier state, the organisms reside in the biliary tree and are excreted in bile. We saw an 80-year-old patient who had his gallbladder removed. He gave his surgeon a history of having had typhoid fever when he was 20 years old but had had excellent health since then. The surgeon sent a swab of the patient's gallbladder for culture, and it grew *S. typhi* 60 years after his original infection!

7. Approximately 20% of *S. typhi* isolates in the United States are resistant to multiple drugs, including the three drugs that have long been considered front-line therapy for typhoid fever: chloramphenicol, ampicillin, and trimethoprim-sulfamethoxazole. Not surprisingly, multidrug-resistant organisms are widespread in countries from which the majority of typhoid strains are imported, countries on the Indian subcontinent and Vietnam. In those countries, antimicrobial agents are freely available over the counter, leading to increased antimicrobial pressure and selection of resistant strains. For example, four of five patients who acquired *S. typhi* in Vietnam were infected with multidrug-resistant isolates. This finding is consistent with the observation that 80% of *S. typhi* isolates in Vietnam are multidrug resistant. A new drug-resistance problem is emerging in *S. typhi* strains from the Indian subcontinent and Vietnam. With the increasing use of fluoroquinolones to treat typhoid fever and other gastrointestinal infections, multidrug-resistant strains are also becoming resistant to the quinolone nalidixic acid, and strains resistant to the fluoroquinolones are also being seen. If fluoroquinolone resistance becomes widespread, there will be no remaining effective oral agent to treat typhoid fever in the developing or industrialized world.

8. Because of the increasing problem with antimicrobial resistance in *S. typhi*, prevention of infection has become of even greater importance. This is particularly true for travelers to the Indian subcontinent and Vietnam, where rates of multidrug-resistant *S. typhi* are high. The simplest way to avoid becoming infected by *S. typhi* is to avoid consuming fecally contaminated food and water. However, this is simply not practical in many parts of the world, where fecal contamination of food and water is the norm. Travelers to areas where sanitary conditions are poor should consider vaccination against *S. typhi*. Three vaccines are currently available and found to be protective. The initial vaccine that was developed against *S. typhi* was a parenteral, killed, whole-cell vaccine. The primary series of this vaccine includes two doses given at 4-week intervals followed by boosters every 3 years. Because of the potential for side reactions to the killed, whole-cell vaccine, including fever and localized swelling and erythema in a significant portion of the vaccinated populations (25 to 40%), the other two *S. typhi* vaccines are preferred. One is derived from the Vi polysaccharide antigen found on the surface of the typhoid bacilli. A single injection of this vaccine is given, with boosters recommended every 2 years. Vaccine efficacy is between 65 and 70%. This vaccine is not recommended for use in children less than 2 years of age. The other is a live, attenuated oral vaccine. It is given as four oral doses over a 1-week period. It elicits primarily a cell-mediated response, and immunity with this vaccine lasts for approximately 7 years. Estimates of vaccine efficacy range from 60 to 99%. Booster doses after the initial series of four oral doses are currently recommended after 5 years. A disadvantage of the oral vaccine is that it must be kept refrigerated, which may be problematic for vaccine campaigns in tropical areas. Currently this vaccine is not recommended for children less than 6 years of age.

There have been no head-to-head comparison of the three vaccines so it is unclear which vaccine is most protective, although vaccine failures have been reported with all three vaccines.

REFERENCES

1. **Ackers, M.-L., N. D. Puhr, R. D. Tauxe, and E. D. Mintz.** 2000. Laboratory-based surveillance of *Salmonella* serotype Typhi infections in the United States: antimicrobial resistance on the rise. *JAMA* **283:**2668–2673.

2. **Farmer P. E., and F. M. Graeme-Cook.** 1999: Case records of the Massachusetts General Hospital (Case 8–1999): a 28-year-old man with gram-negative sepsis of uncertain cause. *N. Engl. J. Med.* **340:**869–876.

3. **Finlay, B. B., and S. Falkow.** 1997. Common themes in microbial pathogenicity revisited. *Microbiol. Mol. Biol. Rev.* **61:**136–169.

4. **Hessel, L., H. Debois, M. Fletcher, and R. Dumas.** 1999. Experience with *Salmonella typhi* Vi capsular polysaccharide vaccine. *Eur. J. Clin. Microbiol. Infect. Dis.* **18:**609–620.

5. **Wain, J., T. T. Hien, P. Connerton, T. Ali, C. M. Parry, N. T. T. Chinh, H. Vinh, C. X. T. Phuong, V. A. Ho, T. S. Diep, J. J. Farrar, N. J. White, and G. Dougan.** 1999. Molecular typing of multiple-antibiotic-resistant, *Salmonella enterica* serovar Typhi from Vietnam. *J. Clin Microbiol.* **37:**2466–2472.

CASE 53

This 71-year-old woman was admitted with a recurrence of her poorly differentiated squamous cell carcinoma of the cervix. She underwent extensive gynecologic surgery (excision of the organs of the anterior pelvis) and was maintained postoperatively on broad-spectrum intravenous antibiotics. The patient had a central venous catheter placed on the day of the surgery.

Beginning 3 days postoperatively, the patient had temperatures of 38.0 to 38.5°C, which persisted without a clear source. On day 8 postoperatively, she had a temperature of 39.2°C. Cultures of blood and of the tip of the central line both grew an agent that was ovoid and reproduced by budding (Fig. 1). Growth of the organism on chocolate agar medium is seen in Fig. 2.

1. What is the differential diagnosis of this patient's infecting organism?

2. The organism was subsequently shown to form germ tubes (Fig. 3). What is the organism?

3. Is this organism part of the normal flora in humans?

4. How did treatment with broad-spectrum antibiotics predispose this patient to infection with this organism?

5. The same organism was present in a positive culture of blood and in a culture of the central venous catheter tip. What does this suggest in terms of the portal of entry of the organism causing the infection?

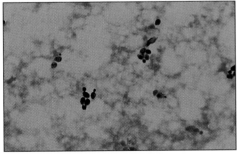

Figure 1

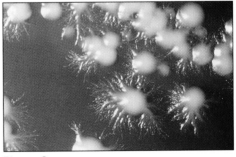

Figure 2

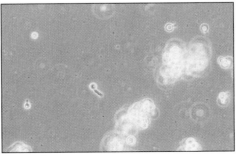

Figure 3

CASE DISCUSSION

CASE 53

1. The presence of ovoid yeast cells that reproduce by budding is consistent with a *Candida* species (such as *Candida albicans, C. tropicalis, C. parapsilosis, C. lusitaniae, C. krusei, C. glabrata,* and *C. guilliermondii*). *Histoplasma capsulatum* is also often ovoid and may demonstrate budding, but is not confused with *Candida* species because of its smaller size. In addition, it grows more slowly than do *Candida* species. Although isolates of *Cryptococcus neoformans* typically demonstrate round yeast cells, this species should also be considered when a yeast is isolated from a blood or cerebrospinal fluid sample.

2. *C. albicans* can be differentiated from other *Candida* species on the basis of its ability to form germ tubes or by demonstrating the presence of specific enzymes. The germ tube test is done by incubating suspected *Candida*-like isolates in animal serum for 1 to 2 hours and examining the organisms microscopically for germ tube formation (Fig. 3). Most but not all *C. albicans* isolates are germ tube positive. The germ tube test is an excellent presumptive test for identification of *C. albicans*. Biochemical tests based upon the detection of the enzymes L-proline aminopeptidase and beta-galactosaminidase in yeast cells in a colorimetric assay are another method to presumptively identify a yeast isolate as *C. albicans*. If both of the enzymes are present, the yeast can be presumptively identified as *C. albicans*. The advantage of this test compared with the germ tube test is that it is faster (only 30 minutes of incubation) and does not require microscopy. *Candida* species which are germ tube negative are identified by using carbohydrate assimilation and biochemical tests as well as by the observation of chlamydospores when the organism is grown on corn meal agar.

3. *Candida* species are a part of the normal flora of the skin, gastrointestinal tract, oropharynx, and vagina.

4. Treatment of patients with broad-spectrum antibacterial agents may allow the growth of yeasts in situations in which they would otherwise be inhibited by the normal bacterial flora. For example, candidal vaginitis is often associated with the prior use of a broad-spectrum antibiotic. The normal bacterial flora may compete with *Candida* species for nutrients or may create an unfavorable environment for *Candida* species growth. When the normal flora is reduced or eliminated by antimicrobial agents, *Candida* is no longer inhibited, and overgrowth of the yeast may occur.

In recent years mucocutaneous candidiasis (often including involvement of the esophagus) has been caused by immunosuppression due to infection with HIV. Unexplained oral thrush is one clue to the presence of an HIV infection.

5. Since *C. albicans* does not normally invade the bloodstream, a breakdown of host defenses must occur to allow a blood-borne infection (candidemia). The two most important defenses in this regard are (i) the presence of an intact barrier (skin) between the blood vessels and the outside and (ii) the presence of an adequate number of functioning neutrophils. With regard to the first defense, since *Candida* species may be present on the skin, intravenous drug users and patients with intravenous devices (such as this patient) are at increased risk of blood-borne infection. Patients lacking the second defense often have cancer and have been left with few neutrophils as a result of chemotherapy. The presence of the same organism in blood and on the central venous catheter tip suggests that this patient may have become infected via the breach in the integrity of the skin at the site of the central line insertion. The blood would then be seeded from the central line. A less likely possibility is that the patient had candidemia from another source and that the central venous catheter was seeded from the blood-borne infection. Treatment of candidemia can be with either amphotericin B or fluconazole. As a result of the increased recognition of drug resistance among *Candida* species, there is now an increased interest in the performance of susceptibility testing for *Candida* isolates. Although *C. albicans*, the most common cause of candidemia, is typically susceptible to fluconazole, resistance has been described. Resistance to fluconazole and other azoles is more common for other *Candida* species, such as *C. krusei*, which is a rather uncommon cause of infection. There are a number of agents that have been recently approved by the Food and Drug Administration or are now in late-stage clinical trials that may also prove to be of use in treating candidemia.

REFERENCES

1. **Pfaller, M. A., R. N. Jones, G. V. Doern, H. S. Sader, S. A. Messer, A. Houston, S. Coffman, and R. J. Hollis.** 2000. Bloodstream infections due to *Candida* species: SENTRY antimicrobial surveillance program in North America and Latin America, 1997–1998. *Antimicrob. Agents Chemother.* **44:**747–751.

2. **Rangel-Frausto, M. S., T. Wiblin, H. M. Blumberg, L. Saiman, J. Patterson, M. Rinaldi, M. Pfaller, J. E. Edwards, Jr., W. Jarvis, J. Dawson, and R. P. Wenzel.** 1999. National epidemiology of mycoses survey (NEMIS): variations in rates of bloodstream infections due to *Candida* species in seven surgical intensive care units and six neonatal intensive care units. *Clin. Infect. Dis.* **29:**253–258.

3. **Rex, J. H., J. E. Bennett, A. M. Sugar, P. G. Pappas, C. M. van der Horst, J. E. Edwards, R. G. Washburn, W. M. Scheld, A. W. Karchmer, A. P. Dine, M. C. Levenstein, and C. D. Webb.** 1994. A randomized trial comparing fluconazole with amphotericin B for the treatment of candidemia in patients without neutropenia. *N. Engl. J. Med.* **331:**1325–1330.

CASE 54 The patient was a 34-year-old male who presented with a 6-week history of acute, intractable lower back and right leg pain. A magnetic resonance imaging (MRI) study was done and showed a large extradural defect at the L4-5 space in the lumbar spine. The MRI study was consistent with a possible herniated disk, and he was admitted for surgery. A frozen section done during the operative procedure showed acute inflammation. A biopsy from the lumbar spine was sent for pathologic testing and culture. The tissue showed acute and chronic inflammation as well as scattered giant cells consistent with granulomatous inflammation. On the basis of the operative findings, purified protein derivative (PPD) and control skin tests were placed immediately postoperatively. The patient was anergic. Laboratory studies were within normal limits except for an elevated erythrocyte sedimentation rate. Cultures of biopsy material and blood cultures obtained postoperatively grew the organism shown in Fig. 1 (Gram stain) and Fig. 2 (growth on sheep blood agar). This organism was rapidly urease positive.

When the identity of the organism infecting this patient was known, a more extensive social and travel history was elicited. It was learned that 11 months earlier he had visited his family in Mexico. During his visit, both his mother and a brother had a febrile illness. He also admitted to consuming goat milk and cheese obtained from his father-in-law, who raised goats.

1. What organism do you think is causing his infection? You should be able to give the species name based on a clue in the case. What is the clue?

2. Why was a PPD test done on this patient? What does anergic mean? What skin test antigens are used to test for anergy?

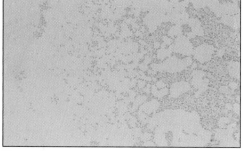

Figure 1

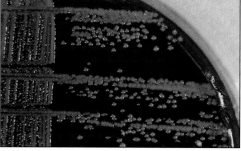

Figure 2

3. In what organs are lesions usually seen with infection due to this organism? Explain the probable steps in the pathogenesis of this patient's infection.

4. What factors concerning the pathogenicity of this organism should be taken into account when deciding on antimicrobial therapy to manage the infection?

5. If this patient had no identifiable risk factor for the organism that was infecting him, what possibility must be considered?

CASE DISCUSSION

1. This patient had a paravertebral abscess due to *Brucella melitensis*. Paravertebral abscesses occur in approximately 20% of patients with brucellosis. *Brucella* spp. are tiny gram-negative rods which will grow only on enriched laboratory medium. *B. melitensis* produces a highly active urease enzyme, so the positive urease test is consistent with this organism. Although the diagnosis of brucellosis was established in this patient on the basis of a positive culture, in many cases the diagnosis is established on the basis of serologic testing. Brucellosis is a zoonotic infection and is usually obtained in one of three ways: ingestion of unpasteurized dairy products, primarily from cows or goats, direct inoculation through cuts or scratches when caring for or attending the birth of (veterinarians, farm workers/herders) or rendering (abattoir workers) cattle (*Brucella abortus*), sheep (*B. melitensis*), pigs (*Brucella suis*), goats (*B. melitensis*), or (rarely) dogs (*Brucella canis*); or via an accident in a microbiology laboratory. The key clue in identifying the organism infecting the patient was his history of consuming dairy products from goats when visiting his family in Mexico. It is estimated that between 15 and 40% of goats in Mexico are infected with *B. melitensis*, and this organism is present in the milk of infected animals. *Brucella* attack rates are much higher in Mexico than the United States, in part because of the consumption of unpasteurized goat dairy products; brucellae are killed by pasteurization. It has been suggested that *B. melitensis* can survive stomach acidity better than other *Brucella* species.

2. The patient's presentation was also consistent with vertebral spondylitis due to *Mycobacterium tuberculosis*. This pathogenic process is also called Pott's disease. Because this patient had a pathologic process consistent with extrapulmonary *M. tuberculosis* infection, he had a skin test to see if he would react to the skin test antigen of *M. tuberculosis* PPD. A positive skin test in this clinical setting would be strong evidence that the patient was infected with *M. tuberculosis*. The patient was found to be anergic, which means that he did not give a positive skin test response to additional antigens that were given concurrently with the PPD. The other antigens were ones most patients would have been exposed to in the past and thus they should have positive skin test reactions to them. Failure to respond to these ubiquitous antigens is evidence of anergy and is frequently seen in immunosuppressed patients, especially HIV-infected patients with low CD4 cell counts. The skin test for *M. tuberculosis* is uninterpretable in anergic individuals. Antigens used for anergy testing include diphtheria and tetanus toxoids, streptokinase-streptodornase (enzymes produced by group A streptococci), and antigens derived from either *Candida* sp. (yeast commonly found on skin and mucous membranes) or mumps virus.

3. The organism is typically found in organs of the reticuloendothelial system (RES), including liver, lungs, spleen, lymph nodes, kidneys, and bone marrow. The localization of brucellae in the RES is due to the ability of this organism to survive and multiply within phagocytes. Survival within phagocytes leads to granuloma formation in infected organs. The steps in the pathogenesis of this patient's infection include ingestion of contaminated milk or cheese; increased organism survival in the acidic conditions found in the stomach; translocation across the gut wall; phagocytosis with intracellular survival and multiplication of the bacteria, followed by lysis of the parasitized phagocytes; dissemination of bacteria to bone marrow in the spine; phagocytosis with intracellular survival, multiplication, and granuloma formulation; and eventually abscess formation in the paravertebral area. The extradural process detected by the MRI study was a granuloma/abscess caused by the localization to the spine of his *Brucella* infection.

4. The ability of brucellae to survive within phagocytes makes antimicrobial therapy difficult. Many antimicrobial agents penetrate poorly if at all into phagocytes. As a result, relapse following antimicrobial therapy is common with brucellosis. Two treatment strategies are important for optimal management. First, therapy should be long-term, with antimicrobial regimens of 6 weeks frequently being recommended. Long-term therapy is necessary because relapse is frequently reported even with this prolonged regimen. Second, the use of antimicrobial agents that are active against brucellae and can penetrate into the phagocyte are desirable attributes. Because of its lipid solubility, rifampin can penetrate into white cells. It is active against brucellae as well. Because of the rapid rate at which most bacteria, including *Brucella* spp., develop resistance to rifampin if used as monotherapy, this agent should be used in combination with other drugs active against brucellae. The combination of rifampin with doxycycline has proved to be effective clinically, although the combination of doxycycline and streptomycin may be less likely to result in a relapse.

5. If no identifiable risk factors were present, the possibility that the patient is the victim of bioterrorism must be considered. These individuals may become ill after exposure directly to the organism or to the actual bioterror target, herd animals such as cattle or sheep who have fallen ill. Unusual numbers of cases of brucellosis in farmers, herders, feedlot workers, and veterinarians would be a cause for concern because infection with *Brucella* species is rare in the United States. This organism is on the list of critical biological agents along with smallpox virus, *Bacillus anthracis*, *Yersinia pestis*, *Francisella tularensis*, and others. In the absence of an identifiable risk factor for acquiring brucellosis, contact should be made with both public health and law enforcement authorities.

REFERENCES

1. **Centers for Disease Control and Prevention.** 2000. Biological and chemical terrorism: strategic plan for preparedness and response. Recommendations of the CDC Strategic Planning Workgroup. *Morb. Mortal. Wkly. Rep.* **49**(RR-4):1–14.

2. **Centers for Disease Control and Prevention.** 2000. Suspected brucellosis case prompts investigation of possible bioterrorism-related activity—New Hampshire and Massachusetts, 1999. *Morb. Mortal. Wkly. Rep.* **49**:509–512.

3. **Corbel, M. J.** 1997. Brucellosis: an overview. *Emerg. Infect. Dis.* **3**:213–221.

4. **Franz, D. R., P. B. Jahrling, A. M. Friedlander, D. J. McClain, D. L. Hoover, W. R. Bryne, J. A. Pavlin, G. W. Christopher, and E. M. Eitzen, Jr.** 1997. Clinical recognition and management of patients exposed to biological warfare agents. *JAMA* **278**:399–411.

CASE 55

A 37-year-old Panamanian man with AIDS and a history of *Pneumocystis carinii* pneumonia, treated syphilis, and a reactive purified protein derivative (PPD) skin test for which he received prophylactic isoniazid for 1 year, was admitted to an outside hospital for persistent fever, weight loss, and pancytopenia. On admission to the outside hospital, the patient had a white blood cell count of 1,700/μl, a hemoglobin level of 8.7 g/dl, and 39,000 platelets per μl. A bone marrow examination demonstrated pancytopenia, granulomas, and the presence of yeast. The patient was begun on intravenous amphotericin B. Desiring to be closer to his family, he left the outside hospital and came to our institution for evaluation.

On examination, he was a cachectic, weak-appearing man in no acute distress. He had oral thrush and 3+ pitting lower extremity edema to his knees. The culture of his bone marrow from the outside hospital was positive for a dimorphic fungus that was subsequently identified both morphologically and by nucleic acid probe. The bone marrow aspirate showing the infecting organism is shown in Fig. 1. Growth of the organism incubated at 30°C is shown in Fig. 2. Microscopic morphology of the organism growing at 30°C is shown in Fig. 3.

1. What are the dimorphic fungi? With which of these agents is he infected? Which of these have an increased rate of dissemination in patients with AIDS?

2. Does the patient's country of origin, Panama, give any additional clues? Are there particular environments in which this organism is more likely to be found?

3. How does this organism appear at body temperature? At room temperature?

4. Clinically, how does disseminated disease with this organism occur in patients who have not recently been exposed to this organism?

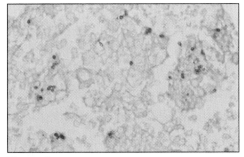

Figure 1

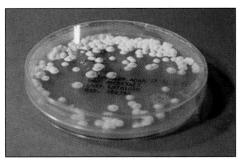

Figure 2

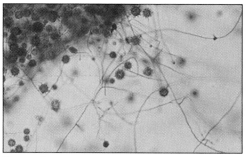

Figure 3

5. What other infections can invade the bone marrow, leading to fever, weight loss, and pancytopenia in patients with AIDS?

6. Culture is a rather slow method (on the order of several weeks) of diagnosing a disseminated infection with this organism. What other methodology is clinically useful in diagnosing patients suspected of having disseminated disease with this organism?

CASE DISCUSSION

CASE 55

1. Dimorphic fungi exist in the mold form at 25 to 30°C and as a yeast or yeastlike form at body temperature. The dimorphic fungi include *Blastomyces dermatitidis*, *Coccidioides immitis*, *Histoplasma capsulatum*, *Paracoccidioides brasiliensis*, *Penicillium marneffei*, and *Sporothrix schenckii*. Of these fungi, *C. immitis*, *H. capsulatum*, and *Penicillium marneffei* all have an increased rate of dissemination in patients with AIDS, and the mortality of infection with *Paracoccidioides brasiliensis* is high. This patient is infected with *H. capsulatum*, the etiologic agent of histoplasmosis. Identification is based on the mold form of the fungus. In Fig. 3, the characteristic tuberculate macroconidia of *H. capsulatum* can be seen.

2. Histoplasmosis was originally described in Panama. Regions where *H. capsulatum* is endemic include many areas in the midwestern and south central United States, especially in the Ohio, Mississippi, and Missouri River valleys. What is not well appreciated, however, is that much of the Caribbean basin is also a region in which this disease is endemic. In fact, of patients admitted to Boston Medical Center Hospital over the past several years who were subsequently found to have histoplasmosis, all were originally from the Caribbean basin, with Puerto Rico the most common site of origin. This, of course, reflects the patient population at this institution, but it also serves to illustrate that the geographic distribution of this disease is not limited to certain regions within the United States.

In nature, although *H. capsulatum* does not infect birds, soil that is rich in bird or bat droppings (near chicken coops, pigeon roosts, caves, starling roosts, etc.) is a rich nutrient source for this organism and often contains it.

3. At body temperature, *H. capsulatum* is found in the yeast form (Fig. 1). Within the body, it is typically small, oval cells with thin walls. At room temperature, conversion to a mold occurs (Fig. 2). The tuberculate macroconidia of the mold form of *H. capsulatum* are shown in Fig. 3.

4. Although most cases of disseminated histoplasmosis complicating AIDS are likely due to exogenous infection, the disease can also occur as a result of reactivation of a latent infection. Like infections with *Mycobacterium tuberculosis*, an acute infection with *H. capsulatum* is usually contained by the body's cell-mediated immunity. An exposure to *H. capsulatum* in an immunocompetent host typically results in an immune defense to wall off the organism. If the host's cell-mediated immunity is compromised, as it can be, for example, by infection with HIV, the balance between the host and the pathogen will be tilted in favor of the pathogen. The host's cell-mediated immunity may no longer be able to keep the *H. capsulatum* in check, and in some cases disseminated disease occurs.

In patients with HIV who have disseminated histoplasmosis, the standard recommendation for therapy is to continue it for life, though in patients who have a good clinical response to highly active antiretroviral therapy, it is unknown if this is required.

5. The other common infectious agents which cause pancytopenia in AIDS patients and invade the bone marrow are the *Mycobacterium avium* complex (MAC), *M. tuberculosis*, and those *Leishmania* species that cause visceral leishmaniasis. Disseminated MAC infections typically occur when the patient's cell-mediated immunity is markedly suppressed, as can be seen by the low number of CD4-positive cells in the patient's blood. Both of these mycobacterial infections can be diagnosed by appropriate mycobacterial culture techniques. Although visceral leishmaniasis has not been reported from Panama, sand fly vectors do exist in parts of the country that could potentially transmit the infection if it was introduced into Panama.

6. A test that detects *H. capsulatum* polysaccharide antigen in urine has been developed that has good sensitivity and good specificity in AIDS patients with disseminated histoplasmosis. The detection of this polysaccharide antigen in urine is much more rapid than culture. In addition, the level of antigen falls in response to successful antifungal therapy and can be used to monitor cases for relapse. The drawback of this test is that it is not done in hospitals and must be sent to a reference laboratory, extending the time until a result is available. Cross-reactions with other dimorphic fungi such as *Blastomyces* and *Coccidioides* spp. may occur. In some cases of disseminated histoplasmosis, the organisms can be seen in smears from peripheral blood or buffy coat.

REFERENCES

1. **Centers for Disease Control and Prevention.** 1995. Histoplasmosis—Kentucky, 1995. *Morb. Mortal. Wkly. Rep.* **44**:701–703.

2. **Levitz, S. M., and E. J. Mark.** 1998. Case records of the Massachusetts General Hospital. Weekly clinicopathological exercises. Case 38-1998. A 19-year-old man with the acquired immunodeficiency syndrome and persistent fever. *N. Engl. J. Med.* **339**:1835–1843.

3. **Singh, V. R., D. K. Smith, J. Lawerence, P. C. Kelly, A. R. Thomas, B. Spitz, and G. A. Sarosi.** 1996. Coccidioidomycosis in patients infected with human immunodeficiency virus: review of 91 cases at a single institution. *Clin. Infect. Dis.* **23**:563–568.

4. **Wheat, L. J., P. A. Connolly-Stringfield, R. L. Baker, M. F. Curfman, M. E. Eads, K. S. Israel, S. A. Norris, D. H. Webb, and M. L. Zeckel.** 1990. Disseminated histoplasmosis in the acquired immune deficiency syndrome: clinical findings, diagnosis and treatment, and review of the literature. *Medicine* **69**:361–374.

CASE 56

The patient was a 7-week-old girl who was admitted with a 1-week history of rhinitis, cough, mild fever, peeling rash on the hands and feet, and a white blood cell count of 35,600/µl. Other than being premature (35 weeks, 2.97 kg), she had an uneventful neonatal period. One week after admission, she was intubated because of respiratory difficulty. The respiratory difficulty resolved, and she was extubated 6 days later. Radiographic findings demonstrated hepatosplenomegaly and extensive periostitis with metaphyseal destructive changes.

↑ WBC

The infant was treated with intravenous penicillin G for 14 days, and laboratory test results indicated that treatment was successful. In addition, both parents were counseled and given appropriate treatment.

1. What infection do you think this child had? (Important clues: rhinitis, peeling rash on hands and feet, and periostitis.)

2. How did this child become infected?

3. Why were the parents treated?

4. Discuss the different stages and natural history of this infection. At what stage is this child?

5. How is the diagnosis of this infection made? What are some of the problems associated with this diagnostic strategy in newborns?

6. What alternative diagnostic strategy(ies) would be attractive for diagnosis of this infection?

CASE DISCUSSION

1. The findings of rhinitis, peeling rash more pronounced on the palms and soles, leukocytosis, hepatosplenomegaly, and periostitis with metaphyseal destruction on radiographic examination are all consistent with the diagnosis of congenital syphilis. Syphilis is caused by the spirochete *Treponema pallidum*. Neonatal infections with this organism are often difficult to diagnose clinically because they mimic other congenital and neonatal infections such as those caused by cytomegalovirus, herpes simplex virus, coxsackievirus, and *Toxoplasma gondii*. In addition, many congenitally infected newborns lack obvious clinical manifestations of syphilis. Some of these will develop manifestations of syphilis while others will remain asymptomatic for life.

2. This patient was infected in utero by the mother. The outcome of the pregnancy in congenital syphilis is related to the stage of maternal syphilis at the time of conception or the stage of gestation at the time of infection. The mother probably acquired a primary infection (see answer to question 4 for further explanation) during pregnancy and transmitted the organism transplacentally. Primary infection of the mother during the first trimester is associated with an especially high rate of fetal mortality. Neonates with congenital syphilis frequently are premature. The role this infection plays in prematurity is difficult to determine, especially since other factors associated with low-birth-weight babies such as a lack of prenatal care, use of recreational drugs, infections with other sexually transmitted diseases, etc., are also associated with prematurity. Infection acquired during passage through a *T. pallidum*-infected birth canal is another possible mode of transmission to neonates but is believed to occur infrequently.

3. For the child to be congenitally infected, the mother must be infected. The presence of congenital syphilis indicates that the mother was unaware of her *T. pallidum* infection and untreated, or she was diagnosed during pregnancy but inadequately treated. Among the 529 cases of congenital syphilis reported in the United States in 2000, 434 (82%) occurred because the mother had no documented treatment or had received inadequate treatment of syphilis before or during pregnancy. The U.S. Centers for Disease Control and Prevention (CDC) recommends serologic testing for syphilis (such as with the rapid plasma reagin test; see answer to question 5 for further information) for all women during the early stages of pregnancy. In areas where the prevalence of the disease is high and among women who are at high risk for syphilis, the CDC recommends testing twice in the third trimester, including once at delivery. In addition, all women who deliver a stillborn infant after 20 weeks of gestation should be tested. Syphilis screening should be offered in emergency departments, jails, prisons, and other settings that provide episodic care to pregnant women at high risk for syphilis. In addition, the natural course of syphilis may be altered by pregnancy, so these women are more likely to be asymptomatic. The great majority of serologically positive pregnant women who deliver either congenitally infected stillborn or live infants do not have a history of primary or secondary syphilis.

As is the case with all sexually transmitted diseases, the sexual partners of this woman should also be treated since they are at risk for infection.

4. The natural history of syphilis (*T. pallidum* infection) may evolve through three stages. The primary stage usually occurs approximately 3 weeks (range 9 to 90 days) after contact (typically sexual) with an infected individual. It is characterized by the development of the primary lesion, the chancre, at the site of infection. Many individuals, especially females, do not seek treatment because the lesions are painless and may not be visible if they are in the vagina or anal region. The primary stage of the disease appears to be the stage during which syphilis is most readily transmitted from mother to fetus. Because the mother may not be aware of her infection, and thus not seek treatment, the fetus is at significant risk for this infection.

Individuals who are untreated progress to secondary syphilis. In this stage, a skin rash that begins on the trunk and spreads to the extremities, including the palms and soles, is seen in 90% of patients. This manifestation is due to dissemination of spirochetes throughout the body. Skin lesions are teeming with spirochetes, so gloves should be worn when examining the skin rash in individuals with suspected secondary syphilis. Constitutional symptoms such as fever, malaise, and arthralgia are common in adults. Adenopathy and hepatitis are frequent. The spirochetes may also invade the central nervous system. The rash and other clinical findings resolve, but in a minority of cases will occur again.

After resolution of the secondary stage of the disease, the infection enters the latent phase. The individuals are asymptomatic but still have evidence of disease based on the presence of specific antibodies to *T. pallidum*.

Latent syphilis develops into late or tertiary syphilis in approximately 30% of untreated individuals. This stage of disease occurs years after resolution of the secondary stage. The most severe manifestations of late syphilis are neurosyphilis and cardiovascular syphilis. Two of the most important findings in neurosyphilis are general paresis and tabes dorsalis. Late syphilis, especially of the cardiovascular system, is encountered infrequently now that penicillin has been established as a safe and effective therapy for syphilis.

Congenital syphilis has features consistent with both the secondary and tertiary stages of disease. Bacteremia with the spirochete is overwhelming in this infection, with a variety of organ systems, including the central nervous system, being infected.

5. Congenital syphilis can be definitively diagnosed in two ways. The spirochetes can be visualized directly in lesions or mucous discharge from the infected neonate or by detecting the organism in the placenta and/or umbilical cord. Scrapings of lesions or mucous discharge are examined using dark-field microscopy. Placental and umbilical cord tissues are typically examined using silver staining. Dark-field microscopy requires a special microscope and an individual skilled in this special microscopic technique. It is used primarily in STD (sexually transmitted disease) clinics, where large numbers of potentially infected individuals are available for study. The definitive diagnosis of congenital syphilis may be made by demonstrating the spirochetes in tissues or fluids obtained from the neonate, but this method is used infrequently.

The diagnosis of congenital syphilis, as well as most other cases of syphilis regardless of disease stage, is made serologically. Screening tests, either the RPR (rapid plasma reagin) or VDRL (Venereal Disease Research Laboratory) test, are used in this diagnosis. These two tests detect antibodies to reagin, an antigen present in patients with syphilis. This antigen is believed to be produced as a result of interactions of *T. pallidum* with the patient's tissue. A positive RPR or VDRL test does not necessarily confirm the diagnosis of *T. pallidum* infection. The reason is that patients with other disease states, both infectious and autoimmune, can have a false-positive RPR or VDRL test. Patients who have positive results in one of these two serologic tests should have a confirmatory test done.

The confirmatory tests for *T. pallidum* diagnosis use the organism or components from its surface as antigens to which the patient's antibodies are tested. Two tests are widely used, FTA-abs (fluorescent treponemal antibody-absorbed) and MHA-TP (microhemagglutination assay–*T. pallidum*). The confirmatory tests remain positive throughout the infected individual's lifetime.

One of the major problems in making a serologic diagnosis of congenital syphilis is that children born to seropositive mothers can be positive in both screening and confirmatory tests without having the disease. The reason for this is the transplacental transfer of *T. pallidum*-specific immunoglobulin G (IgG) antibodies from the infected mother to the fetus. To determine if the child is infected, sequential sera should be titered using the nontreponemal serologic tests, i.e., RPR or VDRL. In this particular child, her initial RPR titer was 1:256 and she had a positive FTA-abs. A second RPR titer done approximately 3 weeks later showed a fourfold rise to 1:1,024, which is consistent with active infection with *T. pallidum*. Approximately 5 weeks after the completion of penicillin therapy, her RPR titer had dropped 32-fold to 1:32, which would be consistent with successful therapy. Her clinical picture and her serologic response are consistent with congenital syphilis.

6. Because this organism cannot be cultivated in vitro and direct examination of tissues for this organism lacks sensitivity, alternative means to detect this organism would be useful. DNA amplification techniques such as PCR (polymerase chain reaction) have been described for the detection of this organism and show promise. However, PCR is mainly restricted to research settings. Another approach is to detect IgM specific for *T. pallidum* in the serum of neonates suspected of having congenital syphilis. IgM detection has two advantages. IgM does not cross the placenta, so the presence of *T. pallidum*-specific IgM antibodies would indicate that the neonate was infected. Second, IgM antibodies are the initial antibodies produced in response to an antigen, so they would be present before IgG antibodies. Two different approaches are used early in the disease course to detect *T. pallidum* IgM-specific antibodies, a Western blot (immunoblot) assay and IgM-FTA-abs. Like DNA amplification, both of these techniques are used mainly in research settings.

REFERENCES

1. **Centers for Disease Control and Prevention.** 2001. Congenital syphilis—United States, 2000. *Morb. Mortal. Wkly. Rep.* **50**:573–577.

2. **Larsen, S. A., B. M. Steiner, and A. H. Rudolph.** 1995. Laboratory diagnosis and interpretation of tests for syphilis. *Clin. Microbiol. Rev.* **8**:1–21.

3. **Stoll, B. J.** 1994. Congenital syphilis: evaluation and management of neonates born to mothers with reactive serologic tests for syphilis. *Pediatr. Infect. Dis. J.* **13**:845–853.

4. **Wicher, V., and K. Wicher.** 2001. Pathogenesis of maternal-fetal syphilis revisited. *Clin. Infect. Dis.* **33**:354–363.

CASE 57 The patient was an 18-year-old female who presented to the ear, nose, and throat clinic complaining of hoarseness and difficulty in swallowing. She had a 1-week history of sore throat, fever, easy fatigability, and myalgia. Her examination was significant for enlarged tonsils touching at the midline with exudate present. Bilateral tender anterior and posterior cervical lymphadenopathy, as well as splenomegaly, were present. Her complete blood count (CBC) showed a hematocrit of 44% and a white blood cell (WBC) count of 7,000/μl with 40% neutrophils, 28% lymphocytes, 12% atypical lymphocytes, and 20% monocytes. Liver function tests showed an aspartate aminotransferase (AST) level of 155 U/liter, an alanine aminotransferase (ALT) level of 208 U/liter, and an alkaline phosphatase level of 189 U/liter. Electrolytes were normal. Lateral neck radiographs showed a clear airway; the chest radiograph was negative. She was admitted to the hospital. A throat culture was sent to rule out gonococcal infection and beta-hemolytic streptococci. Viral serologic tests were ordered. She was treated with intravenous hydration, clindamycin, and steroids. On hospital day 2, a Monospot test result was positive. The clindamycin therapy was stopped, and oral prednisone was given. Her condition showed some improvement, with decreased tonsillar size evident on examination by hospital day 5.

1. What was the differential diagnosis? What viral serologic tests should have been ordered given her physical examination and laboratory findings?

2. What was the etiology of her illness? How was her diagnosis confirmed?

3. Briefly describe the epidemiology of infections with this etiologic agent.

4. Why was this patient given corticosteroids?

5. What complications can the agent causing her infection produce in immunosuppressed hosts? What is our current understanding of how this occurs?

CASE DISCUSSION

CASE 57

1. Her clinical presentation of fever, pharyngitis, cervical lymphadenopathy, and splenomegaly is consistent with infectious mononucleosis. Infectious mononucleosis is typically due to either Epstein-Barr virus (EBV) or cytomegalovirus (CMV). Acute HIV infection can also present as an infectious mononucleosis-like syndrome. Given her significant pharyngitis, her physician also considered group A streptococcal infection and, because she was sexually active, gonococcal pharyngitis. Her enlarged tonsils raised the possibility of retropharyngeal abscess, often a surgical emergency. Given her significant lymphadenopathy, acute toxoplasmosis would also need to be included in the differential diagnosis. Her fever and elevated liver enzymes might suggest she has an acute hepatic infection.

The serologic tests that should be done would be a test for heterophile antibodies such as the Monospot test (see answer to question 2 for further details), CMV immunoglobulin M (IgM) antibodies to diagnose acute CMV-associated mononucleosis, and HIV antibody enzyme immunoassay (EIA). IgM antibodies for hepatitis A, surface antigen and antibodies to core antigen of hepatitis B virus, and hepatitis C virus antibodies might also be sought, although their diagnostic value in this particular clinical setting likely would be minimal.

2. This patient had infectious mononucleosis due to EBV. This diagnosis was based on her clinical presentation, her physical findings, a WBC count and differential count consistent with infectious mononucleosis, and a positive Monospot test result. Her diagnosis was confirmed by specific EBV serologic tests. The positive Monospot test result indicated that the patient had heterophile antibodies. Heterophile antibodies, detected by an agglutination reaction typically to antigens of sheep or horse erythrocytes coated on latex beads, represent a nonspecific activation of B cells. These antibodies are present in 90% of EBV-infected patients at some time during acute illness. The heterophile antibody test is helpful when positive. The test is highly sensitive in adolescents and adults, although less so in children younger than 4 years.

False-positive heterophile tests are unusual and are typically seen in patients with lymphoma or hepatitis. To confirm acute EBV infection, EBV-specific antibodies can be sought. Detection of IgM antibodies to EBV viral capsid antigen (VCA) is the most accurate serologic test for diagnosis of acute EBV infection.

If the heterophile antibody test is negative and the clinical suspicion for EBV-induced mononucleosis is high, specific EBV serologic tests should be done since a small number of EBV infections in adults can be heterophile negative. CMV can also cause a heterophile-negative acute mononucleosis syndrome, so CMV serologic tests should also be done in heterophile-negative individuals.

3. EBV is often referred to as the "kissing disease" because it is typically acquired by intimate oral contact such as kissing. EBV replicates in oropharyngeal epithelial cells and B cells. The virus is shed in saliva by most seropositive individuals. Following acute infection, B cells become latently infected with the viral genome, forming a circular episome in the nucleus. The virus is believed to persist within infected resting B cells.

In infants and children, EBV infections are typically either asymptomatic or very mild and do not attract clinical attention. In the small subset of individuals who are not infected early in life, EBV infections can be more severe. Disease incidence is highest in adolescents and young adults (15 to 24 years), such as the patient discussed in this case. These individuals may have significant morbidity, with a spontaneously resolving illness lasting 2 to 3 weeks being the norm. Acute EBV infection is a frequent reason for hospitalizations among college students and an important cause of lost training days in military recruits. Deaths from acute EBV infections are rare and are typically due to airway obstruction caused by swelling of lymphoidal tissue, splenic rupture, or encephalitis.

EBV has been associated with an unusual form of lymphoma found primarily in equatorial Africa, Burkitt's lymphoma. It has also been associated with nasopharyngeal carcinoma. In both of these malignancies, EBV DNA can be found in the tumor cells. EBV has been associated with malignancy in HIV-infected patients and a lymphoproliferative disorder in transplant recipients. These latter two disorders are discussed in the answer to question 5.

4. The use of steroids in uncomplicated cases of mononucleosis is controversial. However, in patients with significant tonsillar enlargement where concern exists that airway obstruction may occur, patients should be hospitalized and corticosteroid use is indicated, as it has been shown to quickly reduce this enlargement. Steroid use is also indicated to treat EBV-induced hemolytic anemia and thrombocytopenia. Currently, no antiviral agents have been shown to be effective in the therapy of acute infectious mononucleosis.

5. Reactivation of latent EBV infection can occur in immunosuppressed patient populations with defects in cell-mediated immunity, especially in HIV-infected patients with low CD4 counts and in patients receiving organ transplant. In individuals with AIDS, this reactivation can result in an oral hairy leukoplakia that causes a white corrugated oral lesion typically seen on the side of the tongue. This is a nonmalignant condition where replicating virus can be seen. Non-Hodgkin's lymphoma is a common malignancy found in HIV-infected individuals.

In both individuals with AIDS and transplant recipients, a lymphoproliferative disorder can occur. EBV-seronegative transplant recipients are at much greater risk for this disorder than are seropositive individuals. There is a spectrum of disease associated with this disorder ranging from a mononucleosis-type illness to lymphomas in various locations throughout the body. This disorder will often spontaneously resolve with reduction of immunosuppressive therapy or improvement in immune status in HIV-infected individuals.

EBV is specific for B cells and can infect these cells latently, with the virus forming a circular episome in resting B cells. A transmembrane protein called latent membrane protein 1 (LMP1) can be expressed by latently infected cells. This protein has been found on the surface of cells from patients with posttransplant lymphoproliferative disorder and AIDS-associated lymphomas. In these two disease states, LMP1 has been shown to bind a cytoplasmic signal-transducing molecule called TRAF. TRAF activates another protein, NF-κB transcription factor, which enters the nucleus and causes B cells to proliferate.

REFERENCES

1. **Auwaerter, P. G.** 1999. Infectious mononucleosis in middle age. *JAMA* **281:**454–459.

2. **Cohen, J. I.** 2000. Epstein-Barr virus infection. *N. Engl. J. Med.* **343:**481–492.

3. **Liebowitz, D.** 1998. Epstein-Barr virus and a cellular signaling pathway in lymphomas from immunosuppressed patients. *N. Engl. J. Med.* **338:**1413–1421.

CASE 58

The patient was a 41-year-old man who had returned from central Africa 12 days prior to his admission. He had been in Africa for approximately 7 months, working as a civil engineer on road-building projects in Rwanda, Zaire, and Uganda.

Seven days before admission, the patient noted the acute onset of fevers with chills as well as cough and myalgias. For the next 3 days he had episodes of fevers and chills approximately three times a day. After 3 days of fever, he sought care at a local urgent care center. He told the physician that he had just returned from Africa and had not taken malaria prophylaxis. He had a white blood cell count of 2,200/µl and a platelet count of 102,000/µl. No hematocrit was reported. A purified protein derivative (PPD) test was placed. The patient was diagnosed as having bronchitis, treated with azithromycin, and instructed to return in 48 hours to have his PPD test read.

On his return, the patient was found to be hypotensive and had mental status changes. He was noted to be icteric. He was referred to the emergency room of the local hospital to "rule out hepatitis." A diagnosis was made in the hematology laboratory (Fig. 1). At the time of his diagnosis, he was noted to have a hematocrit of 25%. After 24 hours of antimicrobial and aggressive fluid therapy, he had a cardiopulmonary arrest. He was resuscitated. In the next 24 hours, he developed acute respiratory distress syndrome and became progressively hypotensive.

He was transferred to a tertiary care hospital. On arrival, he was intubated and comatose. On physical examination, he was grossly icteric and tachycardic and was noted to be oozing bright red blood from his mouth and catheter sites. His laboratory tests were significant for a hematocrit of 23%, platelets of 39,000/µl, prolonged bleeding times, and a 3+ hemoglobin level and 25 to 50 red blood cells in his urine. He was given fresh-frozen plasma, platelets, and cryoprecipitate for his disseminated intravascular coagulation. Exchange transfusion was begun, and the patient received 7 units of packed red blood cells. His condition continued to deteriorate, and he had a cardiac arrest from which he could not be resuscitated.

1. With what organism was this patient infected? (You should be able to identify this organism to the species level.)

2. What factors put this individual at risk for infection with this organism? What factor contributed to his fatal outcome?

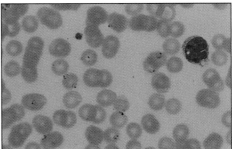

Figure 1

3. Briefly describe the pathogenesis of infection with this organism.

4. Why do you think exchange transfusion was used as a therapeutic strategy in this patient?

5. Discuss the problems associated with chemotherapy for infection with this organism.

6. The number of cases of infection with this organism is increasing in the United States. Explain why this is occurring.

7. Describe preventive strategies for this organism.

CASE DISCUSSION

58

1. Figure 1 demonstrates the delicate ring forms which are consistent with trophozoites of *Plasmodium falciparum*. Also note the red blood cells infected with multiple ring forms. This finding is most frequently observed in, but is not limited to, *P. falciparum*. *P. falciparum* is the most common cause of malaria in sub-Saharan Africa. It also is the species of human malaria most frequently associated with mortality (see answer to question 3 for further details). Other species of malaria include *Plasmodium vivax*, *Plasmodium ovale*, and *Plasmodium malariae*. Other life-threatening infections in travelers to Africa are less common but include typhoid fever, meningococcal disease, and (rarely) viral agents that cause hemorrhagic fever.

2. There are several factors which put this patient at increased risk for developing malaria. First, he traveled to a malarious region. Second, he failed to take appropriate prophylaxis. Approximately 85% of the patients from the industrialized North—the United States, Canada, and Europe—who develop malaria either fail to take prophylaxis (60%) or take the wrong prophylaxis (25%). Third, because the patient was from an area where malaria is not endemic, he did not have any immunity to malaria. In comparison, individuals who live in areas where malaria is endemic have partial immunity if they survive the initial bouts of malaria when they are young. (Malaria is a major cause of death in children less than 5 years of age.) This is due to frequent exposure to the parasite. Fourth, because the patient was working on road projects, it is likely that he spent at least some time in rural areas where he was more likely to have been exposed to the infected vector of the parasite, the female *Anopheles* sp. mosquito. This patient's fatal outcome was due to many factors: his delay in seeking medical attention when he first became symptomatic, the failure of his physician to make the proper diagnosis and to institute antimalarial agents on initial presentation, and the aggressive nature of this parasitic organism, due in part to the patient's lack of immunity to the organism. This outcome should have been prevented. The important lesson of this case is that a travel history to an unusual locale should expand your differential diagnosis to include organisms you would not normally encounter in the United States, Canada, or Europe. In particular, any patient with fever, regardless of other symptoms, who has been to sub-Saharan Africa, the Indian subcontinent, or Central America in the prior 8 weeks and failed to take malaria prophylaxis should be considered to have malaria until laboratory tests for the parasite prove otherwise. This patient's peripheral hematology smear was loaded with the ring forms due to the high level of parasitemia (estimated initially as 20%) and the hematology laboratory quickly notified the physician caring for the patient after his hospitalization. Interestingly, the physician who saw the patient initially considered tuberculosis as a potential cause of his illness. Critical review of the patient's history suggests that malaria would have been much more likely.

3. This patient had several complications associated with *P. falciparum*-induced malaria, including significant anemia, pulmonary failure, and two complications classically associated with this species: acute renal failure (blackwater fever) and cerebral malaria. Acute renal failure was manifested in this patient by the presence of red blood cells and hemoglobin in his urine. This can cause discoloration of urine, from which blackwater fever got its name. The clinical manifestation of cerebral malaria seen in this patient included altered mental status changes progressing to coma and death.

The pathogenesis of *P. falciparum* begins with the bite of an infected female *Anopheles* mosquito. The sporozoite stage of the parasite is present in the salivary gland of the mosquito and enters the bite wound via the mosquito's saliva. From there, the sporozoites are carried by the bloodstream to infect hepatocytes. The parasites multiply in the hepatocytes and are released into the bloodstream as the merozoite phase. The merozoite phase then infects erythrocytes. As the *P. falciparum* parasites mature in red blood cells (RBCs), knobs develop on the surface of these cells. A malaria-derived protein found on the surface of the infected RBCs (*P. falciparum* erythrocyte membrane protein-1 [PfEMP1]) is believed to play a central role in the pathogenesis of malaria. PfEMP1 mediates the adherence of infected RBCs to endothelial cells in the microvasculature of the brain, kidney, lung, and other organs. In addition, it is thought to be responsible for the rosetting of uninfected RBCs to infected ones. The end result of infected RBCs binding to the endothelial cells and the rosetting of uninfected RBCs is a process known as sequestration. Sequestration is believed to cause obstruction of the microvasculature in the central nervous system and kidneys. This obstruction along with the production of increasing levels of cytokines induced by the infection are thought to play central roles in cerebral malaria, a frequently fatal manifestation of *P. falciparum* infection. Consequences of this blockage include decreases in glucose and oxygen levels in the brain tissue. Similar microvascular blockage in the kidney, coupled with high levels of circulating hemoglobin, is thought to cause the renal failure seen in *P. falciparum*-infected patients.

4. Exchange transfusion is a desperate measure used in only the most severe cases of malaria (usually caused by *P. falciparum*). Typically, exchange transfusions are done in nonimmune individuals such as this patient who have high-level parasitemia (10 to 15% or higher). A physician who reviewed this patient's initial hematology smear estimated the parasitemia to be 20%. The idea of exchange transfusion is to eliminate damaged RBCs and a large percentage of parasites while restoring the RBC volume to normal.

5. Chloroquine is the treatment of choice for malaria because the drug is cheap, safe, and generally effective. However, chloroquine-resistant *P. falciparum* is endemic in sub-Saharan Africa. Therefore, this patient should be treated as if he were infected with a chloroquine-resistant organism. A variety of drugs are available to treat resistant *P. falciparum* infections, including quinine, quinidine, mefloquine, halofantrine, doxycycline, atovaquone-proguanil, and pyrimethamine-sulfadoxine. The pros and cons of the different treatment regimens are beyond the scope of this book and will not be considered here. The interested student should consult the Centers for Disease Control and Prevention (CDC) website (http://www.cdc.gov) or call the CDC National Center for Infectious Disease, Division of Parasitic Diseases, at 770-488-7788.

6. Several factors are involved in the increasing numbers of cases of malaria that are being seen in the United States, but two stand out. One is the increasing travel of non-immune individuals to malarious regions, especially in sub-Saharan Africa, the Indian subcontinent, and Central America. Those who become infected while visiting these regions almost always acquire malaria because they either fail to take prophylaxis, take the wrong prophylaxis, or are not compliant with appropriate prophylaxis. In 1995, the last year for which comprehensive data are available, 56% of cases occurred in U.S. citizens who traveled to malarious regions of the world. The second important factor is an increase of infections in individuals who migrate to or visit the United States from malarious regions, especially Africa and the Indian subcontinent. This patient population is responsible for approximately 43% of cases seen in the United States. There is typically a small number of cases (less than 1%) transmitted in the United States annually. Most are due either to vertical transmission from mother to child (the mother is typically an immigrant from a region where malaria is endemic) or from blood transfusion. A very small number of cases are actually acquired in the United States from infected *Anopheles* mosquitoes. Two different scenarios have been suggested. One is that the individual acquires malaria via a mosquito that has taken a blood meal from an infected individual who had recently entered the United States from a malarious region. An example of this mode of transmission was recently reported from Long Island in New York State. This type of transmission only occurs if environmental conditions, i.e., temperature, humidity, and breeding grounds for the *Anopheles* mosquito, are favorable for the malaria developmental cycle in the mosquito. The second scenario is that infected mosquitoes are carried by jet planes from tropical regions. Studies have shown that viable female *Anopheles* mosquitoes can be found on jets that have flown from tropical regions to the industrialized North. Individuals living near international airports in both England and the United States have developed malaria without having traveled to malarious regions. In the United States, malaria was found along much of the mid-Atlantic seaboard and the Gulf of Mexico until the 1940s, suggesting that environmental conditions are in place which would allow malaria transmission in these regions. The vector for malaria is widespread throughout the United States, so there is, in theory at least, the potential for endemic malaria returning to the United States. Given these factors, the importance of maintaining a public health infrastructure to prevent the return of endemic malaria must be emphasized.

7. Prevention of malaria is dependent upon chemoprophylaxis as well as measures to prevent bites by the vector of malaria, the *Anopheles* mosquito. For *P. falciparum* in areas in which chloroquine resistance is documented, mefloquine is the current prophylactic agent of choice, while doxycycline can be used as an alternative in older children and adults who cannot tolerate mefloquine. These drugs must be taken before leaving home, to ensure that the agent can be tolerated; throughout the stay in an area where malaria is endemic; and for 4 weeks after leaving that area, so that infections acquired near the departure time will be prevented. Recommendations for malaria chemoprophylaxis can change. The most current recommendations can be obtained by calling the CDC Malaria Hotline at 888-232-3228.

The *Anopheles* mosquito feeds primarily at night, making evening and nighttime precautions particularly important. Measures to prevent exposure to this mosquito include staying in well-screened areas in the evening and night; using mosquito netting, preferably impregnated with insecticides, when sleeping; wearing clothing that covers the trunk and extremities; using insect repellent containing N,N-diethyl-*m*-toluamide (DEET); and spraying sleeping areas with insecticides.

Finally, attempts are ongoing to develop a malaria vaccine. Progress in developing such a vaccine has been slow, in part because of the exceedingly complex life cycle of the parasite. A number of different vaccine strategies have been proposed. One promising approach is a recombinant vaccine derived from the major surface protein of the infective stage of the parasite, the sporozoite. The target antigen is a polypeptide of circumsporozoite protein that has been fused with HbsAg. This vaccine has been referred to as the RTS,S vaccine. In a small number of nonimmune patients, the vaccine has been shown to be safe and efficacious. One of the challenges in developing this vaccine is the need to induce immunity such that the parasite is eliminated before it enters the hepatocyte. Once the sporozoite enters the hepatocyte and develops into merozoites, it undergoes significant antigen variation, making it unlikely that the immune response generated by this vaccine will be effective at the next stage of the disease.

REFERENCES

1. **Centers for Disease Control and Prevention.** 1999. Malaria surveillance—United States, 1995. *Morb. Mortal. Wkly. Rep.* **48**(SS-1):1–23.

2. **Centers for Disease Control and Prevention.** 2000. Probable locally acquired mosquito-transmitted *Plasmodium vivax* infection—Suffolk County, New York, 1999. *Morb. Mortal. Wkly. Rep.* **49**:495–498.

3. **Chen, Q., M. Schlichtherle, and M. Wahlgren.** 2000. Molecular aspects of severe malaria. *Clin. Microbiol. Rev.* **13**:439–450.

4. **Phillips, R. S.** 2001. Current status of malaria and potential for control. *Clin. Microbiol. Rev.* **14**:208–226.

5. **Stoute, J. A., M. Slaoui, D. G. Heppner, P. Momin, K. E. Kester, P. Desmons, B. T. Wellde, N. Garcon, U. Krzych, M. Marchand, W. R. Ballou, and J. D. Cohen for the RTS,S Malaria Vaccine Evaluation Group.** 1997. A preliminary evaluation of a recombinant circumsporozoite protein vaccine against *Plasmodium falciparum* malaria. *N. Engl. J. Med.* **336**:86–91.

6. **Zucker, J. R.** 1996. Changing patterns of autochthonous malaria transmission in the United States: a review of recent outbreaks. *Emerg. Infect. Dis.* **2**:37–43.

with several newly recognized pathogens. The initial outbreak of the hantavirus pulmonary syndrome occurred in the Four Corners Region of New Mexico, Colorado, Utah, and Arizona and is believed to be due to increasing contact between humans and the deer mouse, the reservoir of the Sin Nombre virus. Ehrlichiosis, which is spread by ticks, is being seen with increasing frequency because humans are more frequently entering environments (either to live or for recreational purposes) where their exposure to infected ticks is increased. Increases in yellow fever and dengue fever in tropical regions may be due in part to clearing of forest for farmland as well as a failure of mosquito control. The impact of global warming on mosquito-borne infections is an active area of research.

Recognition of West Nile virus in the metropolitan New York region reminded us that the combination of worldwide jet travel and the presence of appropriate insect vectors can result in seemingly exotic tropical infectious agents' establishing new environmental niches in the industrialized world, including the United States, Canada, and northern Europe. In just three years this viral agent has spread throughout the continental United States despite strong public health efforts to eradicate it through vector control measures.

New variants of the influenza A virus may have emerged through recombination events between avian and human viral strains. One new variant, H5N1, was responsible for several deaths in Hong Kong in 1999. Prompt destruction of a large number of fowl stopped the Hong Kong outbreak, but whether this influenza strain emerges in human populations remains to be determined. However, studies suggest that this influenza strain was not spread from person to person.

Organisms may also become more virulent for reasons that are not currently well understood. Recent outbreaks of necrotizing fasciitis due to group A streptococci have been sensationalized in the press as being caused by "flesh-eating bacteria." Careful epidemiologic study has shown that certain comparatively rare serotypes producing specific virulence factors have caused these outbreaks. With the use of powerful molecular techniques, studies of isolates recovered from widely dispersed geographic locales have shown that these isolates are genetically closely related, giving rise to the idea of a highly virulent "superbug." What is particularly remarkable is that these outbreaks have been short-lived, with these virulent organisms seemingly "disappearing" after a few weeks or months.

The early 1980s were a time of great optimism for the control of infectious diseases. This optimism has been replaced by the realization that infectious diseases will not disappear any time soon. However, recent advances in the war on AIDS sparked by the development of several increasingly potent antiviral agents, coupled with some encouraging results of experimental HIV vaccines, have given new hope that HIV infection might be controlled. For now, this disease continues unabated in most with several newly recognized pathogens. The initial outbreak of the hantavirus pulmonary syndrome occurred in the Four Corners Region of New Mexico, Colorado, Utah, and Arizona and is believed to be due to increasing contact between humans and the deer mouse, the reservoir of the Sin Nombre virus. Ehrlichiosis, which is spread by ticks, is being seen with increasing frequency because humans are more frequently entering environments (either to live or for recreational purposes) where their exposure to infected ticks is increased. Increases in yellow fever and dengue fever in tropical regions may be due in part to clearing of forest for farmland as well as a failure of mosquito control. The impact of global warming on mosquito-borne infections is an active area of research.

Recognition of West Nile virus in the metropolitan New York region reminded us that the combination of worldwide jet travel and the presence of appropriate insect vectors can result in seemingly exotic tropical infectious agents' establishing new environmental niches in the industrialized world, including the United States, Canada, and northern Europe. In just three years this viral agent has spread throughout the continental United States despite strong public health efforts to eradicate it through vector control measures.

New variants of the influenza A virus may have emerged through recombination events between avian and human viral strains. One new variant, H5N1, was responsible for several deaths in Hong Kong in 1999. Prompt destruction of a large number of fowl stopped the Hong Kong outbreak, but whether this influenza strain emerges in human populations remains to be determined. However, studies suggest that this influenza strain was not spread from person to person.

Organisms may also become more virulent for reasons that are not currently well understood. Recent outbreaks of necrotizing fasciitis due to group A streptococci have been sensationalized in the press as being caused by "flesh-eating bacteria." Careful epidemiologic study has shown that certain comparatively rare serotypes producing specific virulence factors have caused these outbreaks. With the use of powerful molecular techniques, studies of isolates recovered from widely dispersed geographic locales have shown that these isolates are genetically closely related, giving rise to the idea of a highly virulent "superbug." What is particularly remarkable is that these outbreaks have been short-lived, with these virulent organisms seemingly "disappearing" after a few weeks or months.

The early 1980s were a time of great optimism for the control of infectious diseases. This optimism has been replaced by the realization that infectious diseases will not disappear any time soon. However, recent advances in the war on AIDS sparked by the development of several increasingly potent antiviral agents, coupled with some encouraging results of experimental HIV vaccines, have given new hope that HIV infection might be controlled. For now, this disease continues unabated in most exception of vancomycin became a common cause of nosocomial infection. These organisms, called methicillin-resistant *S. aureus* (MRSA), along with another important nosocomial pathogen, *Clostridium difficile*, were treated with vancomycin. The resulting antimicrobial pressure is believed to have made a major contribution to the emergence of another nosocomial pathogen, vancomycin-resistant enterococci (VRE). Bacteremia due to this organism has a high mortality rate. In 1997, strains of *S. aureus* with reduced susceptibility to vancomycin were also described. Only a handful of cases due to the so-called VISA or GISA strains (for vancomycin-intermediate *S. aureus* or glycopeptide-intermediate *S. aureus*) have been reported, indicating that VISA strains are not yet clinically significant, though a patient infected with a vancomycin-resistant *S. aureus* (VRSA) strain was reported in 2002. New antimicrobial agents, linezolid and quinupristin-dalfopristin (Synercid), have been developed to treat infections from MRSA, VISA, and VRE. Unfortunately, drug resistance to these new antimicrobials appears to be emerging in VRE. Multidrug-resistant organisms, such as aminoglycoside-resistant *Pseudomonas aeruginosa*, have long been recognized as important nosocomial pathogens. Recently, selected enteric gram-negative bacilli, such as strains of *Klebsiella pneumoniae* and *Escherichia coli*, have acquired genes that encode "extended-spectrum β-lactamases." Organisms producing these enzymes are resistant to many but not all of the currently available beta-lactam antibiotics.

The recognition and reemergence of a number of enteric pathogens has caused a reexamination of our food and water supply. Diarrhea due to *E. coli* O157:H7 has been clearly associated with the consumption of undercooked ground beef. Diarrhea after the consumption of imported fruit has been shown to be caused by *Cyclospora* spp., a parasite. A major diarrheal disease outbreak in Milwaukee, Wis., in 1993, in which more than 400,000 individuals became ill, was due to the failure of the municipal water system to filter *Cryptosporidium parvum* from the water supply. In Peru, the consumption of a marinated fish dish, cerviche, contaminated with *Vibrio cholerae* was responsible for initiating a major outbreak of cholera in South and Central America; hundreds of thousands of individuals became ill, some seriously and some fatally. The death toll has been even higher among refugees infected in camps in Rwanda.

The recognition of a variant of Creutzfeldt-Jakob disease in the United Kingdom and France, which has been associated with the consumption of beef from herds with bovine spongiform encephalopathy (BSE), has caused a reexamination of feeding practices of domestic animals. BSE is thought to be a disease caused by a prion. The prolonged incubation period for prion diseases means that all the human cases of this disease have not yet been recognized. The final toll of this disease is not yet known nor are accurate estimates possible.

Changes in lifestyle in the developed world and pressures due to population growth in the developing world may be bringing humans into ever-increasing contact

INTRODUCTION TO SECTION VII

In 1992, the Institute of Medicine published a monograph on the worldwide problem of emerging and reemerging infectious diseases. The concerns raised in this document were in stark contrast to the optimism that the medical community felt in the early 1980s when the development of new vaccines and ever more powerful antibiotics and antiviral agents led to the belief that infectious diseases were being brought under control in the developed world. This optimism faded with the emergence of HIV and the many opportunistic infections associated with the immunodeficiency caused by this virus, including meningitis caused by *Cryptococcus neoformans*, encephalitis due to *Toxoplasma gondii*, bacteremia caused by *Mycobacterium avium* complex, *Bartonella henselae*, and *Rhodococcus equi*, diarrhea due to *Cryptosporidium parvum*, and especially pneumonia due to *Pneumocystis carinii*. As a consequence of the combination of the dismantling of much of the public health infrastructure in the United States and the large number of immunocompromised patients with HIV infection, multidrug-resistant *Mycobacterium tuberculosis* became a major problem during the late 1980s and early 1990s. It remains a problem in, for example, prisons in the former Soviet Union.

A new infectious disease problem emerged in October 2001. Bioterrorism moved from the pages of best-selling novels to a grim reality in the United States. *Bacillus anthracis* spores used to intentionally sicken and, in five instances, kill individuals generally failed to infect the bioterrorists' targets, highly visible members of the media and government. Rather, they infected people who handled contaminated mail, the vehicle used to transmit this agent. Fears of anthrax and other agents that the Centers for Disease Control and Prevention (CDC) listed as category A organisms became part of the public discourse in our society. Most ominously, the specter of the return of smallpox, a disease defeated by humanity, has caused great concern among public health officials as well as the general public. This has resulted in a robust response from the scientific community led by the CDC and supported by the U.S. government to ensure adequate smallpox vaccine supplies in the United States.

The emergence of a number of bacteria resistant to multiple antimicrobial agents has caused serious concern in the medical community. This emergence is due in large part to indiscriminate and often inappropriate use of these "wonder drugs." Young children with otitis media due to *Streptococcus pneumoniae* are infected increasingly with strains resistant to all currently available oral antimicrobial agents, making management of these children's infections difficult. This problem has resulted in the development of a 7-valent pneumococcal vaccine that has been shown to be efficacious in preventing invasive pneumococcal disease in children less than 2 years of age. This is now one of the recommended childhood vaccines. Nosocomial infections due to multidrug-resistant organisms have become a problem of increasing seriousness. In the early 1990s, *Staphylococcus aureus* strains resistant to all antimicrobial agents with the

SEVEN

Emerging and Reemerging Infectious Diseases

the immune response generated by this vaccine will be effective at the next stage of the disease.

REFERENCES

1. **Centers for Disease Control and Prevention.** 1999. Malaria surveillance—United States, 1995. *Morb. Mortal. Wkly. Rep.* **48**(SS-1):1–23.

2. **Centers for Disease Control and Prevention.** 2000. Probable locally acquired mosquito-transmitted *Plasmodium vivax* infection—Suffolk County, New York, 1999. *Morb. Mortal. Wkly. Rep.* **49**:495–498.

3. **Chen, Q., M. Schlichtherle, and M. Wahlgren.** 2000. Molecular aspects of severe malaria. *Clin. Microbiol. Rev.* **13**:439–450.

4. **Phillips, R. S.** 2001. Current status of malaria and potential for control. *Clin. Microbiol. Rev.* **14**:208–226.

5. **Stoute, J. A., M. Slaoui, D. G. Heppner, P. Momin, K. E. Kester, P. Desmons, B. T. Wellde, N. Garcon, U. Kraych, M. Marchand, W. R. Ballou, and J. D. Cohen for the RTS,S Malaria Vaccine Evaluation Group.** 1997. A preliminary evaluation of a recombinant circumsporozoite protein vaccine against *Plasmodium falciparum* malaria. *N. Engl. J. Med.* **336**:86–91.

6. **Zucker, J. R.** 1996. Changing patterns of autochthonous malaria transmission in the United States: a review of recent outbreaks. *Emerg. Infect. Dis.* **2**:37–43.

regions of the world, with rapid increases in case numbers in the Indian subcontinent and China. *Haemophilus influenzae* type b invasive disease has been brought under control in the developed world by the widespread use of a conjugated vaccine. Hope exists that the new 7-valent conjugated *S. pneumoniae* vaccine will yield similar results. Strategies are urgently needed to make access to both of these vaccines global. Cases of many viral illnesses are in decline in the developed world, again because of the development of new vaccines. New classes of antimicrobial agents and vaccines are in development. Perhaps we have finally reached the point where we finally recognize that we must be more judicious in our use of antimicrobials or lose their value. New infectious diseases will continue to be recognized, but so too will new strategies designed to prevent or diagnose and cure them. We must also hope that humans will not pervert the great advances in understanding of molecular biology of microorganisms to create chimeric "superbugs" which are both highly virulent and highly drug resistant and can be used as weapons against fellow humans.

TABLE 7 EMERGING AND REEMERGING INFECTIOUS DISEASES

ORGANISM	IMPORTANT CHARACTERISTICS	PATIENT POPULATION	DISEASE MANIFESTATION
Bacteria			
Bartonella henselae	Fastidious, gram-negative bacillus	Children; AIDS	Cat scratch disease; bacillary angiomatosis, hepatic peliosis, bacteremia
Corynebacterium jeikeium	Gram-positive bacillus; susceptible only to vancomycin	Nosocomial, immunocompromised	Line-related sepsis
Ehrlichia spp.	Morulae seen in cytoplasm; organism does not Gram stain or grow on artificial media	Individuals bitten by infected tick	Monocytic and granulocytic ehrlichiosis, i.e., "Rocky Mountain spotless fever"
Enterococcus faecium, vancomycin resistant	Frequently resistant to first-line antimicrobials	Nosocomial	Urinary tract, wound, and bloodstream infections
Escherichia coli, enterohemorrhagic	Produces Shga toxins I and II	Individuals who eat raw or inadequately cooked hamburgers	Bloody diarrhea, hemolytic uremic syndrome
Gram-negative bacilli, drug resistant	Extended-spectrum β-lactamase production; AmpC; other mechanisms of resistance	Nosocomial	Urinary tract, wound, pulmonary, and bloodstream infections
Group A streptococci (*Streptococcus pyogenes*)	Superantigen producing, M1 and M3 serotypes	Children and adults, especially with varicella or immunocompromised	Necrotizing fasciitis, toxic shock syndrome
Helicobacter pylori	Helical, curved, gram-negative bacillus	Adults	Gastric and duodenal ulcers
Mycobacterium tuberculosis, multidrug resistant	Acid-fast bacillus, resistant to two or more first-line antituberculous drugs	AIDS patients and their caregivers, homeless, prison inmates	Tuberculosis; rapid, fatal disease course in AIDS/HIV-infected individuals
Rhodococcus equi	Partially acid-fast, club-shaped bacillus	AIDS	Bacteremia, pneumonia
Staphylococcus aureus, glycopeptide intermediate and resistant	Rare strains identified	Nosocomial	Wound, pulmonary, and bloodstream infections

Organism	Characteristics	Epidemiology	Disease
Parasites			
Angiostrongylus cantonensis	Rat lungworm	Ingestion of snails or snail-contaminated vegetables	Eosinophilic meningitis
Cyclospora spp.	8–10-μm cysts	Travelers to countries with poor hygiene; individuals eating imported food	Diarrhea
Viruses			
Dengue viruses	Enveloped, ssRNA[a]	Travelers bitten by infected *Aedes* mosquitoes, especially in Caribbean basin	"Breakbone" fever, severe joint and muscle pain with rash; hemorrhagic fever; shock can occur
Ebola virus	Enveloped, ssRNA	Epidemics in Africa; health care workers exposed to blood of infected patients	Hemorrhagic fever
Hantaviruses	Enveloped, circular ssRNA	Children and adults in Argentina, southwestern U.S., Korea	Adult respiratory distress syndrome, pneumonia, hemorrhagic fever
Human immunodeficiency virus-1 (HIV-1)	Enveloped ssRNA retrovirus	Infants of HIV-infected mothers, sexually active adolescents and adults, injection drug users	Acute phase, mononucleosis-like infection; late phase, profound immunosuppression, multiple opportunistic infections
Influenza A H5N1	Enveloped segmented ssRNA virus	Human cases with high mortality rate associated with fowl in Hong Kong	Severe respiratory disease
Marburg virus	Enveloped, ssRNA	Epidemics in Africa; health care workers exposed to blood of infected patients	Hemorrhagic fever
Nipah virus	Enveloped, ssRNA	Malaysia and Singapore associated with pigs; natural host is a bat	Epidemic of encephalitis with 40% mortality rate
West Nile virus	Enveloped, ssRNA	Introduction of mosquito-borne virus into North America; found in parts of Europe, Asia, Middle East, Africa	Ranges from asymptomatic infection to fatal encephalitis (usually in elderly)
Prions			
Variant Creutzfeldt-Jakob disease	Prion	Seen in United Kingdom and associated with outbreak of bovine spongiform encephalopathy, primarily in young adults	Fatal spongiform encephalopathy

[a] ssRNA, single-stranded RNA.

CASE 59

This 53-year-old man with a medical history of Guillain-Barré syndrome was ventilator dependent because of persistent weakness. He had a history of several prior episodes of bloodstream infection due to organisms including methicillin-resistant *Staphylococcus aureus*, vancomycin-resistant *Enterococcus faecium*, and *Pseudomonas aeruginosa* and had received multiple courses of antibiotics. At a rehabilitation hospital, the patient complained of shortness of breath and chest discomfort. On examination he was found to be hypotensive and tachycardic. Pulse oximetry revealed desaturation, and the patient was transferred to a tertiary care facility where he was placed in the medical intensive care unit.

On admission, his vital signs were notable for a temperature of 39.3°C, a heart rate of 130 beats/min, and a systolic blood pressure of 44 to 80 mm Hg while receiving pressors (intravenous medications that increase blood pressure via action on the cardiovascular system). Examination revealed an awake ventilated man, with cool extremities, an indwelling Foley catheter, and a sweet-smelling blue-green sacral decubitus wound.

Laboratory studies were remarkable for acidosis (arterial blood gas pH, 7.26), an elevated anion gap of 19, a creatinine level of 2.7 mg/dl, and a white blood cell count of 6,000/μl with a differential count of 48% polymorphonuclear leukocytes and 45% band forms.

Two sets of blood cultures were obtained. The patient was begun on intravenous antibiotics. The blood cultures were subsequently positive for growth of gram-negative bacilli in the aerobic bottle of each set.

1. What is commonly considered a "set" of blood cultures? Why is more than one set of blood cultures obtained on patients with suspected bacteremia?

2. The combination of his clinical findings and the presence of bacteremia caused by gram-negative organisms is consistent with what clinical entity?

3. Of what clinical significance is the physical finding of a "sweet-smelling blue-green sacral decubitus wound"? If this is the source of the patient's bacteremia, what is the most likely organism causing his infection?

4. What are other sources of bacteremia caused by gram-negative organisms?

5. The patient had recently received many courses of intravenous antibiotics for serious infections. Of what concern is this?

CASE DISCUSSION

CASE 59

1. A set of blood cultures consists of two bottles that are inoculated with the aseptically collected blood of a patient. Most commonly, one of the bottles is incubated under conditions and with media that will support the growth of aerobic bacteria, and one bottle is incubated under conditions and with media that will support the growth of anaerobic bacteria. It is worth noting that many human bacterial pathogens are able to grow under both aerobic and anaerobic conditions and, therefore, may grow in either (or both of) the aerobically and anaerobically incubated bottles. Since anaerobic bacteria vary in their ability to grow in the presence of oxygen (aerotolerance), it is even possible to recover some anaerobes from aerobically incubated bottles. In recent years, as the number of obligate anaerobes causing bacteremia has been found to be significantly less than the number of bacteria that are able to grow under aerobic conditions, some authorities have questioned the wisdom of including an anaerobe bottle in blood culture sets. It has been suggested that the yield of blood cultures can be improved by using two aerobic blood culture bottles in each set, reserving anaerobe bottles for those situations in which there is a high clinical suspicion of an anaerobic infection, such as in surgical patients. This is a question that is not yet settled. The best practice will likely vary from institution to institution based upon the patient population and types of infections found.

The use of two sets of blood cultures is based upon the knowledge that bacteremia is usually intermittent rather than continuous. As a result, a single set of blood cultures is less likely to recover the organism causing bacteremia than are two sets of blood cultures drawn at different times. The optimal time between sets of blood cultures has not been determined in any published studies.

2. This patient has septic shock. Terms such as bacteremia, septicemia, septic shock, and sepsis have been used as synonyms in the past. As the understanding of the pathophysiology of sepsis has increased, there has been a need for more precision in the use of these terms. Bacteremia, the presence of viable bacteria in the blood, has been diagnosed in this patient on the basis of positive blood cultures. He has the "systemic inflammatory response syndrome," as he meets the criteria for this syndrome with two or more of the following: temperature >38°C or <36°C; heart rate >90 beats/min; respiratory rate >20 breaths/min or $PaCO_2$ <32 mm Hg; and white blood cell count >12,000/mm^3, <4,000/mm^3, or >10% immature (band) forms. Since the systemic inflammatory response syndrome occurs in response to an infection, this meets the definition of "sepsis." Given that the sepsis is associated with hypotension, this meets the criteria for "severe sepsis." Finally, given that he has sepsis-induced hypotension despite adequate fluid resuscitation along with the presence of perfusion abnormalities, this meets the criteria for "septic shock."

Severe sepsis is the result of a trigger, such as bacterial endotoxin (i.e., lipopolysaccharide) in gram-negative bacteria or peptidoglycan fragments from gram-positive organisms, causing the stimulation of the inflammatory and procoagulant responses to infection. Production of cytokines, such as interleukin-1 and tumor necrosis factor α (TNF-α), activates the coagulation cascade by stimulating the release of tissue factor. Both the inflammatory cytokines and thrombin, generated by the coagulation cascade, suppress fibrinolysis. Ultimately, there may be organ ischemia and multiorgan dysfunction as a result of endovascular injury due to this chain of events. In one recently published study, the use of activated protein C, which exerts an anti-inflammatory effect by inhibiting the production of interleukin-1 and TNF-α and exerts an antithrombotic effect by inhibiting clotting factors Va and VIIIa, has been shown to reduce mortality in patients with severe sepsis.

3. Characteristic odors on physical examination can give a clue to the organism involved in an infection. Anaerobic infections are commonly referred to as "foul smelling" or "putrid." Thus, a patient with fetid breath and fever should be suspected of having a polymicrobial anaerobic infection such as a lung abscess. In this case, the presence of a sweet-smelling wound and a blue-green sacral decubitus wound are suggestive of a wound infection with *P. aeruginosa*. Although pus can have a green color due to the myeloperoxidase in white blood cells, the blue-green color is likely due to one or more of the pigments produced by *P. aeruginosa*, such as pyoverdin and pyocyanin. The sweet smell is noted in the laboratory as well, and cultures of *P. aeruginosa* have been described as smelling like grapes. In fact, blood cultures from this patient were positive for *P. aeruginosa*.

4. In addition to the wound infection, other potential sources of bacteremia caused by gram-negative organisms include the urinary tract, the lung, the biliary tree, the gastrointestinal tract, and intra-abdominal and pelvic abscesses. Less commonly, bacteremia caused by gram-negative organisms may be the result of contaminated intravascular catheters or contaminated intravenous solutions. Infrequent causes include the injection of illicit drugs (e.g., heroin) contaminated with gram-negative bacteria, purposeful injection of contaminated material as in Munchausen syndrome, septic thrombophlebitis in which a blood clot (usually due to the presence of an intravenous catheter) is infected, contaminated blood products, endocarditis caused by gram-negative organisms, infection due to a zoonotic agent such as *Yersinia pestis* or *Francisella tularensis*, and bacteremia in the hyperinfection syndrome due to *Strongyloides stercoralis*.

This patient was at risk for infection of the urinary tract by gram-negative organisms due to the presence of an indwelling Foley catheter. He was also at risk for pneumonia, as he was chronically ventilated and unable to protect his airway. When history and initial imaging studies do not identify a source for bacteremia due to gram-negative organisms, occult sites of infection should be sought. These include the prostate in men and the pelvis in women, both of which may be infected without localizing symptoms. A significant fraction of patients with bacteremia due to gram-negative organisms will not have an identifiable source of the infection.

5. Therapy with antibiotics results in a change in a person's bacterial flora to include antibiotic-resistant bacteria. In this patient's case, the isolate of *P. aeruginosa* was resistant to nearly all antibiotics tested (Fig. 1). Not only had the patient received multiple courses of antibiotics, a clear risk factor in the development of infections with resistant organisms, but also he had already been infected with methicillin-resistant *S. aureus* and vancomycin-resistant *E. faecium*, two resistant organisms. Nosocomial (hospital-acquired) infections are more likely to be due to antibiotic-resistant organisms than are infections occurring in the community. His multiple hospitalizations (he was in a rehabilitation hospital when this infection occurred), use of antibiotics, inability to clear bacteria from his airway, lack of skin integrity, and the fact that many similar patients are likely to have been in close proximity to him—each with his or her own antibiotic-resistant flora—placed him at a very high risk for acquiring an infection with a resistant organism.

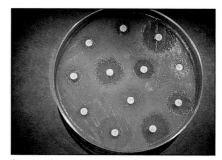

Figure 1

REFERENCES

1. **Bernard, G. R., J. L. Vincent, P. F. Laterre, S. P. LaRosa, J. F. Dhainaut, A. Lopez-Rodriguez, J. S. Steingrub, G. E. Garber, J. D. Helterbrand, E. W. Ely, and C. J. Fisher, Jr.** 2001. Efficacy and safety of recombinant human activated protein C for severe sepsis. *N. Engl. J. Med.* **344:**699–709.

2. **Bone, R. C., C. J. Grodzin, and R. A. Balk.** 1997. Sepsis: a new hypothesis for pathogenesis of the disease process. *Chest* **112:**235–243.

3. **Wheeler, A. P., and G. R. Bernard.** 1999. Treating patients with severe sepsis. *N. Engl. J. Med.* **340:**207–214.

CASE
60

This 63-year-old alcoholic was taken to the emergency room of an outside hospital with obvious gangrene of both feet. He was stuporous. During that evening, he had a seizure and was treated with phenytoin and barbiturates. By the night of transfer he was noted to have opisthotonic posturing and to have developed increasing rigor, respiratory distress, and unresponsiveness. On examination, he had a rectal temperature of 41.7°C, a blood pressure of 70/30 mm Hg, a heart rate of 110 beats/min, and a respiratory rate of 40/min. Examination was notable for marked trismus. The neck was stiff and hyperextended. Necrotic, blackened areas were present over both feet, and several draining ulcers were noted on the heels and toes. Neurologically the patient responded to deep pain with a grimace. On the basis of these findings, specific therapy, in addition to supportive care, was initiated, and the patient ultimately recovered.

1. What is the etiology of his infection? What virulence factor produced by the etiologic agent of his infection was responsible for his trismus?

2. How did this patient become infected with this organism? What was the role of his gangrenous feet in the development of this infection?

3. What is the specific therapy used to treat this infection?

4. How might this infection have been prevented?

CASE DISCUSSION

CASE 60

1. This patient had tetanus, which is caused by *Clostridium tetani*. *C. tetani* is a strictly anaerobic, spore-forming, gram-positive rod. Because of its exquisite sensitivity to oxygen, it is very difficult to recover from clinical specimens, so this diagnosis is generally made on the basis of clinical findings. The organism produces a protein exotoxin, tetanospasmin, that is, along with botulinum neurotoxin, one of the most potent toxins known. The tetanus neurotoxin enters the presynaptic nerve cytosol via receptor-mediated endocytosis and synaptic vesicle reuptake.

The gene encoding the tetanus neurotoxin resides on a plasmid. The neurotoxin is produced under nutritional regulation and is primarily made following cessation of bacterial growth. It is synthesized as a 150-kDa peptide that is modified by posttranslational proteolytic cleavage, forming a molecule with a 50-kDa light chain and a 100-kDa heavy chain that are linked by a disulfide bond. The light chain is the zinc-dependent protease's catalytic domain, the N-terminal portion of the heavy chain is the translocation domain, and the C-terminal portion of the heavy chain is the receptor-binding domain of the toxin. Following the binding of the C-terminal portion of the heavy chain to its receptor on a presynaptic nerve cell and its being taken into the cell, pores are formed that permit the entry of the light chain into the nerve cell. It is transported via retrograde axonal transport to the nerve cell body, reaching nerves in the brain stem and spinal cord. At these sites, it causes the presynaptic blockade of the release of the inhibitory neurotransmitters glycine and GABA (γ-aminobutyric acid) by the cleavage of a protein component of synaptic vesicles. This suppression of inhibitory nerve function results in an increased activation of nerves innervating muscles (such as the masseter), causing muscle spasm. The term tetanus is derived from the Greek word *tetanos*, which means rigid, or to contract.

2. *C. tetani* is typically found as spores in soil. It has a worldwide distribution. These spores can enter wounds, and if the oxidation-reduction potential (a measure of how anaerobic an environment is) is low enough, i.e., anaerobic enough, the spores will germinate and the organism can begin to grow and produce toxin in the wound. Gangrenous tissues create an ideal environment for *C. tetani* to grow. These tissues are composed of devitalized (dead) tissue with compromised blood flow, which creates a very low oxidation potential at that site. If these tissues become contaminated by soil, infection with *C. tetani* may ensue. Since many alcoholics have extremely poor hygiene, soil contamination of the feet is certainly very feasible.

3. Two strategies are required. The patient will require supportive therapy, particularly cleansing of the wound, ventilatory assistance, and the use of muscle relaxants. The other strategy is to use human tetanus immune globulin (TIG). This product, in comparison with the horse product used for diphtheria, is not as likely to cause anaphylaxis or serum sickness. This patient received human TIG and recovered.

4. The same strategy used for prevention of diphtheria is also operative for tetanus. A series of four primary immunizations starting at 2 months of age, followed by appropriate booster vaccinations at 4 to 6 years, at 11 to 12 years, and in adults at 10-year intervals, is highly effective in preventing tetanus. Nevertheless, in the United States from 1995 to 1997, an average of approximately 40 cases of tetanus per year were reported to the Centers for Disease Control and Prevention. Thirty-five percent of tetanus cases are seen in individuals older than 60 years, and it is speculated that these individuals have not received booster immunization. An increasing proportion of tetanus cases reported, 11% of all cases, were in injection-drug users with no known acute injury. Clinically, the severity of disease was found to be directly related to previous vaccination status, as was the case-fatality ratio, with no deaths occurring in patients who had previously received three or more doses.

In developing countries, tetanus is an important problem. In this setting, immunization of children may not occur, leaving large populations susceptible to the disease. An additional problem in these countries is neonatal tetanus, which may occur due to a contaminated umbilical stump. This infection is an important, preventable cause of neonatal death in the developing world. Prevention of this condition is possible with the use of vaccination during pregnancy.

REFERENCES

1. **Centers for Disease Control and Prevention.** 1996. Progress toward elimination of neonatal tetanus—Egypt, 1988–1994. *Morb. Mortal. Wkly. Rep.* **45:**89–92.

2. **Centers for Disease Control and Prevention.** 1998. Tetanus surveillance—United States, 1995–1997. *Morb. Mortal. Wkly. Rep.* **47**(SS-2):1–13.

3. **Galazka, A., and F. Gasse.** 1995. The present status of tetanus and tetanus vaccination. *Curr. Top. Microbiol. Immunol.* **195:**31–53.

4. **Weinstein, L.** 1973. Tetanus. *N. Engl. J. Med.* **289:**1293–1296.

C A S E
61

The patient was a 21-year-old male who presented with complaints of nausea, vomiting, diffuse body aches, productive cough, fever, and loose, watery diarrhea. He had not urinated in the prior 24 hours and complained of dizziness on standing. He denied headache or abdominal pain. His past medical history was significant for sore throat 3 weeks previously. He also reported that he had a urinary tract infection "a month or two" ago. He received intravenous fluids, and a fecal specimen was sent for culture and ova and parasites. The stool specimen was positive for *Entamoeba histolytica* (or *Entamoeba dispar*). At that point, one day into his hospitalization, the infectious disease (ID) service was consulted.

The patient was from North Carolina, had no recent travel history, drank alcohol, and denied sexual contact. On physical examination by the ID fellow, the patient was found to have a fever of 39.1°C, a heart rate of 104/min, and blood pressure of 134/84 mm Hg. His physical examination revealed enlarged tonsils but no cervical, axillary, or inguinal adenopathy. No hepatosplenomegaly was noted. Bowel sounds were normal. Chest was clear on auscultation and by chest radiograph. Laboratory studies were significant for a white blood cell count of 2,200 with 57% polymorphonuclear leukocytes, 33% lymphocytes, and 6% atypical lymphocytes. Aspartate aminotransferase (AST) was 650 U/liter, alanine aminotransferase (ALT) was 830 U/liter, and lactate dehydrogenase (LDH) was 1,000 U/liter. Hepatitis A, B, and C virus and HIV serologic test results were negative. The etiologic agent of his primary illness was detected by culture, positive antigen test, and PCR (polymerase chain reaction). *E. histolytica* was a secondary infection.

[handwritten margin notes: ↑AST ↑ALT]

1. This patient has two infections. One, with *E. histolytica*, explains the occurrence of nausea, vomiting, diarrhea, and dehydration. However, infection with this organism does not explain his sore throat 3 weeks prior to his hospitalization, the tonsillar enlargement, his systemic complaints of fever, nausea, vomiting, and diffuse body aches, or the laboratory findings of atypical lymphocytes and hepatitis. With what syndrome seen in young adults are these findings consistent? What are the viral etiologies of this condition?

2. What is his specific diagnosis? Explain why serologic testing for antibody was negative at this stage of his infection, but the pathogen was identified by antigen testing, PCR, and culture. What clues were present in his history that suggested the possibility of this pathogen?

3. What populations are at increased risk for infection with this agent?

4. Describe the pathogenesis of his infection which resulted in the mono-like illness. What is the natural history of this infection?

5. His *E. histolytica* infection was successfully treated. How should this patient be managed for the other infection?

CASE DISCUSSION

61

1. The patient has an infectious mononucleosis-like syndrome. Epstein-Barr virus, cytomegalovirus, and HIV can cause a mono-like illness.

2. His diagnosis is acute, primary HIV infection. During this early stage, there is a massive burst of viral replication. Plasma viral RNA, antigen, and replicating virus can be detected before the host's antibody response becomes positive. The antibody response can take weeks to occur following initial infection.

Two clues in this patient's case history suggest that he might be infected with HIV. First, he is infected with *E. histolytica,* an intestinal parasite that is quite rare in non-migrant farmworkers or in persons who have not traveled to developing countries where the disease is endemic. However, this parasite can be spread by anal intercourse or oral-anal contact. Second, this individual gave a vague history of a recent "urinary tract" infection. Urinary tract infections are quite rare in 21-year-old males. However, urinary tract infection symptoms, such as dysuria and urgency, are common in persons with sexually transmitted diseases (STDs). It is highly likely that the history this individual gave of a recent urinary tract infection actually represented an STD. It should be noted that the diagnosis of primary HIV infection is often missed. Patients with primary HIV infection may present to physicians who are not specialists in managing HIV-infected patients. All physicians need to be highly suspicious of primary HIV infection in any person presenting with a mono-like illness. The index of suspicion should increase if the patient has a history of an STD.

On closer questioning, the patient admitted to participating in anonymous anal intercourse in the previous month.

3. A variety of populations are at increased risk for acquiring HIV infections. HIV is spread primarily in three ways: sexually, including vaginal and anal intercourse and oral-genital contact; by contact with infected blood; or vertically, from mother to fetus. In the United States, homosexual or bisexual males have the highest rates of infection, followed by intravenous drug users and their sexual partners. Infection rates are lowest among individuals who are infected through heterosexual activity, but this at-risk population in the United States is developing an increasing incidence of HIV infection. This is in sharp contrast to the spread of HIV in other parts of the world, where the major mode of transmission is via heterosexual activity. Although the spread of HIV has slowed in the United States, HIV infection is becoming an increasing public health problem in India, China, Russia, and many countries throughout Southeast Asia. In Russia and China, the increasing prevalence of HIV is largely driven by very high rates of infection in intravenous drug users. AIDS continues to be a devastating disease in sub-Saharan Africa, where 14 million people have died of AIDS as of 2001.

Approximately 9% of adults in Africa between 15 and 50 years of age are infected with HIV. Worldwide, HIV disease and AIDS have surpassed tuberculosis and malaria as the leading fatal infectious diseases.

At one time, HIV-contaminated blood product transfusions were an important mode of transmission. Serologic tests and aggressive donor screening have led to a greatly reduced rate of transfusion-associated HIV infection in the developed world. In areas of the world where screening of blood for HIV antibodies is prohibitively expensive, transfusion-associated HIV infection continues to be a major health problem.

4. Upon infection, large amounts of virus are produced which transiently peak and rapidly decline. At this point, the host's immune response provides some control, which can be measured, in part, by detectable anti-HIV antibodies. The clinical manifestations of acute HIV infection can include gastrointestinal, dermatologic, neurologic, and lymphoid tissue involvement arising from the widespread dissemination of HIV infection. Some individuals can be asymptomatic during the acute stage of HIV infection.

An individual's course of HIV infection and disease is variable and is largely determined by complex interactions between the virus and the host's immune system. The rate at which immune dysfunction occurs determines the rate of disease progression. Advanced HIV disease and AIDS result in complications such as opportunistic infections and malignancies.

Even early in infection, very large amounts of virus are present in lymphoid tissues and in plasma (up to 10^6 particles per ml). An estimated 10^{10} viral particles are generated daily. Within 1 year after seroconversion, each infected individual establishes a "steady-state" level of HIV RNA which largely determines the rate at which CD4$^+$ T lymphocytes subsequently decline. Concurrent use of plasma HIV RNA measurements and CD4 cell counts can predict disease progression.

5. The patient should be counseled about his HIV infection status, what clinical course to expect, and what behaviors should be modified to avoid transmitting his infection to others. He should be considered for combination antiretroviral therapy. Since 1996, combination antiretroviral therapy involving three or more anti-HIV medications, including a protease inhibitor or a non-nucleoside reverse transcriptase inhibitor plus two nucleoside reverse transcriptase inhibitors, has become the standard of care in the industrialized world. These regimens have resulted in a substantial reduction in HIV-related deaths in populations that have access to these drugs. In 1998, a major reason that AIDS was removed from the list of the 15 leading causes of death in the United States was the widespread use of highly active antiretroviral therapy.

Greater viral load reductions and CD4$^+$ cell count increases can result from combination antiretroviral therapy. At the acute stage of HIV infection, the viral population is relatively homogeneous and thus less variable. Treating with more than one antiretroviral agent significantly lessens the risk that drug-induced resistance will develop.

Antiretroviral drug resistance is becoming a critical issue. Some drug-resistant viral isolates can be transmitted. A slowly increasing proportion of patients with primary HIV infection have acquired HIV that is already resistant to one or more classes of antiretroviral drugs. Patient response to therapy is measured by determining the level of HIV viral RNA present in the blood by means of a quantitative PCR or similar test. This is called "viral load testing." The goal of therapy is to make viral RNA undetectable in the blood for as long as possible. Viral loads are monitored at approximately 3-month intervals. If there is a sudden increase in viral load levels, there are two common explanations. One is that the patient may have stopped taking his or her medicine. Although current HIV antiretroviral regimens are very effective, they are complicated to take and can have unpleasant and (rarely) fatal side effects. The second explanation is that the isolate may have developed resistance to one or more of the agents with which the patient is being treated. The HIV genome is highly mutable, as are most RNA viruses, and mutations leading to resistance are frequent. Antiretroviral resistance testing can potentially benefit patients by allowing individualized treatment regimens based upon the analysis of the genotype or phenotype of the patient's viral isolate. This testing is quite expensive.

Because of the cost of antiretroviral therapy, the vast majority of HIV-infected individuals in the world today do not have access to these life-saving drugs. For them, HIV infection is still a death sentence. Hope for control of this disease rests largely with behavior modification and with the development of a vaccine. At the present time, HIV vaccine development has been difficult because of the mutability of the virus and our limited understanding of what constitutes protective immunity. A key feature of any successful HIV vaccine will be the capacity to induce mucosal immunity to prevent the acquisition of HIV via sexual activity.

REFERENCES

1. **Ho, D. D., A. V. Neumann, A. S. Perelson, W. Chen, J. M. Leonard, and M. Markowitz.** 1995. Rapid turnover of plasma virions and CD4 lymphocytes in HIV-1 infection. *Nature* **373:**123–126.

2. **Mellors, J. W., A. Munoz, J. V. Giorgi, J. B. Margolick, C. J. Tassoni, P. Gupta, L. A. Kingsley, J. A. Todd, A. J. Saah, R. Detels, J. P. Phair, and C. R. Rinaldo, Jr.** 1997. Plasma viral load and CD4$^+$ lymphocytes as prognostic markers of HIV-1 infection. *Ann. Intern. Med.* **126:**946–954.

3. **Saag, M. S., M. Holodnix, D. R. Kuritzkes, W. A. O'Brien, R. Coombs, M. E. Poscher, D. M. Jacobsen, G. M. Shaw, D. D. Richman, and P. A. Volberding.** 1996. HIV viral load markers in clinical practice. *Nat. Med.* **2:**625–629.

4. **Schacker, T., A. C. Collier, J. Hughes, T. Shea, and L. Corey.** 1996. Clinical and epidemiologic features of primary HIV infection. *Ann. Intern. Med.* **125:**257–264.

5. **U.S. Department of Health and Human Services.** 2001. Guidelines for the use of antiretroviral agents in HIV-infected adults and adolescents (http://hisatis.org/trtgdlns.html).

CASE 62

The patient was a 32-year-old woman with a history of poly-substance abuse, including intravenous drug use, who presented to the emergency room of an outside hospital with a 3-day history of fevers, abdominal pain, nausea, vomiting, headaches, and myalgias. She also had a toothache for which she had taken approximately 50 acetaminophen tablets over a 24-hour period. She was transferred to our institution for management of her acetaminophen overdose.

On physical examination, she was alert and her vital signs were within normal limits. Significant physical findings included right upper quadrant pain on palpation with guarding. She was mildly jaundiced. Laboratory findings were significant for the following: blood urea nitrogen (BUN), 83 mg/dl; creatinine, 10.1 mg/dl; aspartate aminotransferase (AST), 791 U/liter; alanine aminotransferase (ALT), 1,398 U/liter; γ-glutamyltransferase (GGT), 166 U/liter; and lactate dehydrogenase (LDH), 826 U/liter. She also had an acetaminophen level of 15 µg/ml on admission.

Results of hepatitis serologic tests were as follows: hepatitis A immunoglobulin M (IgM), nonreactive; hepatitis A total antibody, nonreactive; hepatitis B surface antigen, negative; hepatitis B surface antibody, negative; hepatitis B core IgM, nonreactive; hepatitis C antibody, positive.

Her hospital course was significant for acute renal and liver failure with coagulopathy and encephalopathic changes. After dialysis in the intensive care unit, her renal and liver function improved and her encephalopathy resolved. After transfer to the medical ward, four abscessed teeth were extracted.

↑ creat.
↑ AST
↑ ALT

1. With what agent was this woman infected? What role did it have in her clinical presentation?

2. How is this disease typically diagnosed?

3. Describe the natural history of infection with this agent.

4. How is this agent typically transmitted? How did she likely get infected?

5. For what other infectious agents would she be at increased risk?

6. Why is it important to determine the genotype of the agent with which she is infected?

7. How can this disease be prevented?

CASE DISCUSSION

62 **1.** Serologic studies, her liver enzyme levels, and her clinical presentation suggest that she had an acute hepatitis C virus (HCV) infection. HCV is a single-stranded RNA virus that belongs to the flavivirus family, which includes both the yellow fever and dengue viruses. It was the first infectious agent detected by the technique of molecular cloning. Although most cases (~90%) of acute HCV infection are asymptomatic, the clinical presentation of nausea, vomiting, right upper quadrant abdominal pain with guarding, and jaundice, along with the finding of a chemical hepatitis as evidenced by highly elevated liver enzymes, are all consistent with viral hepatitis. Serologic test results that were negative for both hepatitis A virus (HAV) and hepatitis B virus (HBV) but positive for HCV antibodies support the diagnosis of an HCV infection. She also had a positive HCV RNA level (249,000 IU/ml), further supporting the HCV diagnosis.

Her clinical presentation of HCV is greatly confounded by her drug overdose with acetaminophen, a hepatotoxic drug. However, many aspects of her clinical history consistent with acute HCV infection predate her drug overdose. Her acute renal and liver failure were likely due to a combination of the drug overdose and her HCV infection. It is possible that she would not have sought medical care if she had not taken a drug overdose. Therefore, her HCV diagnosis was somewhat serendipitous due to an astute clinician's doing a careful work-up of her acute liver failure.

2. The strategy for diagnosing HCV is to detect HCV-specific antibodies by enzyme immunoassay (EIA) followed by a confirmatory test. The EIA test is highly sensitive, detecting ≥97% of HCV-infected patients and becoming positive 4 to 10 weeks postinfection. It cannot distinguish between acute, chronic, or resolved infections. The confirmatory test is necessary because of a fairly high false-positive rate with the EIA test. Two approaches can be used, a recombinant protein immunoblot assay (RIBA) or the determination of the presence of HCV RNA by a molecular amplification technique such as polymerase chain reaction (PCR). Detection of HCV RNA may be a useful diagnostic tool in acute infections before the patient develops an antibody response. However, since the vast majority of patients present for medical care in the chronic phase of HCV infection, when they are seropositive, detection of HCV RNA is rarely used as a primary diagnostic tool.

3. Approximately 15 to 25% of patients with HCV have an asymptomatic or mild infection that resolves spontaneously. The other 75 to 85% progress to develop chronic infection that may evolve over decades from chronic active hepatitis to cirrhosis with liver failure and/or hepatocellular carcinoma. Both chronic alcohol abuse and HIV infection may accelerate the natural history of liver disease. HCV infection is the leading indication for liver transplant in the United States. Fulminant, acute hepatitis with liver failure and death is a very rare manifestation of HCV infection.

4. There are four major modes of transmission recognized for HCV: (i) sharing of needles, syringes, or drug-preparing utensils during the injection of illicit drugs; (ii) from blood transfusions, including the use of clotting factors in hemophiliacs; (iii) sexually; and (iv) needle-stick injuries or mucous membrane exposure of blood from an HCV-infected individual. The major mode of HCV transmission in the United States and the industrialized world is intravenous drug use. Acute infection occurs most commonly in the 20- to 39-year age group. Up to 90% of intravenous drug users are HCV positive. Given that this patients was an intravenous drug user, she was mostly likely infected by sharing drug paraphernalia with an HCV-infected individual. Prior to 1990 and the development of screening tests for HCV, blood transfusion was a major mode of transmission. This form of hepatitis was called non-A, non-B hepatitis. With the introduction of screening, the risk of HCV transmission by blood transfusion has dropped and is now estimated in the United States to be less than 1 in 103,000 units transfused. HCV can be spread sexually, but its transmission is inefficient compared with HIV and other sexually transmitted diseases, including hepatitis B. As with HIV, sexual transmission is more efficient from males to females, and individuals with multiple partners are at greater risk of infection.

Exposure of health care workers to the blood of HCV-infected individuals via needle-stick or sharps injuries or mucous membrane exposures carries a 3% risk of infection, which is 10-fold less than the risk for acquiring HBV infection but 10-fold greater than the risk for acquiring HIV infection.

Vertical transmission of HCV from an infected mother occurs infrequently and is more likely from mothers who are concurrently infected with HIV. The virus does not appear to be transmitted via breast milk.

5. Her intravenous drug use puts her at risk for a number of blood-borne pathogens, including HIV, HBV, hepatitis D virus (which must occur concurrently with HBV), and severe acute or even fulminant HAV infection if the patient has been chronically infected with HCV. Intravenous drug users are also at increased risk for bacterial and fungal endocarditis, especially due to *Staphylococcus aureus*, beta-hemolytic streptococci, *Pseudomonas aeruginosa*, *Aspergillus* sp., and *Candida* sp. They may also suffer complications associated with endocarditis, including endophthalmitis and brain and liver abscesses.

Because intravenous drug users may trade sex for drugs, sexually transmitted diseases other than HIV should be considered. The patient presented here was found to be seropositive for syphilis, which often causes hepatitis during the secondary stage, and was treated for that infection as well.

6. Genotyping is important for disease prognosis and to predict response to therapy. Six different genotypes, 1 through 6, have been described for HCV. There are also subtypes of the six different genotypes. The nucleic acid sequence of the six genotypes may vary by as much as 33%. High mutation rates resulting in variation in the viral envelope proteins may be important in escaping the immune system, allowing chronic infection. In the United States and Europe, genotype 1 is most common, followed by types 2 and 3. Genotype 1b appears to be the most pathogenic of the genotypes, with more rapid progression of chronic active hepatitis to cirrhosis and a greater likelihood of requiring liver transplantation compared with patients infected with other genotypes. HCV patients who develop hepatocellular carcinoma are often infected either with genotype 1b or with a combination of genotypes including 1b.

Genotyping data have greater value in treatment management. The therapy of choice for chronic active HCV infection is interferon alfa-2b and ribavirin. The attachment of polyethylene glycol to the interferon alfa, called pegylated interferon, results in a higher response rate than does monotherapy with interferon. Trials of the combination of pegylated interferon and ribavirin are now in progress. Patients with genotype 1 receive a longer treatment regimen than those infected with genotype 2 or 3. Patients with genotype 2 and 3 are treated for 24 weeks. At that time, if their HCV RNA level is negative, they are considered cured. Patients infected with genotype 1 are tested at 24 weeks, and if negative for HCV RNA at that time, they receive an additional 24 weeks of therapy. If patients infected with any three of these genotypes are still positive at 24 weeks for HCV RNA, they have not responded to treatment and cannot be cured with currently available regimens. Therapy in genotype 1 patients would be discontinued at week 24 if the patient were HCV RNA positive at that time. Patients with genotype 1 have a lower cure rate despite extended duration of therapy, as compared with patients with genotype 2 and 3.

A recent study of acute HCV infection suggests that treatment with interferon alfa-2b for 24 weeks has a much higher cure rate than treatment of chronic active infection. In that study over 90% of patients had negative HCV RNA levels at the conclusion of therapy and 24 weeks later. It is important to note that 60% of patients in that study were infected with genotype 1. Longer follow-up is needed to determine how durable this response will be.

7. Unlike the other two common agents of hepatitis, HAV and HBV, there is no vaccine for HCV, so behavior modification of infected individuals is key. The first important step on the road to prevention is that individuals need to know they are infected and are a risk to others. This seems simple enough. However, this is a chronic infection with which viremic, i.e., infectious, individuals may be asymptomatically infected for years before showing symptoms. Many chronically infected patients do not know that they are infected. Therefore, screening for HCV infection of at-risk populations, such as intravenous drug users and individuals who seek care at clinics for sexually transmitted diseases, is important. Behaviors that prevent the spread of HIV also prevent the spread of HCV because their modes of transmission are similar, although HIV is spread more readily sexually while HCV is spread more efficiently by intravenous drug use. Because a common mode of HCV transmission is intravenous drug use, encouraging behaviors which prevent the sharing of "works" (i.e., needles, syringes, water, and utensils used to prepare the drugs for injection) so that infected individuals will not expose noninfected individuals to contaminated blood is essential. This can be very difficult in a culture where the sharing of works is ingrained.

As previously mentioned, the sexual transmission of HCV is inefficient but does occur, especially in individuals who have multiple partners. Condom use is to be very much encouraged in HCV-positive individuals as a means of preventing the spread of this agent to their partners, even in monogamous relationships where transmission is quite low.

Instituting programs among health care workers to prevent needle-stick injuries and mucous membrane exposures to blood will prevent not only HCV infections but also HIV and HBV infections. Gamma globulin does not prevent acquisition of HCV following needle sticks. It is not known whether treatment with interferon alfa-2b will prevent transmission following needle sticks, although recent data suggest that it can be used to effectively treat acute HCV infection, at least in the short term.

REFERENCES

1. **Centers for Disease Control and Prevention.** 1998. Recommendations for the prevention and control of hepatitis C virus (HCV) infection and HCV-related chronic disease. *Morb. Mortal. Wkly. Rep.* **47**(RR-19):1–39.

2. **Jaeckel, E., M. Cornberg, H. Wedemeyer, T. Santantonio, J. Mayer, M. Zankel, G. Pastore, M. Dietrich, C. Trautwein, and M. Manns for the German Acute Hepatitis C Therapy Group.** 2001. Treatment of acute hepatitis C with interferon alfa-2b. *N. Engl. J. Med.* **345**:1452–1457.

3. **Lauer, G. M., and B. D. Walker.** 2001. Hepatitis C virus infection. *N. Engl. J. Med.* **345**:41–52.

4. **Zein, N. N.** 2000. Clinical significance of hepatitis C virus genotypes. *Clin. Microbiol. Rev.* **13**:223–235.

CASE 63 The patient was an 8-year-old male with a 2-day history of diarrhea. He presented with worsening diarrhea (14 movements that day) which had become bloody. He also complained of pain on defecation. He had vomited once. He had attended a cookout 6 days previously. He claimed that his mother made him eat a hamburger that was "pink inside" even though "he did not like it." His physical examination was benign except for obvious dehydration. His laboratory findings were significant for a white blood cell count of 13,100/μl with 9,700 neutrophils/μl, a methylene blue stain of feces that showed abundant polymorphonuclear cells, and a positive stool guaiac. He was treated with trimethoprim-sulfamethoxazole and intravenous fluid therapy for dehydration. He quickly improved and was discharged within 24 hours. Culture of his stool specimen on MacConkey-sorbitol agar is shown in Fig. 1.

1. What is the most likely etiologic agent of his infection? What two important clues are found in this case that helped you determine the etiology of his infection?

2. What are the major virulence factors produced by this organism? How do they act and what are their roles in the pathogenesis of disease?

3. Why are these organisms so difficult to detect in feces? Think about one of the major virulence factors produced by this organism and how it is encoded genetically.

4. Besides cultures, what other methods may prove useful for detecting this organism? Explain how these methods could be used to detect this organism. Why might such methods be of value in studying outbreaks of disease with this organism?

5. How is the organism usually spread? How can infection with this organism be prevented?

6. Was using antibiotic therapy in this patient an appropriate clinical decision?

7. What are sequelae associated with this infection? What organ and cell types are specifically targeted? What is the outcome of these sequelae?

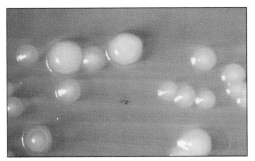

Figure 1

CASE DISCUSSION

1. This patient has <u>bloody diarrhea</u> due to enterohemorrhagic *Escherichia coli*, or <u>EHEC.</u> The specific serotype of *E. coli* with which he is infected is <u>O157:H7.</u> This is the most common of the approximately 50 serotypes of *E. coli* that have been recognized as causing this disease. The two important clues in this case were bloody diarrhea, which is found in approximately 90% of patients with EHEC-associated diarrhea, and the ingestion of a <u>hamburger that was "pink inside"</u> (see answer to question 5 for further details). <u>Abdominal pain,</u> characterized as "pain on defecation" in this patient, would also be consistent with this infection.

2. EHEC possesses two major virulence factors which play a role in the diarrheal disease process. One, a transmembrane bacterial protein called <u>intimin, mediates pedestal formation</u> on the apical surface of enterocytes. At the point of bacterial attachment, there is an accumulation of actin in the cytoskeleton that may explain this unique cell structure. This accumulation of actin is an end result of a signal transduction pathway induced by proteins secreted by the attached bacteria. Through a not-yet-defined sequence of events, this intimate binding results in changes in the enterocyte which cause diarrhea. The organism also produces a family of toxins, encoded by a lysogenic phage, referred to as Shiga toxin (STx). There are at least two different Shiga toxins. STx-1 is immunologically and biochemically identical to the Shiga toxin (Stx) produced by *Shigella dysenteriae*. STx-2 is actually believed to be a family of toxins that are immunologically distinct from STx-1 but are immunologically and biochemically related to each other. Strains may produce one or both toxins. Strains producing STx-2 alone or in combination with STx-1 are believed to be more virulent. STx-1 and 2 have the same mechanism of action, i.e., inhibition of protein synthesis by enzymatic inactivation of the 80S ribosome. STx-1 and 2 are A-B toxins. The B toxin binds to a specific receptor, a cell surface glycolipid called globotriaosylceramide. The toxin-glycolipid complex is internalized by cells, and the enzymatically active A subunit can then bind to and inactivate the 80S ribosome, inhibiting protein synthesis. In addition to inhibiting protein synthesis, there is a growing body of evidence that these toxins can cause apoptotic changes in endothelial cells as well. It appears that these toxins target endothelial cells and can damage intestinal blood vessels, resulting in the bloody diarrhea seen in these patients. These toxins are thought to play a prominent role in the complications of this infection discussed in the answer to question 7.

3. *E. coli* is very common in the gastrointestinal flora, and most individuals are colonized with nonpathogenic strains. Diarrheagenic strains of *E. coli* possess both chromosomal and extrachromosomal genes which encode virulence factors. STx-1 and 2 are encoded by genes contained in a lysogenic phage. The <u>genes for effacement</u> are encoded on a pathogenicity island on the chromosome called LEE (Locus for Enterocyte Effacement) whereas genes for the early stage of attachment mediated by bundle-forming pili are encoded on the EAF (EPEC Adherence Factor) plasmid. It is often difficult to distinguish pathogenic from nonpathogenic *E. coli* strains unless methods are used to detect virulence factors or the genes encoding them. Fortunately, O157:H7, the most common cause of EHEC, can be distinguished from over 90% of all other *E. coli* strains by its inability to ferment sorbitol. Almost all other *E. coli* isolates can ferment this sugar. To detect *E. coli* O157:H7 in the presence of nonpathogenic *E. coli* and other related enteric bacteria, stool specimens are plated on a differential and selective medium which contains sorbitol and a pH indicator to show whether sorbitol has been fermented. The isolates that fail to ferment sorbitol are screened serologically to determine if they are *E. coli* O157:H7 (Fig. 1). This culture technique is widely available but may need to be specifically requested if this organism is being sought in feces. Detection rates are much higher in patients with bloody diarrhea than in those with non-bloody diarrhea. It is recommended that *E. coli* O157:H7 be sought in the stools of all patients with bloody diarrhea.

4. Besides culture, EHEC can be detected either by demonstrating toxin production or by finding genes encoding the toxins. Detection of toxin production can be done directly on stool specimens or clinical isolates. Two approaches for toxin detection are widely used. One is based on the ability of STx to cause cytotoxic changes in a continuous cell line called Vero (thus the name verotoxin). Cytotoxicity assays are done as follows. Two aliquots of stool or culture filtrate are prepared. One is mixed with an equal volume of buffer, and the other is mixed with an equal volume of antiserum specific for STx. After a brief incubation period, the two samples are added to Vero cell monolayers and observed for up to 72 hours. The following chart explains the interpretation of the test.

FILTRATE AND BUFFER	FILTRATE AND ANTISERUM	TEST INTERPRETATION
No cytotoxicity	No cytotoxicity	No toxin present
Cytotoxicity	No cytotoxicity	Toxin present
Cytotoxicity	Cytotoxicity	Uninterpretable
No cytotoxicity	Cytotoxicity	Technical error

Another approach to toxin detection is to use an enzyme immunoassay (EIA) for the immunologic detection of toxin. Tests for STx are commercially available. They are not as sensitive as the bioassay in cell monolayers but may be sensitive enough to detect clinical disease.

In addition to toxin detection, detection of genes coding for STx production may also be used. Detection of toxin genes can be done directly in stool or on isolates suspected of causing disease. The most common technique used to detect these STx genes is PCR (polymerase chain reaction).

It is estimated that approximately 80% of episodes of enterohemorrhagic colitis (the disease this child had) are due to the O157:H7 serotype. Other *E. coli* serotypes that have been shown to produce STx, particularly O26 and O111, have been associated with both outbreaks and sporadic cases of bloody diarrhea. Isolates of these serotypes usually are sorbitol positive, so culture on sorbitol-MacConkey plates will be of no diagnostic value for these organisms. When looking for non-O157 serotypes that produce STx, methods that detect either the toxin or toxin gene must be employed.

5. The major reservoir for *E. coli* O157:H7 is cattle. Consumption of ground beef, eaten either raw or as undercooked hamburgers, appears to be the typical way in which this organism is spread. Several large outbreaks of this disease have been associated with the consumption of undercooked hamburgers prepared by fast-food restaurant chains. Why is ground beef most frequently associated with outbreaks of this disease? Fast-food restaurant chains want a consistent product, so they often buy ground beef in huge lots containing thousands of pounds of meat. To produce these large lots, multiple animals, frequently from different herds from different locales, find their way into a single lot. Cattle can carry *E. coli* O157:H7 as part of their normal fecal flora, with carriage rates being higher during warm weather months. During rendering, the carcass of an *E. coli* O157:H7-infected animal may become contaminated with fecal flora containing this organism. During production of these multi-animal lots of beef, the entire lot may become contaminated by this single animal. The process of grinding will introduce the organism throughout the meat. When beef patties are formed, organisms may be at the center of the patty. If the interior temperature of the patty fails to reach 150°F (ca. 67°C), the organism will survive cooking. Rare to medium-rare hamburgers ("pink inside") may not reach this temperature, resulting in the survival of some percentage of the EHEC. As with *Shigella* spp., it appears that a low inoculum size can produce EHEC infection, so the small number of organisms that survive in the interior of undercooked hamburgers may be sufficient to be a hazard. Other sources of EHEC organisms have included unpasteurized cow's milk and cheese, water fecally contaminated by cattle, unpasteurized cider made of apples which had fallen in a cattle grazing area and were not washed before cider preparation, and radish and alfalfa sprouts for which cow manure was used as a fertilizer. Sausages made from both cattle and pigs have been a source of this organism as well. Person-to-person spread has now been well documented with this organism in day care centers, fresh water lakes, and water parks.

Preventive measures include avoiding the consumption of undercooked or raw hamburgers and unpasteurized dairy products. A simple rule is to cook hamburgers until the juices run clear and the meat is not pink inside. By law in several states, restaurants may not sell rare to medium-rare hamburgers; they must be thoroughly cooked.

6. There is conflicting evidence, some of which supports the contention that antibiotic therapy is contraindicated in patients with EHEC infections. The reason for this is that hemolytic uremic syndrome (HUS) (see answer to question 7 for further details) may be seen more frequently in EHEC-infected patients treated with antibiotics than in those who are untreated. The reason is that several different antibiotics have been shown to increase toxin production/release by this organism. In particular, the antibiotic trimethoprim-sulfamethoxazole, with which this patient was treated, has been shown in some studies to be associated with HUS. Fortunately, this patient did not develop this syndrome.

7. The major sequela associated with this disease is the hemolytic uremic syndrome (HUS). The syndrome may also be called thrombotic thrombocytopenic purpura (TTP) in adults. HUS is seen primarily in children, and it is estimated to occur in approximately 10% of individuals following diarrheal disease with EHEC. The pathophysiology of HUS is due to the action of STx on endothelial cells, particularly those in the kidney. HUS is characterized by thrombocytopenia, microangiopathic hemolytic anemia, and acute renal failure. STx-induced changes in the endothelial cells result in increased binding of platelets to the wall of renal microvasculature, which contributes to this triad of events. Therapy for this disease is primarily supportive and may include erythrocyte transfusions and dialysis. The mortality of this disease is estimated at 5%; an additional 5 to 10% of patients have some degree of chronic kidney failure. HUS is the most common cause of kidney failure in children.

REFERENCES
1. **Gilligan, P. H.** 1999. *Escherichia coli*: EAEC, EHEC, EIEC, ETEC. *Clin. Lab. Med.* **19:**505–521.

2. **Kiyokawa, N., T. Taguchi, T. Mori, H. Uchida, N. Sato, T. Takeda, and J. Fujimoro.** 1998. Induction of apoptosis in normal renal tubular epithelial cells by *Escherichia coli* Shiga toxins 1 and 2. *J. Infect. Dis.* **178:**178–184.

3. **Nataro, J. P., and J. B. Kaper.** 1998. Diarrheagenic *Escherichia coli*. *Clin. Microbiol. Rev.* **11:**142–201.

4. **Safdar, S., A. Said, R. E. Gangnon, and D. G. Maki.** 2002. Risk of hemolytic uremic syndrome after antibiotic treatment of *Escherichia coli* O157:H7 enteritis: a meta-analysis. *JAMA* **288:**996–1001.

5. **Wong, C. S., S. Jelacic, R. L. Habeeb, S. L. Watkins, and P. I. Tarr.** 2000. The risk of hemolytic-uremic syndrome after antibiotic treatment of *Escherichia coli* O157:H7 infections. *N. Engl. J. Med.* **342:**1930–1936.

CASE 64

A 30-year-old woman presented to the clinic with fever, backache, and headache of 2 days' duration. She complained bitterly about intense myalgias in the upper arms and pain on moving her eyes. She had just returned from a trip to El Salvador, where she had extensive exposure to mosquitoes. On physical examination, she appeared uncomfortable but not toxic. A blanching, erythematous rash was present on the face, arms, trunk, and thighs (Fig. 1). There was no enanthem, murmur, or splenomegaly. Her white blood cell count was 1,600/μl with a normal differential, platelet count was 140,000/μl, and hemoglobin was 17.5 g/dl. Convalescent-phase antibodies to a mosquito-borne viral disease were diagnostic.

1. What viral disease did the patient acquire? By what type of mosquitoes is this virus transmitted? Are any of these mosquitoes found in the United States?

2. Severe disease, which can be fatal, can occur with this infection. What are the clinical manifestations of severe disease as a result of infection with this virus?

3. There are four distinct serotypes of this virus. Would you predict that infection with one serotype would confer immunity to the other serotypes?

4. In what geographic range is this disease endemic? What means can be taken by travelers to endemic regions to prevent infection?

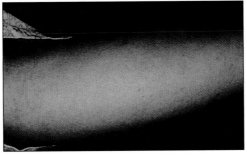

Figure 1

CASE DISCUSSION

1. This patient had convalescent-phase antibodies that were diagnostic of infection with one of the four serotypes of dengue virus. The dengue viruses are members of the flavivirus family, which includes the yellow fever virus and viruses which can cause mosquito-borne encephalitis. Flaviviruses are enveloped, single-stranded RNA viruses which replicate in the cytoplasm of the infected cell. Dengue virus replicates primarily in monocytes or macrophage-type cells. It is transmitted most commonly by the highly domesticated urban mosquito *Aedes aegypti*. Other mosquitoes within this genus can also serve as vectors of dengue.

The potential exists for epidemic spread of dengue virus within the United States. There have been small numbers of autochthonous cases in Texas during the 1980s (due to dengue 1) and 1990s (due to dengue 2) and 119 confirmed cases in Hawaii during 2001 and 2002 (due to dengue 1). *A. aegypti* has been found in the United States for more than 200 years, and can be found during the summer months in Gulf Coast states. In addition, *Aedes albopictus* (the Asian tiger mosquito), which transmits dengue in Asia, was introduced into the United States several years ago in the water found in old automobile tires imported from Asia. Since that time, this mosquito has greatly expanded its range and is now found in at least 26 states. Although documentation of cases of dengue transmitted by this mosquito in the United States is lacking, the potential for spread is real.

2. Dengue is normally manifested by fever, chills, headache, myalgias, and bone or joint pain. A rash may be seen, as in this case. The disease is typically painful (hence the term "breakbone fever") but is usually self-limiting. Severe manifestations of dengue include dengue hemorrhagic fever (DHF) and dengue shock syndrome (DSS).

DHF results in a low platelet count and hemoconcentration, which usually develops on days 3 to 7 of illness at the time of defervescence. Patients may have hemorrhagic manifestations, such as bleeding from mucous membranes, sites of previous needle sticks, and the gastrointestinal tract. A fraction of patients with DHF become hypotensive or have a narrowing of the pulse pressure and have DSS. Patients with DHF/DSS may have abdominal pain and pleural effusions due to marked capillary leakage. Gastrointestinal hemorrhages are often found at autopsy.

Because of their anticoagulant properties, aspirin and other nonsteroidal anti-inflammatory agents are contraindicated in patients with dengue fever. Acetaminophen is recommended for the management of pain and fever.

3. Unfortunately, it is possible to be infected by each of the four serotypes of dengue virus. The immunity that is acquired after infection with one of the dengue virus serotypes confers lasting immunity only to that serotype. In fact, there is evidence that

DHF and DSS are due, in part, to prior immunity to a dengue serotype different from the one infecting the patient. The pathogenesis of this process is thought to be caused by immune complexes that contain both the preexisting antibody to a different serotype (which is not protective) and the virus. These immune complexes are taken into host cells more easily than is the virus without the antibody. An alternative hypothesis is that DHF may be due to increases in the virulence of the infecting viral strain.

4. The geographic range of dengue virus corresponds with the range of the mosquito vector *A. aegypti*. This includes most tropical areas in both hemispheres as well as some temperate areas. More than half of the world's population live in areas at risk for dengue infection. It is estimated that worldwide there are 50 to 100 million cases of dengue fever each year. There has been a dramatic increase in the range of *A. aegypti* in the Western Hemisphere during the past 2 decades and a corresponding increase in cases of dengue fever.

A. aegypti is a species that is well adapted to urban environments and is often found in or near human dwellings. *A. aegypti* normally bites during the day, with peak biting activity in the early morning and late afternoon, but may feed at any time during the day. Thus, to prevent dengue infection, mosquito repellent such as *N,N*-diethyl-*m*-toluamide (DEET) must be worn during the daytime when the mosquitoes are most active. It is important to recognize that DEET may need to be reapplied fairly frequently due to evaporation. In addition, wearing protective clothing and remaining in well-screened or air-conditioned areas are recommended. *A. aegypti* may be present in geographic areas in which malaria is endemic. *Anopheles* spp., which transmit malaria, most commonly bite at night. There is currently no available vaccine against dengue. As a result, prevention of dengue infection can be accomplished only by preventing the bite of the vector mosquito.

REFERENCES

1. **Centers for Disease Control and Prevention.** 2000. Imported dengue—United States, 1997 and 1998. *Morb. Mortal. Wkly. Rep.* **49:**248–253.

2. **Gill, J., L. M. Stark, and G. G. Clark.** 2000. Dengue surveillance in Florida, 1997–98. *Emerg. Infect. Dis.* **6:**30–35.

3. **Rigau-Perez, J. G., G. G. Clark, D. J. Gubler, P. Reiter, E. J. Sanders, and A. V. Vorndam.** 1998. Dengue and dengue haemorrhagic fever. *Lancet* **352:**971–977.

CASE 65

This 60-year-old woman with a medical history of a gastric ulcer had recently noted symptoms of dyspepsia. She characterized her discomfort as a pressure in the upper abdominal area that radiated to her chest and neck. She underwent an upper gastrointestinal series that showed radiologic findings compatible with a thickened fold within the stomach.

An outpatient esophagogastroduodenoscopy (EGD) was performed. A biopsy of the antral portion of the stomach was consistent with moderate gastritis. No tumor was seen. In addition, the biopsy demonstrated 3+ to 4+ of a bacterial organism (Fig. 1).

1. What bacterium has been associated with chronic gastritis?

2. What clinical syndromes, other than chronic gastritis, have been linked to this organism?

3. Other than by histopathologic examination of a biopsy specimen, how can infection with this organism be diagnosed?

4. What special properties of this organism allow it to live in the rather inhospitable environment of the human stomach?

5. What is the epidemiology of infection with this organism?

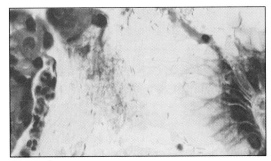

Figure 1

CASE DISCUSSION

CASE 65

1. The curved or helical gram-negative rod *Helicobacter* (formerly *Campylobacter*) *pylori* has been associated with chronic gastritis. The photomicrograph (Fig. 1) is consistent with *H. pylori* in an area of chronic gastritis.

2. There are data to support links between *H. pylori* and chronic gastritis and peptic ulcer disease, including both gastric and duodenal ulcers. In addition, there is epidemiologic evidence that both adenocarcinoma of the stomach and high-grade gastric B-cell lymphoma are associated with *H. pylori* infection. As you might imagine, it would be difficult to employ Koch's postulates in demonstrating that *H. pylori* is an etiologic agent of these two malignancies. There have been a number of studies that suggest that *H. pylori* infection is associated with an increased risk of coronary artery disease. This is far from clear, and the data are conflicting.

3. The isolation of *H. pylori* in a culture of a gastric biopsy specimen is the "gold standard" for establishing the diagnosis of infection with this organism. Unfortunately, this method, which requires endoscopic biopsy and gastric biopsy culture, is not routinely performed by all clinical microbiology laboratories. Culture is estimated to be relatively insensitive, with approximately 70% of cultures being positive in patients with *H. pylori*-associated disease. Another method, also requiring endoscopic biopsy, is the *Campylobacter*-like organism (CLO) test, which relies on the presence of the bacterial enzyme urease (see answer to question 4). In this test, gastric biopsy tissue is added to a tube that contains urea agar. The urease activity of tissue containing *H. pylori* is demonstrated by the change in color of the indicator present in the urea agar (Fig. 2). Noninvasive tests include the demonstration of urease production following the ingestion of ^{13}C- or ^{14}C-labeled urea by the detection of labeled carbon dioxide in expired air, serologic tests to demonstrate the presence of IgG antibody to *H. pylori*, and most recently the use of an *H. pylori* stool antigen enzyme immunoassay. Because of its noninvasive nature, *H. pylori* serology is being used with increasing frequency in an attempt to

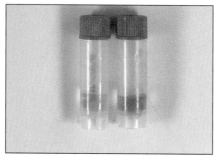

Figure 2

diagnose this infection. Serologic results must be interpreted cautiously. Seropositivity increases with age, presumably because of increased exposure to the organism. Patients can be seropositive without clinical evidence of gastritis or peptic ulcer disease.

4. The ability of *H. pylori* to survive and multiply in the ecologic niche of the acidic stomach is quite unusual. Although the details have not been worked out, there are a number of interesting adaptations that are important in allowing colonization and subsequent multiplication of the organism. The enzyme urease, which may represent as much as 6% of the protein synthesis of *H. pylori*, is active at the low pH of gastric juice. It has been established by studies with mutant *H. pylori* that lack urease that this enzyme is essential for gastric colonization. Urease catalyzes the hydrolysis of urea, resulting in the production of ammonia, which is believed to raise the pH of the microenvironment, resulting in improved bacterial survival. It is a subject of active debate whether or not this mechanism is the means by which urease contributes to the colonization of the stomach. Another factor of importance is the presence of flagella, conferring motility to *H. pylori*. The flagella enable *H. pylori* to move through the thick mucous coat that is present in the stomach.

5. Studies on the time of acquisition of *H. pylori* infection have generally reported seroprevalence data. The prevalence of infection due to *H. pylori* varies by geographic area, age, race, ethnicity, and socioeconomic status. *H. pylori* is typically acquired during early childhood in developing countries. In developed countries, such as the United States, the infection is not typically acquired during childhood, and the incidence is lower. It is important to recognize that although approximately 50% of Americans have been infected with this organism by age 60, most infections are asymptomatic. In the United States, whites have a lower seroprevalence rate than do either African Americans or Latinos. Although other routes of transmission may occur, it is likely that the most common means of transmission are via either the oral-oral route or the fecal-oral route.

REFERENCES

1. **Brown, L. M.** 2000. *Helicobacter pylori*: epidemiology and routes of transmission. *Epidemiol. Rev.* **22**:283–297.

2. **Lassen, A. T., F. M. Pedersen, P. Bytzer, and O. B. Schaffalitzky de Muckadell.** 2000. *Helicobacter pylori* test-and-eradicate versus prompt endoscopy for management of dyspeptic patients: a randomised trial. *Lancet* **356**:455–460.

3. **Uemura, N., S. Okamoto, S. Yamamoto, N. Matsumura, S. Yamaguchi, M. Yamakido, K. Taniyama, N. Sasaki, and R. J. Schlemper.** 2001. *Helicobacter pylori* infection and the development of gastric cancer. *N. Engl. J. Med.* **345**:784–789.

4. **Vaira, D., P. Malfertheiner, F. Megraud, A. T. Axon, M. Deltenre, A. M. Hirschl, G. Gasbarrini, C. O'Morain, J. M. Garcia, M. Quina, G. N. Tytgat, and the HpSA European Study Group.** 1999. Diagnosis of *Helicobacter pylori* infection with a new non-invasive antigen-based assay. *Lancet* **354**:30–33.

CASE 66

The patient was a 7-year-old female who was admitted to an outside hospital 4 days previously with complaints of fever, malaise, headache, and myalgias. She was referred to our institution because of respiratory distress (respiratory rate of 60 to 70/min) and a chest radiograph consistent with right pleural effusion and pulmonary edema, hepatitis (aspartate aminotransferase [AST], 173 U/liter; alanine aminotransferase [ALT], 128 U/liter; lactate dehydrogenase [LDH], 572 U/liter), and pancytopenia (white blood cell count, 4,000/μl; hemoglobin, 10.3 g/dl; platelets, 116,000/μl). She had a sick sibling and father, both with low-grade fevers. Her family owned exotic animals, including reptiles.

At our institution she underwent elective intubation, bronchoscopy, bone marrow aspiration, and spinal tap. The bronchoscopy revealed mild, nonspecific bronchitis. Cultures for bacteria—including *Chlamydia pneumoniae* and acid-fast bacilli—and fungal and viral cultures were all negative, as was an examination for *Pneumocystis carinii*. Bone marrow examination was consistent with viral suppression. Cerebrospinal fluid (CSF) examination revealed 54 white blood cells per μl (90% neutrophils), 40 red blood cells per μl, a protein level of 110 mg/dl, and a glucose level of 39 mg/dl. Bacterial, fungal, and viral cultures of CSF were negative. Blood cultures done at admission were negative. HIV, cytomegalovirus, hepatitis A, B, and C viruses, psittacosis, *Brucella* sp., *Francisella tularensis*, Epstein-Barr virus, *Leptospira* sp., *Bartonella* sp., and *Ehrlichia* sp. serologies, and antinuclear antibodies were all determined. HIV serology results were problematic: enzyme immunoassays were positive or indeterminate three of four times. Western blot (immunoblot), p24 antigen, and HIV RNA PCR (polymerase chain reaction) results were all negative. A diagnostic serology was reported on hospital day 10. A review of her hospital course showed that the patient's pulmonary edema cleared on hospital day 2, and she was extubated on hospital day 3. Fever persisted until hospital day 14. Liver enzyme levels began improving on hospital day 7. Her antibiotics were discontinued on hospital day 15, and she was discharged. A blood smear from a patient infected with the same organism as this patient is shown in Fig. 1.

1. With what organism do you think this patient was infected? What clinical features did this patient have which are typical for this disease?

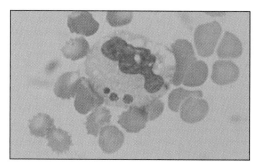

Figure 1

2. Describe the two forms of disease caused by this genus.

3. With what other infectious disease process is this organism frequently confused?

4. Why is this diagnosis usually made serologically?

5. How is this organism spread?

6. Give a possible explanation for why this organism is being seen with increasing frequency.

CASE DISCUSSION

CASE 66

1. This patient was infected with *Ehrlichia chaffeensis*. Serologic studies revealed that this patient had a high initial titer to *E. chaffeensis*, and a convalescent-phase serum showed an eightfold rise in titer to this organism. All other serologies were negative.

Patients with *Ehrlichia* infection generally present with a febrile illness and complaints of headache, malaise, and myalgias, signs and symptoms that can be seen in a variety of systemic infections. All were seen in this patient. Patients with ehrlichiosis usually have elevated liver enzymes and leukopenia. They may also have thrombocytopenia and anemia. These were also seen in our patient, although her leukopenia and thrombocytopenia were mild. Her bone marrow examination was characterized as indicating "viral suppression." Bone marrow hypoplasia has been reported with this infection and is probably due in part to infection of precursor cells by ehrlichiae. Pneumonitis due to *Ehrlichia* infection has also been reported and may explain her respiratory failure, which required a brief period of intubation.

One of the diagnoses that were considered in this patient was HIV infection. Her screening serology was very problematic, being positive or indeterminate three times. However, her Western blot assay (to confirm the screening test results) was consistently negative, as were tests used to directly detect the virus. She is an example of a patient with false-positive screening results, reemphasizing the importance of confirming screening test results.

2. *E. chaffeensis* infects mononuclear cells, mainly monocytes and macrophages. Infection with *E. chaffeensis* (named for Ft. Chaffee, Ark., the geographic site where human infection was first recognized) is found primarily in the south central to southeastern United States. A second form of the disease results from infection of granulocytes and is called human granulocytic ehrlichiosis (HGE). Two agents have been recognized. One is closely related genetically to *Ehrlichia phagocytophila* and *Ehrlichia equi*. A second species, *Ehrlichia ewingii*, previously thought only to infect dogs, has recently been found as a cause of HGE. Human granulocytic ehrlichiosis has been found in the upper Midwest, upstate New York, the New England states, and coastal regions of the Pacific Northwest.

3. Ehrlichiosis has been referred to as "Rocky Mountain spotless fever." *Ehrlichia* spp. belong in the same family (*Rickettsiaceae*) as *Rickettsia rickettsii*, the agent of Rocky Mountain spotted fever. Clinically, the diseases have similar presentations, although a rash is infrequent (spotless) in ehrlichiosis but common (spotted) in *R. rickettsii* infection. Most *Ehrlichia* infections are mild and self-limited. Morbidity and mortality is much higher with *R. rickettsii* than with *Ehrlichia* infection. In North Carolina where *R. rickettsii* and *E. chaffeensis* infection are both endemic, recent studies suggest that in patients with unexplained fever and a history of a tick bite (the vector for both infections), *Ehrlichia* infections are more frequent. Likewise, in areas of the United States where *Borrelia burgdorferi* and HGE are endemic, the clinical presentation of patients infected with either of these two organisms is similar and they also can be confused.

4. A characteristic of members of the family *Rickettsiaceae* is that they cannot grow on artificial media. Rather, they must be grown in other living cells, such as continuous cell lines. Although *E. chaffeensis* can be grown in macrophage-derived cell lines, cultivation of *Ehrlichia* spp. from patients is beyond the capability of almost all hospital and reference laboratories.

The organism produces characteristic cytoplasmic inclusions (morulae; see Fig. 1), which can be seen by microscopic examination of peripheral blood or bone marrow smears. Observation of morulae occurs in approximately two-thirds of patients with human granulocytic ehrlichiosis but only 30% of patients with human mononuclear ehrlichiosis. The test is most likely to be positive only in the acute stage of the febrile illness. PCR has been used successfully to detect this organism in the blood of infected individuals, but this test is only available in a few research and reference laboratories.

Because this organism cannot be grown on artificial media, direct examination is relatively insensitive, and PCR is not widely available; serology is the method of choice for diagnosis. Sera should be obtained during the acute phase of illness and 2 to 4 weeks (convalescent phase) later. A fourfold rise in titer between the acute- and convalescent-phase sera is diagnostic. A major drawback of serology is that the diagnosis often is made retrospectively.

This patient had a high acute-phase titer to *Ehrlichia* sp. and negative titers to a variety of other agents for which she was tested, suggesting that she was infected with *Ehrlichia*. A subsequent eightfold rise in titer of the convalescent-phase sera confirmed her diagnosis.

5. Ehrlichiosis is a zoonotic infection which is spread by ticks. The disease incidence is highest in June through August, which corresponds with the time of maximal tick feeding. Deer, field mice, and other small mammals are the natural hosts for *Ehrlichia* and act as a reservoir for these organisms. Humans are accidental hosts of this organism. Animal studies suggest that transmission of *Ehrlichia* can occur in the first 24 hours of tick attachment. For two other common tick-borne diseases, Rocky Mountain spotted fever and Lyme disease, ticks typically must be attached for more than 24 hours for the infecting agent to be transferred. The major vector of *E. chaffeensis* is the Lone Star tick (*Amblyomma americanum*), while the major vector of the agent of human granulocytic ehrlichiosis is the deer tick (*Ixodes scapularis*). Because *I. scapularis* is also a vector for *B. burgdorferi* (the etiologic agent of Lyme disease), dual infections with *B. burgdorferi* and HGE have been reported.

6. There are several possible explanations for why *Ehrlichia* spp. are being increasingly recognized. First, disease caused by these organisms has only recently been described (in 1986) in the United States. Improved diagnostic tools for detecting *Ehrlichia* spp. are now widely available. Increased surveillance efforts by the Centers for Disease Control and Prevention may contribute to increasing case recognition. Understanding of the clinical disease caused by this organism is evolving. It is highly likely that illness due to this organism has been, and continues to be, underdiagnosed or misdiagnosed as infection due to other agents such as *Rickettsia*. Second, exposure to ticks, the vector of *Ehrlichia*, may be increasing. The reasons are varied. With increased homebuilding and outdoor activities in forested areas, humans have increased exposure to ticks. Populations of the tick's animal hosts such as deer and field mice may be increasing because of reforestation in some areas and elimination of natural predators. Probably both of these factors, along with others, have played a role in the emergence of *Ehrlichia* spp. as more frequently recognized human pathogens.

REFERENCES

1. **Bakken, J. S., and J. S. Dumler.** 2000. Human granulocytic ehrlichiosis. *Clin. Infect. Dis.* **31:**554–560.

2. **Belongia, E. A., K. D. Reed, P. D. Mitchell, P.-H. Chyou, N. Mueller-Rizner, M. F. Finkel, and M. E. Schriefer.** 1999. Clinical and epidemiological features of early Lyme disease and human granulocytic ehrlichiosis in Wisconsin. *J. Infect. Dis.* **29:**1472–1477.

3. **Buller, R. S., M. Arens, P. Himel, C. D. Paddock, J. W. Sumner, Y. Rikihisa, A. Unver, M. Gaudreault-Keener, F. A. Manian, A. M. Liddell, N. Schmulewitz, and G. A. Storch.** 1999. *Ehrlichia ewingii*, a newly recognized agent of human ehrlichiosis. *N. Engl. J. Med.* **341:**148–155.

4. **Carpenter, C. F., T. K. Gandhi, L. Kuo Kong, G. R. Corey, S.-M. Chen, D. H. Walker, J. S. Dumler, E. Breitschwerdt, B. Hegarty, and D. J. Sexton.** 1999. The incidence of ehrlichial and rickettsial infections in patients with unexplained fever and recent history of tick bite in central North Carolina. *J. Infect. Dis.* **180:**900–903.

5. **des Vignes, F., J. Piesman, R. Heffernan, T. L. Schulze, K. C. Stafford III, and D. Fish.** 2001. Effect of tick removal on transmission of *Borrelia burgdorferi* and *Ehrlichia phagocytophila* by *Ixodes scapularis* nymphs. *J. Infect. Dis.* **183:**773–778.

6. **McQuiston, J. H., C. D. Paddock, R. C. Holman, and J. E. Childs.** 1999. The human ehrlichioses in the United States. *Emerg. Infect. Dis.* **5:**635–642.

7. **Parola, P., and D. Raoult.** 2001. Ticks and tickborne bacterial diseases in humans: an emerging disease threat. *Clin. Infect. Dis.* **32:**897–928.

8. **Standaert, S. M., T. Yu, M. A. Scott, J. E. Childs, C. D. Paddock, W. L. Nicholson, J. Singleton, Jr., and M. J. Blaser.** 2000. Primary isolation of *Ehrlichia chaffeensis* from patients with febrile illnesses: clinical and molecular characteristics. *J. Infect. Dis.* **181:**1082–1088.

CASE 67

A 12-year-old boy experienced the acute onset of periumbilical pain. The pain moved to the left upper quadrant. The patient also complained of cough, mild sore throat, backache, headache, and feeling "hot" and tired. He also had watery eyes. Twelve hours after the onset of these symptoms, he was seen in a local clinic in a small rural New Mexico town. His temperature was 39.4°C. He was given antipyretics, reported to have the "flu," and sent home. He returned to the clinic 4 hours later with vomiting and shortness of breath.

His physical exam showed no lymphadenopathy or organomegaly. Cardiopulmonary exam was consistent with a pulmonary process. His oxygen saturation was 89% on room air; no arterial blood gas concentrations were obtained. His chest X ray showed findings consistent with pulmonary edema. He received oxygen and supportive care. On laboratory testing, his chemistries were normal but his complete blood cell count was significant for a platelet count of 85,000/μl. His platelet count fell to a nadir of 57,000/μl over the course of 3 days before returning to normal.

The patient's history was significant for living in a rural area. Many mice were recently seen around his home which he said he had "shot" with a squirt gun.

His diagnosis was confirmed by enzyme immunoassay and Western blot (immunoblot) as well as PCR (polymerase chain reaction) on serum.

1. What is this child's diagnosis?

2. What is our current understanding of the pathogenesis of infection with this agent?

3. What is the reservoir in the United States for this infection? What environmental factors impact on this reservoir?

4. How is this infection acquired? Is there any evidence of person-to-person transmission?

5. What is the geographic distribution of this disease?

CASE DISCUSSION

CASE **67**

1. This patient was diagnosed with the hantavirus pulmonary syndrome (HPS). This syndrome is caused, in the United States, by the Sin Nombre virus, an enveloped, negative sense RNA virus belonging to the *Hantavirus* genus of the family *Bunyaviridae*. This disease was first recognized in 1993 in the Four Corners region (the area where the borders of Arizona, New Mexico, Colorado, and New Mexico meet) of the southwestern United States, where a cluster of cases was reported mostly in healthy young adults. Retrospective review has uncovered cases of HPS in the United States as far back as 1959. This patient's presentation was in many ways typical of HPS. He had the typical HPS prodrome of fever, myalgia, headache, and cough, which was followed by rapid onset of pulmonary edema. Patients typically have thrombocytopenia, oxygen saturation <90% (as were seen in this patient), and hemo-concentration as evidenced by high hematocrits (not seen in this patient). Mortality is estimated to be 45%, with most patients dying from cardiopulmonary failure within a few days of disease onset. Treatment, frequently including intubation and ventilation, is mainly supportive, as no effective antiviral agent has been identified to date. A double-blinded study of the antiviral agent ribavirin is ongoing.

It would be important for the physicians caring for this patient to consider the possibility of plague due to infection with *Yersinia pestis*, given that he had known rodent exposure in New Mexico, the state with the greatest number of cases of this disease.

2. Sin Nombre virus has been shown to specifically infect the endothelial cells in the microvasculature of the lung. Virus-induced changes in the endothelium are believed to be responsible for the leaking of fluid into the lung, resulting in the pulmonary edema characteristic of this disease. The mechanism by which this occurs is not understood, but recent data suggest that it may be an immune-mediated phenomena since high levels of various cytokines, some of which are vasoactive, can be found in the lungs of patients with HPS.

3. The deer mouse *Peromyscus maniculatus* is the principal reservoir for Sin Nombre virus. Field studies in the Four Corners region have shown that up to 30% of mice may be infected. The enzootic hantavirus can cause persistent asymptomatic infection in this mouse species.

Two outbreaks of Sin Nombre virus infection have been reported in the Four Corners region, one in 1993–1994, and the other in 1998–1999. In the winter months prior to each outbreak, there was increased rainfall due to an El Niño weather pattern (change in upper-level wind patterns resulting in increased rainfall to semi-arid and arid regions in the southwestern United States). This increased rainfall resulted in large increases in mouse populations because of favorable environmental conditions, the most important being abundant food supplies. Increases in mouse populations led to greater contact between this infected reservoir and humans. Following the 1993–1994 outbreak, mouse populations declined as environmental conditions became less favorable. With this decline in mouse populations, there was also a marked decrease in the number of HPS cases.

4. Humans are an accidental host of hantavirus, acquiring this infection following exposure to mouse excreta or saliva, most often by the aerosol route. Individuals are typically infected in rural areas where infected mice live in and around dwellings. In this case, the history of the child's squirting mice with his squirt gun near his home indicates the likelihood of exposure to infected mouse excreta. Individuals who become infected with Sin Nombre virus typically either live in or are working in or around dwellings infested with infected mice. A small percentage of individuals have also become infected after camping.

Judging from the lack of secondary spread of this virus to health care workers caring for infected individuals, and the absence of nosocomial spread of the virus, there is currently no evidence to suggest that person-to-person spread occurs with Sin Nombre virus, though there is evidence for person-to-person spread in infections caused by Andes virus, a related hantavirus found in South America.

5. HPS appears to be present throughout the Americas with cases now reported not only from the western United States but also western Canada, Mexico, Panama, Brazil, and Argentina. In all instances reported so far, the reservoir for the virus responsible for this syndrome is a rodent.

Other hantaviruses, seen mostly in Europe and Asia, cause epidemic and sporadic disease manifested as hemorrhagic fever and renal failure. Rodents are the reservoir for these hantaviruses as well.

REFERENCES

1. **Hjelle, B., and G. E. Glass.** 2000. Outbreak of hantavirus infections in the Four Corners region of the United States in wake of the 1997–1998 El Niño–southern oscillation. *J. Infect. Dis.* **181:**1569–1573.

2. **Khan, A. S., R. F. Khabbaz, L. R. Armstrong, R. C. Holman, S. P. Bauer, J. Graber, J. Strine, G. Miller, S. Reef, J. Tappero, R. E. Rollin, S. T. Nichol, S. R. Zaki, R. T. Bryan, L. E. Chapman, C. J. Peters, and T. G. Ksiazek.** 1996. Hantavirus pulmonary syndrome: the first 100 cases. *J. Infect. Dis.* **173:**1297–1303.

3. **Mills, J. N., T. G. Ksiazek, C. J. Peters, and J. E. Childs.** 1999. Long-term studies of hantavirus reservoir populations in the southwestern United States: a synthesis. *Emerg. Infect. Dis.* **5:**135–142.

4. **Mori, M., A. L. Rothman, I. Kurane, J. M. Montoya, K. B. Nolte, J. E. Morman, D. C. Waite, F. T. Koster, and F. A. Ennis.** 1999. High levels of cytokine-producing cells in the lung tissues of patients with fatal hantavirus pulmonary syndrome. *J. Infect. Dis.* **179:**295–302.

5. **Schmaljohn, C., and B. Hjelle.** 1997. Hantaviruses: a global disease problem. *Emerg. Infect. Dis.* **3:**95–104.

C A S E 68 The patient was a 54-year-old male retired U.S. Air Force pilot. He was in excellent health until he developed a chronic neurologic disorder that led to his death 6 months after the onset of symptoms. The initial manifestations were nonspecific: forgetfulness, subtle behavioral changes, headaches, and fatigue. Disequilibrium, a wide-based gait, and double vision developed, and he sought medical attention. There was no history of alcohol use or a tick bite. He was never febrile. He was treated with penicillin and tetracycline for presumed Lyme disease or Rocky Mountain spotted fever (RMSF). His symptoms persisted. Serology tests for *Borrelia burgdorferi* and RMSF were negative. Evaluation by a neurologist revealed findings consistent with cerebellar and brainstem degeneration. Head computed tomogram (CT) scan revealed mild cerebral and cerebellar atrophy but no mass lesions. His social history was significant for having been an air force pilot for 23 years. He spent several years in Guam and Vietnam. His eating habits were conventional. He had never been transfused, and he had no known risk factors for HIV.

Routine hematology and chemistry laboratory tests were normal; HIV and VDRL (Venereal Disease Research Laboratory) tests were negative. Lumbar puncture revealed a normal opening pressure, and all tests on cerebrospinal fluid, including cultures, were within normal limits. No brain biopsy was done.

He had a progressive decline in his neurologic status and died. An autopsy was done and the neuropathologic exam was pathognomonic for this patient's diagnosis. Extensive spongiform changes were prominent in the cerebrum, cerebellum, and basal ganglia (Fig. 1).

1. What viruses can cause chronic neurologic disorders?

2. The etiology in this particular case is an unconventional agent that causes a noninflammatory subacute degenerative neurologic disorder. What is the name of the etiologic agent? What is the disease in humans called?

3. A similar disease occurs in sheep. What is it?

4. What is the disease called in cattle? Describe an ongoing epidemic in cattle that has been associated with a variant of this disease in humans.

5. Name three possible modes of person-to-person transmission of this agent.

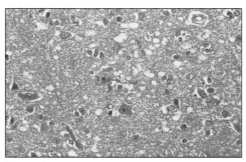

Figure 1

CASE DISCUSSION

1. The following viruses are associated with neurologic conditions resulting from chronic infection:

VIRUS	CLINICAL CONDITION
Measles	SSPE (subacute sclerosing panencephalitis)
Papovaviruses (JC virus)	Progressive multifocal leukoencephalopathy
HIV	AIDS dementia complex (HIV can also cause aseptic meningitis, peripheral neuropathies, and myelopathy)
Human T-cell lymphotropic virus (HTLV-1)	HTLV-1-associated myelopathy (formerly known as tropical spastic paraparesis)

2. The patient had Creutzfeldt-Jakob disease (CJD), which is a spongiform encephalopathy most likely caused by a prion. A prion is a proteinaceous infectious particle with little or no nucleic acid. It is a transmissible agent with unconventional biochemical and biophysical properties. It is not yet entirely clear how it replicates or induces the neuropathologic characteristics seen in CJD. The definitive diagnosis of CJD can only be made on the basis of a brain biopsy or postmortem histologic examination of brain tissue. A noninvasive test that looks for the presence of the 14-3-3 protein has been associated with false-positive results.

3. A similar disease that can occur in sheep is called scrapie. The natural mode of transmission in sheep is unknown. Similar diseases in other animals include transmissible mink encephalopathy and chronic wasting disease of elk and deer.

4. Bovine spongiform encephalopathy (BSE) is a prion disease of cattle that was first investigated in the United Kingdom in 1985 and peaked in 1992, with more than 37,000 cases identified in that year. By the end of 2000, more than 180,000 BSE cases had been identified in the United Kingdom. The BSE epidemic may have been initiated by a change in the processing of sheep- and cattle-derived meat and bone meal protein supplements fed to cattle. This processing change was the removal of a solvent-extraction step that previously might have sterilized the material from this transmissible agent. Smaller numbers of cattle with BSE have been identified in a number of other European countries, including Ireland, Portugal, France, Belgium, Italy, Spain, Denmark, the Netherlands, Germany, Liechtenstein, Greece, Switzerland, and the Czech Republic. Cases of BSE in imported animals have been reported from Canada, the Falkland Islands, and Oman.

As of early 2002, there have been over 100 human deaths from definite or probable infection with a variant form of CJD in the United Kingdom. This variant form has been diagnosed in patients, many of whom are less than 45 years old, an age group in which CJD very rarely occurs. Usually CJD affects middle-aged or elderly individuals. These recently described variant CJD cases are possibly linked to consumption of BSE-contaminated beef, thus constituting a "species jump" of this transmissible agent. This is not yet proven, but the cases are linked by a geographical and temporal association. Recent intensified surveillance in the United States has not detected any evidence of either BSE or this new CJD variant.

Although most patients with variant CJD are in their 20s and 30s, there was a report of a patient in his 70s with variant CJD. It is possible that this disease is underdiagnosed in the elderly, who are at risk for a number of illnesses that are characterized by dementia, such as Alzheimer's disease. One additional concern has been raised by studies in mice showing that the incubation time varies greatly among different mouse strains and that certain mouse prion susceptibility alleles exist. If this proves to be true for human susceptibility to prion diseases as well, it is speculated that the cases in the United Kingdom that have been identified to date represent only those cases in people with a genetic predisposition to a short incubation period. If this extension to human disease from the mouse model is valid, what has occurred in humans in the United Kingdom is only the tip of the iceberg. Time will tell if this is the case.

In an effort to control this epidemic, many countries have instituted a ban on the import of beef from countries in which BSE has been identified. Other public health measures have been implemented, such as a ban on the use of meat and bone meal in animal feed. Concerns have also been expressed about the use of cattle-derived material in human medicines and cosmetics.

5. This condition was first described in New Guinea among a tribe that practiced cannibalism. In this specific epidemiologic situation, this condition was called kuru. Kuru has decreased since the discovery of this association with cannibalism and a subsequent change in this tribe's cultural practices. Sporadic cases occurring elsewhere are called CJD.

A second mode of transmission is by corneal transplantation. Rabies also has been transmitted by corneal transplantation.

In 1985, three deaths due to CJD were reported among hypopituitary patients. These patients had received pituitary-derived human growth hormone. The process of pooling cadaveric pituitary glands to extract human growth hormone presumably included some tissues infected with the CJD transmissible agent.

Rare cases of CJD have also been attributed to the use of contaminated electroencephalogram electrodes and dura mater grafts. Prions are remarkably resistant to many methods of sterilization, including autoclaving.

This patient had a sporadic case of CJD. It is unclear how the patient acquired this condition.

REFERENCES

1. **Anderson, R. M., C. A. Donnelly, N. M. Ferguson, M. E. J. Woolhouse, C. J. Watt, H. J. Uddy, S. MaWhinny, S. P. Dunstan, T. R. E. Southwood, J. W. Wilesmith, J. B. M. Ryan, L. J. Hoinville, J. E. Hillerton, A. R. Austin, and G. A. H. Wells.** 1996. Transmission dynamics and epidemiology of BSE in British cattle. *Nature* (London) **382:**779–788.

2. **Brown, P., M. Preece, J. P. Brandel, T. Sato, L. McShane, I. Zerr, A. Fletcher, R. G. Will, M. Pocchiari, N. R. Cashman, J. H. d'Aignaux, L. Cervenakova, J. Fradkin, L. B. Schonberger, and S. J. Collins.** 2000. Iatrogenic Creutzfeldt-Jakob disease at the millennium. *Neurology* **55:**1075–1081.

3. **Chapman, T., D. W. McKeel, Jr., and J. C. Morris.** 2000. Misleading results with the 14-3-3 assay for the diagnosis of Creutzfeldt-Jakob disease. *Neurology* **55:**1396–1397.

4. **Lloyd, S. E., O. N. Onwuazor, J. A. Beck, G. Mallinson, M. Farrall, P. Targonski, J. Collinge, and E. M. Fisher.** 2001. Identification of multiple quantitative trait loci linked to prion disease incubation period in mice. *Proc. Natl. Acad. Sci. USA* **98:**6279–6283.

5. **Prusiner, S. B.** 1995. The prion diseases. *Sci. Am.* **272:**48–57.

6. **Will, R. G., J. W. Ironside, M. Zeidler, S. N. Cousens, K. Estibeiro, A. Alperovitch, S. Poser, M. Pocchiari, A. Hofman, and P. G. Smith.** 1996. A new variant of Creutzfeldt-Jakob disease in the UK. *Lancet* **347:**921–925.

GLOSSARY

abscess   A cavity of liquefactive necrosis within solid tissue as a result of a localized infection; this can be the result of an acute or a chronic process.

accidental host   A host that harbors a parasite but is not the normal host for that parasite species.

acid-fast   Pertaining to a group of organisms that resist decolorization by acid-alcohol; typically associated with *Mycobacterium* and *Nocardia* spp.

acidosis   Pathological condition in which the arterial pH drops below the normal value.

acute phase   The initial period of an infection that precedes the chronic phase.

acute specimen   A specimen that is collected from the patient during the initial or acute illness.

adenopathy   An enlargement of a lymph node or lymph nodes in response to some stimulus such as inflammation or infection; can occur singly or in multiple nodes; also referred to as lymphadenopathy.

adnexa   An appendage to an organ or structure.

adrenalitis   Inflammation of the adrenal gland.

aerobic   Pertaining to a microorganism that must grow in the presence of oxygen (a "strict aerobe") or may grow in the presence of oxygen (a "facultative aerobe").

aerosol   A collection of solid or liquid particles suspended in a gas, such as a liquid that is dispersed in fine droplets through the air.

afebrile   Relating to the absence of fever.

agglutination   The interaction between a particulate antigen and antibodies specific for the particular antigen; the antigen-antibody complex leads to the aggregation or clumping of the antigen-containing material.

AIDS   Acquired immunodeficiency syndrome, a disease caused by one of the HIV (human immunodeficiency virus) retroviruses; the condition is characterized by the depletion of lymphocytes with subsequent failure of the immune system.

allele   One of a series of two or more different genes that occupy the same location on a homologous chromosome.

alpha-hemolytic   Pertaining to a reaction seen typically on agar medium containing sheep red blood cells in which the area surrounding a colony has a green hue; the most medically important alpha-hemolytic organism is *Streptococcus pneumoniae*.

alveolus   An air sac in the lung consisting of a single layer of cells surrounded by a network of capillaries also consisting of a single cell layer; gas exchange occurs here.

amastigote   A stage of the life cycle for protozoan species within the genus *Leishmania*; the nonflagellated amastigotes survive within macrophages of the host.

ameboma A nodular inflammatory lesion, usually in the wall of the colon, that may develop in chronic amebiasis.

anaerobic Pertaining to a microorganism that must grow in the absence of oxygen (a "strict anaerobe") or may grow in the absence of oxygen (a "facultative anaerobe").

anaphylactic Referring to a severe allergic reaction; caused by mast cells degranulating or by the activation of the complement cascade.

anemia A condition in which the number of functional red blood cells is decreased; this is often associated with symptoms such as pallor, fatigue, shortness of breath, and lethargy.

anergy A lack of the ability of the immune system to respond with a delayed-type hypersensitivity reaction to commonly and previously encountered antigens such as mumps and candida; often seen in patients with AIDS.

anicteric Without icterus (jaundice).

annular rash A circular rash.

anorexia Decreased appetite.

anthropophilic A parasite or fungus with a preference for humans.

antigen A substance that is capable of triggering an immune response.

antigenic drift The gradual change in a microorganism's genetic composition after successive generations of reproduction.

antigenic shift A drastic change in a microorganism's genetic composition with one reproductive cycle; typically associated with influenza A viruses in which a new strain is unrecognizable to a previously immune host.

antigenic variation A sudden or gradual change in the expression of an antigen by a microbe; the variation may result in a drastic change in the pathogenicity of the organism or in the host's ability to mount an immune response; antigenic shift and antigenic drift are types of antigenic variation.

antiphagocytic Inhibiting the ability of the phagocyte to ingest bacteria, foreign materials, or other cellular debris.

antipyretics Fever-reducing agents such as aspirin, acetaminophen, and ibuprofen.

aplastic anemia Decrease in the numbers of all elements in the blood due to the death of their precursor cells in the bone marrow, where the cells usually mature; often associated with specific chemicals or drugs that are toxic to these cells.

arthralgia Severe joint pain, usually characterized as noninflammatory.

arthritis Inflammation or infection of a joint, leading to decreased and painful mobility of the affected joint.

arthroconidium An asexual spore that is budded from the body of some species of fungi.

ascariasis Disease caused by intestinal nematodes of the genus *Ascaris*, usually the giant roundworm species *Ascaris lumbricoides*.

asplenia Absence of the spleen, either congenitally (at birth) or later, often seen in persons with long-standing sickling disease. This condition makes the individual susceptible to infections by certain bacteria and parasites.

asterixis An involuntary jerking motion that occurs with various toxic or metabolic encephalopathies, including hepatitic encephalopathy and uremia; most easily demonstrated when the patient is asked to extend his or her arms with the hands pointed up and the fingers extended.

ataxia The inability to coordinate muscle activity during voluntary movements of the head, limbs, or trunk; often associated with cerebellar dysfunction.

atrophy A process characterized by wasting of specific tissues, organs, or the entire body that can result from a variety of causes.

attenuation Decreased virulence of a pathogen through either natural or experimental means.

auramine-rhodamine A stain that fluoresces under a UV microscope; used for visualizing acid-fast organisms.

auscultation A method based on sounds or sound changes, used during a physical examination to gather data on internal organs such as the heart, lungs, and liver; the most common method involves the use of a stethoscope.

autochthonous Refers to a process that originated in the place where it is found.

autoinfection Process whereby the host is reinfected by the same organism after the organism has undergone a replication cycle.

avirulent Microorganism with limited pathogenic potential.

axillus Armpit; the area between the upper arm and chest wall where the two join.

bacteremia The presence of viable bacteria in the bloodstream; may be a transient phenomenon associated with dental care or may be due to a bacterial infection in which bacteria have entered the bloodstream.

bacteriocins Proteins produced by certain bacteria that have a lethal effect on related bacteria; although the effect has a more narrow range than the effect of antibiotics, it is more potent.

bacteriuria The presence of bacteria in the urine.

bibasilar Pertaining to the bases of both lungs.

bilateral Pertaining to both sides of a symmetrically shaped tissue, organ, or the entire body; for example, the right and left lungs are bilateral organs.

biliary tree System of ducts through which bile is transported.

biopsy A procedure that involves the removal of specific tissues from patients for the purpose of diagnosis.

blanching A temporary lightening of the skin or mucous membranes after direct pressure is applied.

bronchiolitis Inflammation of the bronchioles (conducting airways of less than 1 mm).

bronchitis Inflammation or infection of the airways.

bronchoalveolar lavage (BAL) The instillation of saline into the airways of the lungs so that samples can be removed and the washings (fluids) can be analyzed for malignancy, inflammation, or infection; also done during bronchoscopy.

bronchoscopy   The use of a flexible hollow tube to look directly at the trachea, bronchi, and larger airways in the lungs; it is also possible to obtain samples (biopsies, fluids, brushings) through this device.

bronchospasm   Episodic constriction of smooth muscles lining the bronchi in response to some kind of irritant or stimulus.

cachexia   General weight loss or wasting due to a disease process or emotional imbalance.

calcification   Focal area of increased deposition of calcium compounds.

capsule   A polysaccharide (except for *Bacillus anthracis*, in which it is poly-D-glutamic acid) covering of certain bacteria and yeasts that makes them resistant to phagocytosis by white blood cells; often utilized in vaccines as the antigen.

catheterization   The placement of a catheter, usually through the urethra, into the bladder in order to drain urine from the bladder.

CD4-positive (CD4⁺) cells   Subset of T lymphocytes that are characterized by the presence of CD4 receptors on their cell membrane surfaces; they assist in turning on the immune response by activating other T and B lymphocytes; also called T-helper cells. Less than 200 cells per µl is seen in adult AIDS patients.

cell-mediated immunity   An immune response that is activated and carried out by T lymphocytes; the counterpart to humoral immunity.

cellulitis   Inflammation or infection of the skin and tissues beneath the skin.

cerebellar ataxia   A loss of motor coordination as a result of damage to the cerebellum or its pathways; may be manifested as a loss of balance in the entire body when moving or by unsteady movement of the arms or legs.

cerebellar   Pertaining to the cerebellum (the portion of the brain concerning the coordination of complex movements and balance).

cerebritis   A focal infection or inflammation of the brain tissue.

cerebrospinal fluid   Fluid surrounding the brain and spinal cord.

cervical   Pertaining to the neck or a necklike portion of an organ.

cervicitis   Inflammation of the mucous membranes or deeper structures of the cervix.

cesarean section   A nonvaginal delivery whereby the fetus is surgically removed following an incision through the lower abdominal wall and uterus.

cestodes   The class of helminths that includes the tapeworms (e.g., *Taenia* sp.); these flatworms have a rounded head (scolex) followed by a chain of multiple segments (proglottids).

chancre   The primary lesion of syphilis; typically dull red, hard, and insensitive with a center that erodes and ulcerates, then heals slowly over a period of 4 to 6 weeks.

chemoprophylaxis   The use of chemicals such as antibiotics to prevent the occurrence of disease.

chemotherapy   The use of drugs or chemical substances to treat disease.

chimeric   Combining traits from two different species to create an organism with unique characteristics; in the context of bioterrorism, adding new virulence factors to a microorganism. This term is also used in the context of genetically engineered antibodies in which different parts of the antibody come from different animals.

cholangitis   Inflammation or infection of the bile ducts.

cholecystectomy   Surgical removal of the gall bladder.

cholecystitis   Inflammation of the gall bladder.

chorioamnionitis   Inflammation or infection of the chorion, amnion, amnionic fluid, and often the placental villi and decidua as well.

chorioretinitis   Inflammation or infection of the light-detecting layer (retina) and the underlying vascular tissue (choroid) beneath it in the back of the eye, which can lead to progressive impairment of vision.

chronic obstructive pulmonary disease (COPD)   A group of slowly progressive lung disorders that includes emphysema and chronic bronchitis; results in the marked decrease in airflow through the lungs with the main impact on the ability to exhale; the vast majority of cases are associated with cigarette smoke.

ciliostasis   Pathological process, usually induced by viral infections, that causes changes in the cilia such that they do not beat.

cirrhosis   Destruction of a tissue or organ with loss of normal structure which is replaced with scar tissue; common in association with alcoholism, where it involves the liver.

cohort   A designated group of individuals to be examined for a particular trait, exposure, or outcome.

colitis   Inflammation of the colon.

complement   Component of the immune system that consists of a set of proteins in the serum; the proteins are activated in a cascade fashion and assist in the process of phagocytosis, direct microbe destruction, and phagocyte recruitment.

compliance   The degree to which a patient follows the prescribed regiment of treatment.

confirmatory test   A diagnostic test that is conducted after a positive screening test to confirm the illness or trait; a good confirmatory test should have a low rate of false-positive results.

conjugate vaccine   Type of bacterial vaccine in which a portion of the bacterial capsule is attached to an inactive toxin or protein to enhance the immune response to the capsule antigen.

conjunctivitis   Inflammation of the tissue protecting the front of the eye; often due to infection.

contact dermatitis   Inflammation of the skin caused by superficial contact with an allergen or chemical.

continuous bacteremia   The persistent presence of bacteria in the circulation.

convalescence   The period of time after a disease process has ended but before the return of optimal health.

convalescent specimen A specimen that is collected from the patient either in the late stages of an acute infectious process or when it has ended.

coryza Acute inflammation or infection of the nasal membranes, leading to a thin watery discharge from the nose, as seen with the common cold.

costal Pertaining to the ribs.

costovertebral angle The area in the back where the last ribs join to their respective vertebrae.

crepitance The crackling sound (resembling the sound that occurs when rubbing hair between the fingers) that is heard in certain disease states.

croup The difficult, noisy respirations and hoarse cough that characterize laryngeal problems in children; this is often caused by parainfluenza virus.

cushingoid body habitus An increase in adipose tissue (fat) in certain areas of the body, legs, and trunk. Purplish stripes, especially on the abdomen (striae), are also associated with Cushing's disease.

cyst An abnormal membranous sac containing gas, fluid, or semisolid material. Also a dormant form of the protozoan life cycle that is capable of resisting destruction by heat or dehydration.

cysticercus Larva of *Taenia solium* that is present in the tissue of the hosts.

cystitis Inflammation or infection of the urinary bladder; also called urinary tract infection (UTI). Associated with symptoms including painful urination, increased urination, and/or malodorous urine.

cystocele A hernia of the bladder; the hernia usually protrudes into the vagina.

cytokine One of a diverse group of chemicals produced by various cell types that regulate the immune response.

cytologic Relating to the study of cytology (the anatomy, physiology, pathology, and chemistry of the cell).

cytolysin A substance or antibody that is capable of directly destroying a cell.

cytopathic Changes in intracellular structures due to disease or toxins, usually leading to the death of the cell.

cytotoxic Destructive or damaging to a cell.

debride To remove devitalized tissue and/or foreign material from a wound.

decubitus The position of a patient when lying down; lateral decubitus specifically refers to the patient's lying on his or her side. A decubitus wound is a wound that occurs as a result of pressure on an area, such as the heels or buttocks, because of the way the patient is lying.

defervescence The disappearance of fever.

definitive host A host that supports the adult, sexually reproducing form of a parasite.

dehydration The loss of water resulting in the increased osmolarity of the body fluids.

dementia A state of decreased mental capacity and orientation without a loss in consciousness; can be caused by toxins or by a gradual loss of normal brain tissue.

dermatome   The area of skin that is served by one sensory spinal nerve.

dermatophyte   One of the fungal organisms capable of infecting human skin, hair, or nails; includes the genera *Microsporum*, *Trichophyton*, and *Epidermophyton*.

desaturation   The state of having unoccupied binding sites or the process of removing ligands from the binding sites; with hemoglobin, refers to a decrease in the amount of bound oxygen.

differential medium   A type of medium used to determine if an organism demonstrates a specific characteristic.

dimorphic   A characteristic of certain fungi to exist in two distinct forms, such as yeast and mold.

directly observed therapy   A type of therapy in which a responsible individual directly observes the patient taking the prescribed medications; utilized as a way to improve patient compliance.

disequilibrium   Unsteady balance.

disseminated intravascular coagulation (DIC)   The depletion of clotting elements in the blood; caused by many disease processes; diffuse, severe hemorrhaging can occur; without treatment, it is often fatal.

dorsal   Referring to the back or posterior aspect of a tissue or organ.

dorsal root ganglion   Group of nerve cell bodies outside the spinal cord that convey sensory impulses to the brain.

dysentery   A condition marked by frequent watery stools usually containing blood and mucus; often accompanied by pain, fever, and dehydration; typically the result of amebic, bacillary, helminthic, or viral infections.

dyspepsia   Gastric indigestion (upset stomach) due to alterations of gastric function that are caused by various disorders of the stomach.

dyspnea   The sensation of having difficulty in breathing; also called shortness of breath.

dysuria   Difficulty or pain on urination.

echocardiogram   A real-time, noninvasive study that utilizes ultrasound to evaluate the heart tissue, heart valves, great vessels, and the corresponding blood flow.

ectopic pregnancy   Type of pregnancy in which the embryo implants outside the uterus; this usually occurs in the fallopian tube.

eczema   An itchy, scaly, blistery, or raised rash often seen in children and associated with irritation of the skin.

edema   An accumulation of large amounts of extracellular watery fluid in tissues throughout the body.

effusion   The presence of excess fluid in the tissues or a cavity; for example, referred to as pleural effusion when excess fluid is found in the pleural cavity around the lungs, usually as a result of inflammation, infection, or malignancy.

electromyography (EMG)   Test used to represent electrical currents associated with muscles; used in the diagnosis of neuromuscular diseases.

embolism   An object in the bloodstream that becomes dislodged and carried until becoming trapped in a different vessel; may be a thrombus, immune complex of microorganisms, air, foreign object, fat, or amniotic fluid; leads to partial or complete occlusion of the blood vessel.

emesis   Vomiting.

empiric   Type of therapy that is employed when the causative agent or the antibiotic susceptibility of the organism is not definitively known.

empyema   The collection of purulent material in the pleural space.

enanthema   A mucous membrane eruption; typically occurs in relation to the skin eruptions that are symptoms of acute viral or coccal disease (exanthema).

encephalitis   Inflammation of the brain.

encephalopathic   Referring to the pathological condition of encephalitis.

endemic   Describes the presence of a disease that persists in a community or group of people.

endocarditis   Inflammation or infection of the tissue lining the inside of the heart; usually involves the heart valves.

endocytosis   The process of internalizing external substances by fusion with the plasma membrane and formation of an intracellular vesicle; is specifically considered phagocytosis if the internalized substance is solid.

endometritis   Inflammation or infection of the lining of the uterus.

endophthalmitis   Inflammation within the eyeball.

endoscopy   Procedure involving the passing of a flexible hollow tube into the esophagus or rectum for the purpose of visualizing portions of the gastrointestinal tract; also useful for obtaining diagnostic samples.

endothelial   Referring to the cells lining blood vessels.

endotoxin   A lipopolysaccharide on the outer membrane of gram-negative bacteria; has many possible biological effects on the infected host, including activation of the clotting cascade, shock, or death.

endovascular   Within the lumen of the blood vessels.

enteric   Relating to the intestine.

enterocytes   The absorptive epithelial cells that line the lumen of the intestine.

enterotoxin   Exotoxin that causes fluid secretion in the gut.

enzootic   Referring to a temporal pattern of disease occurrence in an animal population that is marked by predictable regularity with little change over time; for example, the pattern of hantavirus infection in deer mice.

enzyme immunoassay (EIA)   A diagnostic test in which a specific antibody is linked to an enzyme; in a positive test, the antibody is attached to the antigen being tested for with a subsequent reaction by the linked enzyme; the enzyme reaction leads to a noticeable change, such as a change in color.

eosinophilia   Increased number of eosinophils (a type of leukocyte) in the blood, often associated with parasitic infections.

epidemic   An unexpectedly large number of cases of a disease or illness in a community.

epididymitis   Inflammation or infection of the epididymis.

epigastric   Relating to the area of the abdomen that lies between the margins of the ribs.

epiglottitis   Inflammation or infection of the flexible flap of tissue that covers the larynx during swallowing.

episome   An extrachromosomal piece of DNA that is capable of being replicated independently of the host chromosome and integrated within the host DNA.

erythema   Reddening, usually of the skin or mucous membranes.

erythema infectiosum   A mild facial rash in children that has a "slapped cheek" appearance; usually caused by parvovirus B19.

erythema migrans   A circular red rash that has a "bull's-eye" appearance due to the repeating red rings within an expanding outer ring; forms at the site of a tick bite and is indicative of Lyme disease.

erythematous   Relating to erythema.

esophagitis   Inflammation of the esophagus.

ethmoid sinus   Air-filled cavity in the ethmoid bone located below the orbit of the eye and beside the nose.

etiology   The cause of a disease or process.

exacerbation   An increase in the symptoms of a disease.

exanthema   A skin rash that occurs as a symptom of acute viral or bacterial infection.

excoriate   To physically scratch or remove the skin.

exocytosis   A process whereby intracellular vesicles fuse with the plasma membrane in order to empty their contents into the extracellular space.

exotoxins   Proteins secreted by bacteria that have toxic effects on mammalian cells.

extensor surface   The surface of a joint involved in extension or straightening of a limb.

extradermatomal   Not confined to one dermatome.

extradural   Referring to the outer side of the dura mater.

extramedullary cranial ganglion   Outside the medulla oblongata (brain stem), in reference to the cranial nerves.

exudate   Fluid resulting from inflammation or infection; contains an increased number of cells and an increased amount of protein and other cellular debris.

false negative   A test result of negative when the true result is positive.

false positive   A test result of positive when the true result is negative.

fastidious   Type of organism with complex nutritional requirements; requires special media for laboratory culture.

febrile   Relating to fever.

fetid   Foul smelling.

filariform   The third larval stage of the intestinal nematodes; this stage is infectious and nonfeeding.

fimbriae   Appendages on the surface of bacteria that allow for attachment and infection; same as pili.

flaccid paralysis   A loss of muscle function with a resulting absence of muscle tone.

flatulence   The presence of excessive gas in the stomach and intestines.

flocculent   The presence of mucus or a mucus-like material in a fluid such as urine.

fluctuance   A wave-like motion felt on palpitation, due to underlying fluid content.

fomite   An object that is capable of transmitting an organism from one location to another; for example, toys, clothing, utensils, or hairbrushes.

fontanelles   Soft area between the cranial bones of an infant's skull, indicative of areas not yet ossified.

frozen section   A tissue sample typically used during a surgical procedure to allow rapid decisions during the operation.

fulminant   Referring to the sudden occurrence of an intense or severe process.

fungemia   The presence of fungi in the bloodstream.

gait   Manner of walking.

ganglion   A group of nerve cell bodies located in the peripheral nervous system.

gangrene   Necrosis due to any cause (for example, obstructed, diminished, or lost blood flow); may be localized or widespread; may be dry or wet.

gastritis   Inflammation of the stomach, usually involving only the lining inside the organ.

gastroenteritis   Inflammation of the mucous membrane lining of both the stomach and the intestine.

gene   A complete unit of genetic information that encodes a single protein.

geophilic   Pertaining to an organism with a preference for the soil.

gestation   The interval between fertilization of a female and birth of the offspring.

glomerulonephritis   Bilateral inflammatory changes of the glomeruli that are the result of renal disease.

Gram stain   A sequence of dyes and solvents applied to bacteria to enable viewing under the microscope; organisms stain either gram positive or gram negative.

granuloma   A collection of leukocytes, macrophages, and specialized cells of the reticuloendothelial system surrounding a focal area of chronic inflammation or infection; usually forms a nodular mass.

granulomatous   Pertaining to or resembling a granuloma.

guaiac A reagent used to test for occult blood, usually in the feces.

Guillain-Barré syndrome Inflammation of peripheral nerves leading to increasing weakness or paralysis; most often occurs in more distal areas before affecting more proximal portions of the body; may progress to the point where the patient requires support on a ventilator because of weakness of the respiratory muscles.

HAART Highly active antiretroviral therapy, used to treat HIV infection; includes several medications with different mechanisms of action.

hallucinosis A syndrome characterized by hallucinations that are caused by organic substances such as drugs and alcohol; may occur during withdrawal from drugs or alcohol.

hematemesis Vomiting blood.

hematocrit Amount of red blood cells (erythrocytes) in a given volume of blood; usually expressed as a percentage or fraction.

hematogenous Refers to anything produced by, derived from, or spread through the blood.

hematoma A collection of blood within an extravascular space such as an organ, a tissue area, or a potential space; the blood displays various colors and degrees of organization; a bruise.

hematuria The presence of blood in the urine.

hemoconcentration An increase in the number of red blood cells per unit volume of plasma.

hemoglobinopathies Disorders or diseases that are the result of abnormalities of hemoglobin structure; for example, sickle cell disease and thalassemia.

hemolysis Destruction or breakdown of red blood cells.

hemoptysis Coughing up (expectoration) of blood or blood-streaked sputum.

hemorrhage Bleeding; can be either external or internal, e.g., intracranial hemorrhage.

hepatitis Inflammation of hepatic (liver) cells.

hepatomegaly Enlargement of the liver.

hepatosplenomegaly Enlargement of the liver and spleen.

herbivore An animal that obtains its energy and nutrient requirements from plants.

hernia Part of an anatomical structure that has partially expanded out of its normal confined area.

herpangina A disease caused by the coxsackievirus group B; results in fever and ulcers of the palate.

hilum The area of an organ where the nerves and vessels enter and exit; the lungs, kidneys, lymph nodes, ovaries, and spleen all have hilar areas.

horizontal transmission Passage of disease from person to person or by contact with infected materials.

humoral   Referring to substances in the blood; in the immune system, this refers to antibodies to help fight disease rather than the cellular portion which involves leukocytes.

humoral immunity   An immune response that is activated and carried out by B lymphocytes and involving specific antibodies; the counterpart to cell-mediated immunity.

hydrocephalus   Condition characterized by a large accumulation of fluid in the cerebral ventricles; results in increased intracranial pressure and dilatation of the cerebral ventricles. It can also occur secondarily as a result of loss of brain tissue.

hydrops   General term for the accumulation of clear, watery fluid in spaces or cavities throughout the body.

hypercholesterolemia   The presence of abnormally high levels of cholesterol in the blood.

hyperinfection   Infection caused by very large numbers of organisms that result from immunodeficient states.

hypertension   Abnormally high blood pressure.

hyphae   The tube-like structures visible with a microscope that make up filamentous fungi.

hyponatremia   Abnormally low concentrations of sodium in the circulating blood.

hypopituitary   Pertaining to a decrease in the amount of hormones produced by the anterior (forward) portion of the pituitary gland; involves hormones that affect growth, steroid production, thyroid gland function, and reproduction.

hypoplasia   The underdevelopment of tissues or organs as a result of either atrophy or a decrease in the number of cells.

hypotension   Abnormally low blood pressure.

hypotonia   Reduced tone of any structure, especially muscle.

hypoxemia   Low oxygen content in the blood.

icterus   *See* jaundice.

idiopathic cardiomyopathy   Dilatation and weakening of the heart muscle with no known cause.

immune complex   Antigen bound by specific antibodies with or without activated complement; may become insoluble and deposit in the vessel walls, including those of the glomeruli (*see* glomerulonephritis) or tissue.

immunization   The process of exposing an organism to antigen with the intent of generating immunological memory.

immunocompromised   A general term indicating an increased susceptibility to infection; this may be due to defects in cell-mediated immunity, as occurs with AIDS and posttransplant immunosuppressive medications; defects in neutrophil number or function, as occurs following cancer chemotherapy; defects in humoral immunity, as occurs in immunoglobulin deficiencies; or other immune defects such as deficiencies in splenic function.

immunosuppressive   Pertaining to an agent or disease that prevents or interferes with the immune system response.

impetigo   Infection of previously damaged skin by group A streptococci or staphylococci; lesions usually drain honey-colored fluid.

in utero   Inside the womb.

indolent   Nearly or completely painless, sluggish, or inactive; used to describe a disease process.

induced sputum   A clinical sample of secretions from the tracheobronchial tree that is produced by having the patient inhale an irritating aerosol; the presence of macrophages or inflammatory cells suggests an adequate sample whereas the presence of squamous epithelial cells suggests contamination by the upper respiratory tract.

induration   Firmness in soft tissue.

infarct   Tissue death often due to an interruption in the blood supply to that tissue.

infertility   The inability to produce viable offspring; may be due to dysfunctional germ cells or to a structural abnormality.

infiltration   The invasion of the spaces in a tissue by materials not usually found in the tissue, such as tumors, infectious agents, and white blood cells.

inflammation   A series of chemical and physical processes within a tissue in response to toxins, foreign antigens, or injury; involves various cells of the vasculature, connective tissue, and immune system.

intermediate host   A host in which a parasite is capable of completing only part of its life cycle.

interstitial   Pertaining to spaces between the components of a tissue; in the lungs, refers to spaces between the lung parenchyma.

intracranial   Within the skull.

intrapartum   During labor or delivery.

intraperitoneally   Within the peritoneal cavity.

intrauterine   Inside the cavity of the womb, as for a fetus.

intubation   The placement of a tubular device from the oro- or nasopharyngeal cavity into the trachea in order to assist with ventilation.

intussusception   The prolapse of a distal part of an organ back into the proximal portion of the same organ; usually refers to a portion of the intestine where it may result in obstruction.

ischemia   Cellular injury due to a decreased delivery of oxygen to a tissue because of either the impairment of blood flow to the tissue or decreased oxygen content of the blood. This may progress to cellular death.

jaundice   Yellow cast of the skin and mucous membranes due to an increase of bilirubins (bile breakdown products) that occurs when the liver is unable to clear these chemicals from the blood; often due to toxic or infectious hepatitis; also referred to as icterus.

Koplik's spots   Small red spots with a white center that are located on the inside of the cheek; pathognomonic for measles.

lacrimal duct   A small duct that carries fluids (tears) from the lacrimal gland to the surface of the eyeball.

larvae   Developmental stage of the helminths, resembling small immature worms.

laryngitis   Inflammation of the larynx (voice box).

laryngotracheobronchitis   Inflammation of the larynx and larger airways.

latent   Referring to any organism or disease that is quiescent but is capable of being reactivated.

latent infection   Type of infection in which the organism or virus is present but not producing an inflammatory response.

lethargy   Drowsiness or decreased responsiveness.

leukemia   General term used to describe a group of malignancies of either lymphoid or hematopoietic origin; progressive proliferation of abnormal leukocytes can be found in the blood, the hematopoietic tissues, and other organs.

leukoencephalopathy   Encephalitis that is restricted to the white matter of the brain.

leukopenia   Decreased number of white blood cells.

lumbar puncture   A procedure that is used to obtain cerebrospinal fluid for diagnostic purposes, performed by introducing a needle into the lumbar region of the subarachnoid space.

lymphocytosis   An increase in the number of lymphocytes above normal.

lymphoma   Cancer of the lymph cells or their precursors.

lymphoproliferative disorders   Any one of a group of cancers involving cells from which white blood cells or platelets are derived; includes lymphomas, leukemias, and multiple myeloma.

lysogeny A stable and heritable characteristic of bacteria to produce and release bacteriophage as a result of a prophage within the bacterium; the prophage can be either integrated into the bacterial genome or maintained as a plasmid.

macular   Pertaining to lesions that are flat and that are often detected only by a change in color or texture of the lesion compared with surrounding normal tissue.

maculopapular   Pertaining to lesions with properties that are both macular and papular.

malabsorption   Condition in which substances are not absorbed properly; usually occurs in the intestine, where it results in excessive loss of water or nutrients in the stool.

malabsorptive diarrhea   An increase in the total number or volume of stools due to a decrease in the absorption of nutrients (especially fats and fat-soluble vitamins) in the small intestine.

malaise   Generalized feeling of discomfort caused by many disease processes.

melena   Excretion of black, tarry stools containing blood that has been altered by intestinal substances.

meninges   Thin, tough tissue surrounding the brain and spinal cord.

meningitis   Inflammation of the meninges.

> **aseptic**   Low number of white cells in cerebrospinal fluid, predominantly lympho-cytes, most frequently caused by infection by viruses or fungi. Most causes of acute bacterial meningitis, in contrast, have high numbers of white cells and a predominance of neutrophils.

meningoencephalitis   Inflammation of the brain and surrounding membranes (meninges).

metaphyseal   Relating to the conical portion of long bones that lies between the epiphysis and diaphysis.

metastasis   The occurrence of disease at sites distant from and not connected directly with the site where the disease first appeared; this process is seen with malignancies and infections.

MIC   Minimum inhibitory concentration; the lowest concentration of a drug needed to inhibit the growth of a microorganism.

microaerophilic   Pertaining to an atmosphere with a reduced concentration of oxygen; some microorganisms, called microaerophiles, grow best in such an environment.

microvasculature   The part of the circulation that includes the smallest vessels; typically refers to the capillaries.

mitral regurgitation   Any condition of the mitral valve between the left atrium and left ventricle that allows blood to flow back into the atrium when the ventricle contracts; nor-mally, the valve shuts tightly, allowing no flow of blood back into the atrium; often heard as a heart murmur during auscultation.

mitral valve prolapse   A defect in the valve between the left atrium and ventricle caused by a weakening of the tough connective tissue of the valve leaflets, which allows the valve to project back into the left atrium; during normal function, the valve closes tightly during ven-tricular contraction.

morbilliform rash   Rash that resembles the flat to slightly raised (maculopapular) lesions seen in measles.

mucocutaneous   Involving the skin and mucous membranes.

murmur   Normal or abnormal sounds heard on auscultation of the heart or vessels; this is a physical finding that has a variety of causes.

myalgia   Soreness or aching of muscles.

myeloma   Cancer of well-differentiated immunoglobulin-producing cells; progressive pro-liferation of plasma cells is seen with this type of malignancy.

myelopathy   A disease of the neural tissue of the spinal cord.

myocarditis   Inflammation of heart muscle cells.

myonecrosis   Necrosis of muscle tissue.

nares   Nostrils.

nasopharyngeal   Pertaining to the nasal and pharyngeal cavities.

necrosis   Death of cells, tissues, or portions of organs that results from irreversible injury; several different types of necrosis can be distinguished.

necrotizing fasciitis   A destructive soft tissue infection that causes necrosis of the superficial fascia and surrounding tissues; this is often a fulminant process and one that may be difficult to diagnose.

negative-pressure room   A confinement room that has a lower atmospheric pressure than the surrounding rooms; a pressure gradient prevents air from escaping into occupied areas; necessary for the containment of airborne pathogens, such as *Mycobacterium tuberculosis.*

nematode   The class of helminths that includes the roundworms: *Ascaris, Enterobius,* and *Strongyloides* spp.

neonate   A newborn infant.

nephrosis   Degeneration of renal tubular epithelium.

neuralgia   A stabbing pain that follows the course of a nerve.

neuropathy   Diseases or disorders affecting the cranial nerves or the peripheral or autonomic nervous systems.

neutralizing antibody   An antibody that binds to a particle, usually a virus or toxin, to prevent it from being infectious or hazardous.

neutropenia   Abnormally low numbers of neutrophils in the circulating blood.

nocturnal   Referring to night.

nodular   Knotlike or raised solid lesions of the skin or other organs.

nonsuppurative sequelae   Complications, caused by a previous disease status, that do not involve purulence.

nontoxigenic   Not producing a toxin.

normotensive   Normal blood pressure; the usual readings seen in adults are between 90/50 and 150/90 mm Hg.

nosocomial   Any condition resulting from a person's hospital stay; usually used in the context of an infection acquired as a result of hospitalization.

nuchal rigidity   Stiffness of the neck, often associated with meningitis.

occult blood   Blood present in body fluids, such as stool, that cannot be detected with the naked eye; the most commonly used test for occult blood is a guaiac test.

oncospheres   Motile tapeworm embryos that are released from the egg after the egg is ingested by a suitable host.

opisthotonic   Spastic state in which the head and heels are bent backward and the torso extends outward.

opportunistic infection   An infection caused by an organism capable of causing disease only in individuals whose resistance to infection is lowered.

opsonic   Pertaining to an agent (typically an antibody) that, when bound to an antigen such as bacterial proteins, enhances the ingestion of the antigen by white blood cells.

oral rehydration   The reversal of dehydration by drinking a solution of water, electrolytes, and carbohydrates.

organomegaly   Abnormal enlargement of the organs; visceromegaly.

orifice   A normal anatomical opening, such as the mouth.

oropharyngeal   Pertaining to the oral and pharyngeal cavities.

orthostatic hypotension   Decreased blood pressure caused by sitting up or by standing erect; often seen in patients who are dehydrated.

osteomyelitis   Inflammation or infection of bone.

otitis media   Inflammation of the middle ear; visualized with an otoscope, which often shows the presence of fluid or pus behind the eardrum.

ototoxic   Refers to a substance that has a toxic effect on the ear; some antibiotics, for example, have this property.

palatal   Referring to the palate, the bony or soft tissue roof of the oral cavity.

palpation   A technique used during physical examination that involves the use of the hands to feel for organs, abnormal masses, pulses, or vibrations.

pancytopenia   A significant reduction in the number of red blood cells, white blood cells, and platelets in the circulating blood.

pandemic   Relating to a disease that is affecting the population of an extensive region, country, continent, or the entire world.

papular   Pertaining to lesions that are raised and well circumscribed.

paraparesis   Weakness of the legs.

parasite   An organism that lives and derives its nutritional support from a host.

parasitemia   Presence of parasites in the blood.

paravertebral   Beside a vertebra or the vertebral column.

parenchyma   Cells of a gland or organ that are contained within and supported by the surrounding connective tissue network.

paresis   Incomplete paralysis.

paroxysm   The abrupt episodic recurrence of disease or disease symptoms; also, spasms or fits.

pathogen   Any microorganism that causes disease.

pathogenicity island   A small portion of a bacterial chromosome that encodes many of the genes necessary for infection; the physical closeness of genes suggests that the bacteria obtained infectious capability from a single source or process.

pathogenicity locus   The area of the bacterial chromosome that encodes the genes required for infection.

pathognomonic   Symptoms or lesions characteristic of a single disease process, on the basis of which a diagnosis can be determined.

PCP *Pneumocystis carinii* pneumonia. This abbreviation can also refer to primary care physician or to the drug of abuse phencyclidine (phenyl cyclohexyl piperidine ["angel dust"]). Be careful when using abbreviations in medicine.

pelvic inflammatory disease   Inflammation of the female reproductive organs, typically due to infection.

peptidoglycan   The main component of the bacterial cell wall; composed of carbohydrates and amino acids.

perforation   An abnormal opening.

perianal   The area surrounding the anus.

pericarditis   Inflammation of the sac covering the heart.

periorbital   Around the eye socket (orbit).

periosteum   The tough tissue surrounding the surface of any bone.

periostitis   Inflammation of the periosteum (the thick fibrous membrane that covers the surface of the bone, except for the articular cartilage).

peristalsis   The alternating waves of contraction and relaxation within the gastrointestinal tract that propel its contents onward.

peritonitis   Inflammation of the peritoneum (the thin lining of mesothelium and connective tissue that lines the abdominal cavity).

periumbilical   Pertaining to the area surrounding the navel (umbilicus).

periurethral   Pertaining to the area surrounding the urethra.

petechiae   Pinpoint, flat lesions due to hemorrhage of blood into tissues under the skin or mucous membranes.

petechial rash   Small, pinpoint, and flat lesions of the skin and mucous membranes associated with hemorrhage beneath the tissue; similar to purpura, except the lesions seen with purpura are larger.

Peyer's patches   Lymphatic tissues in the walls of the large intestine.

phagocytosis   The process by which cells ingest and digest solid substances such as tissue debris, bacteria, or foreign material.

phalanx   A long bone of the finger.

pharyngitis   Inflammation of the pharynx, the muscular tube connecting the nose and mouth to the esophagus and larynx; sore throat.

phenotypic   Pertaining to the effects of an organism's genes and the environment on its physical appearance, biochemistry, and physiology.

photophobia   Abnormal sensitivity to light.

pica   The desire to eat inappropriate and nonnutritional substances such as soil.

pili   Appendages on the surface of bacteria that allow for attachment and infection; same as fimbriae.

plasmid   A piece of DNA that is physically separate from the host's chromosome and not essential for survival; often a means for bacteria to gain a new function such as antibiotic resistance or toxin production.

pleocytosis   The presence of abnormally large numbers of cells; often refers to the increase in white blood cells that occurs during an infection.

pleuritis   Inflammation of the pleura.

pleurodynia   Episodic pain due to transient spasms of the intercostal muscles (muscles between the ribs that assist in respiration), caused by irritation of the pleura.

pneumonia   Infection of the lung parenchyma.

pneumonitis   Inflammation of the lungs.

poliomyelitis   Inflammation and destruction of the gray matter of the spinal cord, possibly resulting in paralysis; causative agents are the polioviruses.

polymerase chain reaction   A laboratory technique for rapidly replicating DNA strands in a test tube; can be used for diagnosis when a probe specific for an exact sequence of DNA is used to initiate the replication process.

portal circulation   The circulation of blood to the liver from the small intestine, the right half of the colon, and the spleen via the portal vein.

positive predictive value   The probability of having a disease given a positive test result.

post ictal   Referring to a patient's condition following a seizure; many people are less alert, and transient neurologic findings may be observed.

postpartum   The condition or period of time after the delivery of a fetus.

poststreptococcal sequelae   Consequences of infection by group A streptococci that appear after the initial infection; these processes are not a direct result of the organism but rather are due to the host immune response to the organism.

PPD skin test   A diagnostic test in which purified protein derivative is placed just under the skin; this test is used to determine if patients are infected with *Mycobacterium tuberculosis*.

precipitin   An antibody that binds with its antigen and causes the antigen-antibody complex to precipitate from solution.

preeclampsia   A condition of the late stages of pregnancy caused by accumulation of toxins in the blood; characterized by high blood pressure, swelling of the hands and feet, and the presence of proteins in the urine. If seizures occur, it becomes known as eclampsia.

prevalence   The number of people at a given period of time with a given trait or disease divided by the number of people at risk for the trait or disease.

prion   Infectious protein, possibly self-replicating.

proctitis   Inflammation of the rectum.

proctocolitis   Inflammation of the colon (large intestine) and rectum.

prodrome   An early symptom of a disease.

promastigote   A stage in the life cycle of the protozoans in the genus *Leishmania*; a characteristic finding in host macrophages during infection.

prophylactic   Pertaining to agents or procedures that prevent disease processes in suscepti-ble individuals; for example, giving vaccines or antibiotics to prevent an infection before it occurs.

prosthetic   Pertaining to a human-made replacement for a missing or defective body part, such as artificial limbs or heart valves.

Protozoa   A phylum or subkingdom of the animal kingdom; contains many medically important unicellular parasites.

proximal phalanx   The part of a digit (finger or toe) that is closest to its attachment to the body.

pruritus   Itching.

pseudomembrane   A layer of bacteria, fibrin, dead epithelial cells, and blood cells that form a plaque over the surface of the digestive tract.

pseudomembranous colitis   Inflammation of the mucous membranes of the small and large intestines with the formation of false membranes composed of bacteria, white blood cells, fibrin, platelets, and necrotic material; occurs most frequently as a result of the action of exotoxins made by *Clostridium difficile*.

ptosis   Drooping of the upper eyelid.

pulse oximetry   A technique that uses light to rapidly estimate the percentage of hemoglo-bin in the arterial blood which is saturated with oxygen.

punctate   Pertaining to lesions or markings that look like points or dots.

purpura   Purplish lesions of the skin and mucous membranes due to hemorrhage beneath the tissues; usually less than 1 cm in size, lesions may be flat or raised.

purulent   Related to presence, consistency, or formation of pus.

pustular   Pertaining to a skin rash with pus in the lesions.

pyelonephritis   Infection of the kidney with or without a concurrent bladder infection.

pyoderma   An infection of the skin that produces pus; caused by pyogenic bacteria.

pyrogenic   Causing fever.

pyuria   The presence of pus in the urine.

quadriparesis   Weakness of all four extremities.

rales   Abnormal breathing sounds (crackles) heard by auscultation of the lungs during res-piration; classified as dry or moist.

reactivation   The process by which a latent infection can become active after a variable period of dormancy.

rectocolitis   Inflammation of both the rectum and the colon.

rehydration   Restoration of body fluids to normal osmolarity by adding fluids.

renal   Pertaining to the kidney.

reservoir   The host, in nature, of an infectious agent. Some of these infectious agents may cause disease in humans after transmission from the reservoir either directly (e.g., a rabid bat bites a person) or indirectly (e.g., a rodent infected with *Leptospira* urinates in water in which a person later bathes).

restriction fragment length polymorphism (RFLP)   The variation in size of certain DNA fragments that exists between homologous chromosomes; the difference in the fragment lengths is due to alterations at the site of cleavage by restriction enzymes; can be used to determine the genetic variation between organisms of the same species.

reticuloendothelial system (RES)   The specialized white blood cells (especially macrophages) and other cells in the lymph nodes, liver, and spleen.

retinal   Referring to the retina.

retinitis   Inflammation of the light-detecting tissue (retina) in the back of the eye.

Reye's syndrome   An illness that occurs in children with influenza or varicella; can result in encephalitis, liver involvement, coma, and death; strongly associated with the use of aspirin during the initial viral infection.

rhabditiform   The first and second larval stages of the intestinal nematodes; this form is feeding and noninfectious.

rheumatic fever   An autoimmune disease of children and young adults that follows infection by group A streptococci; most severe sequelae are the destruction of heart tissue and scarring of heart valves, both of which can lead to heart failure.

rhinitis   Inflammatory or infectious process involving the mucous membranes lining the nose.

rhinorrhea   Thin, watery discharge from the nose; runny nose.

rhonchi   Coarse, low-pitched sounds associated with the presence of secretions or obstruction of larger airways in the lungs during inspiration and expiration.

rigors   Episodes of rigidity and shaking chills that may be brought on by infections; rigors may precede a fever.

ring-enhancing lesions   Lesions in the brain seen on computed tomography (CT) scans, consisting of lucent (less dense to X rays), rounded masses surrounded by a region of increased density (especially with the use of vascular contrast material); often seen in certain infections of the brain.

RNA   Ribonucleic acid; a polymer of nucleotides that is produced from a DNA template; has several functions in protein synthesis, including directing the sequence of amino acids and serving as a constituent of the ribosome complex.

salpingitis   Inflammation or infection of a tube, usually the fallopian tubes between the ovaries and the uterus.

scarlet fever   An illness caused by infection with group A streptococci; most prominent feature is the formation of numerous bright red spots on the skin and mucous membranes that are subsequently shed; before antibiotics, eruptions led to severe sequelae such as sepsis.

scleral icterus   Yellow discoloration of the sclera due to elevated concentrations of bilirubin.

sclerosing cholangitis   Recurrent or persistent obstructive jaundice due to extensive, destructive fibrosis of the bile ducts; frequently progresses to cirrhosis, liver failure, or portal hypertension; most common in young men.

screening test   An inexpensive, rapid test used for asymptomatic persons for a specific condition or trait; the test should be sensitive in order to avoid false negatives; usually followed by a confirmatory test.

selective medium   Type of medium that inhibits the growth of some organisms while allowing the growth of others; useful when trying to isolate or detect specific organisms.

sensitivity   The ability of a diagnostic test to detect an organism if the organism is present; a very sensitive test should miss only a small number of infections; often a characteristic of screening tests.

sepsis   The presence of the systemic inflammatory response syndrome as a result of an infection.

septate   Having septae; divided into compartments; usually a description of hyphae in fungi.

septic abortion   Infection of the mother after the abortion of a fetus; can involve the uterus and become a widespread systemic infection.

septic embolus   Clot (blood or other occlusive material) carried by the blood which contains infectious agents; often leads to systemic infection; its occurrence can be sudden.

septic shock   Hypotension as a result of sepsis, despite adequate fluid resuscitation, along with the presence of perfusion abnormalities of vital organs. This is a life-threatening condition that results in the increased permeability of the blood vessels, leading to excess fluid escaping into the tissues, abnormalities in cardiac function, and changes in vascular tone.

sequela (pl. sequelae)   A condition that occurs as a result of a disease process.

serology   The study of a patient's blood to determine the presence of specific antibodies to viruses or other microbes; can be used to determine exposure to a pathogen, to follow the course of a current infection by measuring the antibody level present, or to determine vaccination status.

serum sickness   An immune complex disease that occurs approximately 1 to 2 weeks after a drug, foreign serum, or protein has been introduced into the body; local or systemic reactions can occur.

sinusitis   Inflammation of the sinus cavities, often leading to headaches or nasal congestion.

somnolent   Drowsy.

specific gravity   A laboratory measurement that allows for comparison of the density of a liquid with that of distilled water; frequently used to determine the relative concentration or dilution of urine.

specificity   The ability of a diagnostic test to detect the absence of an organism if the organism is absent; a positive test result with a very specific test should "rule in" disease; often a characteristic of confirmatory tests.

splenectomy   Removal of the spleen.

splenomegaly   Enlargement of the spleen.

spongiform   Pertaining to a spongelike appearance; when noted in brain tissue, it is indicative of the presence of Creutzfeldt-Jakob disease or another prion disease.

spores   The product of sexual or asexual reproduction in fungi; a very hardy latent form that certain bacteria can employ in the presence of environmental stress.

stillbirth   The birth of a fetus that is dead before delivery.

stridor   Type of breathing that is characterized by a high-pitched whistling sound; caused by obstruction of the upper airway.

stridorous cough   A characteristic high-pitched cough associated with laryngeal blockage.

stuporous   Being in a state of decreased consciousness that is characterized by a diminished sense of orientation and response to the environment.

subcarinal   Refers to the area below the ridge that separates the right and left main bronchi at their junction with the trachea.

subcostal retractions   Inward movement of the area between the ribs, associated with an increased respiratory effort.

superantigen   An antigen that induces a T-cell response by binding to a T-cell receptor in an area outside of the antigen recognition site; usually results in the activation of many T cells.

superinfection   An infection by organisms because of a previously acquired but ongoing infection (such as a bacterial pneumonia which sometimes occurs during or after a viral pneumonia).

supine   Refers to lying on the back with the face directed upward.

suppurative lymphadenopathy   Enlarged, tender lymph nodes from which pus is draining.

syncope   Transient, brief episodic loss of consciousness (fainting).

synergy   When the actions of two or more processes, structures, or agents are combined to yield a result that is greater than the sum of each individually; for example, antibiotics can be synergistic.

systemic inflammatory response syndrome   A response to a stressor, especially infection, in which two or more of the following criteria are met: temperature >38°C or <36°C; heart rate >90 beats/min; respiratory rate >20 breaths/min or $PaCO_2$ <32 mm Hg; and white blood cell count >12,000/mm^3 or <4,000/mm^3, or >10% immature (band) forms.

tabes dorsalis   A neurologic disease of tertiary syphilis that results in destruction of the sensory or posterior portions of the spinal cord; when the upper spinal cord is involved, parts of the sympathetic nervous system can be involved.

tachycardia   Increased heart rate (>100 beats/min in adults).

tachypnea   Increase in respiratory rate (>20 breaths/min in adults).

tapeworm   Type of flatworm of the class *Cestoda*; infects humans after the larva or eggs are ingested in contaminated food or water; some species may infect only the digestive tract, whereas others may disseminate to the brain, lungs, or liver.

tenosynovitis   Inflammation of the tough sheath surrounding a tendon.

teratogenic   Pertaining to agents that cause the abnormal development of an embryo, commonly resulting in fetal death or birth defects.

thrombocytopenia   A decrease in the numbers of platelets.

thrombosis   Clotting within a blood vessel that may lead to infarction of the tissues supplied by that vessel.

thrush   An infection of the oral cavity that produces a white plaque; caused by the yeast *Candida albicans* in immunocompromised patients or in patients who have received broad-spectrum antibacterial agents.

tonsillitis   Inflammation of one or more tonsils, usually the palatine tonsil.

toxoid   A toxin that has been denatured into a nontoxic form; the toxin retains its antigenic potential and can be used in a vaccination.

trachoma   Infection of the conjunctiva and cornea due to *Chlamydia trachomatis*; it is a major cause of blindness in the developing world.

transbronchial biopsy   During bronchoscopy, the removal of a small piece of lung tissue through the walls of a bronchus (airway).

transient bacteremia   The presence of bacteria in the bloodstream for a brief period of time followed by rapid clearance; patients with transient bacteremia are asymptomatic.

transplacental   Crossing the placenta; pertaining to any substance that passes from the mother to the fetus or vice versa.

trismus   Spasm of the jaw muscles.

trophozoite   The pathogenic or non-cyst stage of certain protozoans.

tropism   Movement of an organism toward or away from a source of light, heat, or some other stimulus.

tuberculoma   A tumor-like mass composed of *Mycobacterium tuberculosis* and the corresponding phagocytes that are unable to digest the bacteria.

tympanic membrane   The eardrum.

ulcer   A lesion on the surface of the skin or mucous membranes caused by superficial loss of tissue; often accompanied by inflammation.

unicellular   Being composed of a single cell.

urethritis   Inflammation of the urethra.

vaccine   Weakened (attenuated) or dead (inactivated) bacteria or virus, inactivated toxin (toxoid), or genetically engineered component of an infectious agent, used to promote immunity against a disease (such as smallpox or diphtheria).

vaginitis   Inflammation of the vagina.

vasculitides   The various forms of vasculitis.

vasculitis   Inflammation of blood vessels leading to lesions on the skin, mucous membranes, or internal organs.

vegetation   A clump of fibrin and platelets on the heart valves; may include microorganisms as a result of endocarditis or may be sterile.

venipuncture   A procedure used to draw blood or inject a solution that involves puncturing a vein.

vertical transmission   Passage of an infectious agent from the mother to a fetus.

vesicular   Pertaining to small, blisterlike lesions filled with clear fluid.

vesiculopustular   Pertaining to blisterlike lesions containing pus.

viral load   A quantitative measurement of the amount of virus present in the bloodstream; used as a means of following the course or severity of a viral infection.

viremia   The presence of viable virus in the bloodstream.

virulence factor   A trait of an infectious agent that gives the organism the capability to cause disease; may be a toxin, a biochemical component of the organism's outer surface, or other factor.

virulent   Refers to the particularly toxic effects of a pathogenic microorganism.

virus   A particle composed of genetic material surrounded by a protein capsule; capable of infecting host cells by inserting the genetic material into the host cell and utilizing the host's machinery and energy to replicate.

viscera   The organs located within the thorax, abdomen, and pelvis.

wheeze   A high-pitched whistling noise made when narrowed airway passages cause difficulty in breathing.

zoonotic   Pertaining to diseases or conditions usually affecting vertebrates other than humans; however, humans can become infected after contact, either direct or indirect, with infected animals, their tissues, or their excrement.

zoophilic   Pertaining to a parasite or fungus with a preference for animals over humans.

zygomycosis   An infection caused by fungi belonging to the zygomycetes; typically seen in either diabetics or immunocompromised hosts.

INDEX